THE THERMOPHYSICS OF GLACIERS

GLACIOLOGY AND QUATERNARY GEOLOGY

A Series of Books

THE THERMOPHYSICS OF GLACIERS

by

I. A. ZOTIKOV
Institute of Geography, U.S.S.R. Academy of Sciences, Moscow, U.S.S.R.

D. REIDEL PUBLISHING COMPANY

A MEMBER OF THE KLUWER ACADEMIC PUBLISHERS GROUP

DORDRECHT / BOSTON / LANCASTER / TOKYO

Library of Congress Cataloging-in-Publication Data

CIP

Zotikov, I. A. (Igor' Alekseevich)
The thermophysics of glaciers.

(Glaciology and quaternary geology)
Translation of: Teplofizika lednikovykh pokrovov.
Includes index.
1. Glaciers. 2. Heat budget (Geophysics)
II. Hardin, R. II. Title. III. Series.
G2405.Z6713 1986 551.3′12 b86–597
ISBN 90–277–2163–7

Published by D. Reidel Publishing Company
P.O. Box 17, 3300 AA Dordrecht, Holland

Sold and distributed in the U.S.A. and Canada
by Kluwer Academic Publishers,
101 Philip Drive, Assinippi Park, Norwell, MA 02061, U.S.A.

In all other countries, sold and distributed
by Kluwer Academic Publishers Group,
P.O. Box 322, 3300 AH Dordrecht, Holland

Originally published in 1982 by Gidrometeoizdat under the title

Теплофизика ледниковых покровов

Translated by Ron Hardin

Printed in The Netherlands

TABLE OF CONTENTS

INTRODUCTION

Glaciers or ice sheets are natural accumulations of ice possessing intrinsic motion, which have appeared on the Earth's land surface as a result of the accumulation and transformation of precipitation [2]. Only a very small portion of the surface of the land is now covered by glaciers, and at low latitudes they are found just at high elevations, on mountain slopes.

However, glaciers are known to play very important roles in shaping the topography of the Earth, in determining its past, present, and future climate, and in creating the state of the World Ocean. Recently there has also been a marked increase in the practical value of our knowledge about glaciers, as a part of the human habitat and as a factor affecting the economy.

Interest in glaciers on other planets is also growing. Voyages of spacecraft to Jupiter, for instance, have shown that some Jovian satellites possess ice sheets tens of kilometers in thickness.

A substantial part of modern glaciology deals with the thermal regime and temperature state of glaciers, and with the thermal interaction between a glacier and its surroundings. The importance of such studies is evident, for the temperature at which a glacier disappears, that is, the melting point of ice, lies within the range of temperature variations typical of our planet. Consequently, the conditions of existence of glaciers, as well as their development and interaction with the climate, are to a considerable extent governed by the thermal state of the glacier, by the conditions of heat exchange in its interior and at its boundaries.

This is why, 30 years ago, G. A. Avsyuk, the founding father of Soviet thermophysical studies of glaciers, wrote: "The temperature state of glaciers and the thermophysical processes taking place in the ice interior constitute one of the most important fields of study in modern glaciology, and in addition to being of paramount theoretical interest they have great practical importance as well." [2]

However, only during recent years have the basic premises of the thermophysics of glaciers been verified by observations. The thermal regime of glaciers was found to be one of the main factors governing the variability and nature of interactions between glaciers and the ocean and atmosphere. Thus this regime must be given prime consideration when formulating a theory of catastrophic glacier movements, including those caused by anthropogenic factors. Consequently, we must know about the thermophysics of glaciers in order to ascertain how the machine we call 'Earth' operates. Such knowledge is also vital when planning and implementing measures for protecting the environment, as well as when constructing models of the Earth's climate, which are becoming increasingly important.

On the other hand, there are some very pressing practical reasons for studying the thermophysics of glaciers, since an increasing number of persons are working in the polar seas, and in polar and mountainous regions, where the ice cover is a major factor. Moreover, it may well be possible to utilize glacier ice in the human economy: via the artificial intensification or retardation of glacier melting; by burying radioactive wastes in the Antarctic Ice Sheet; by transporting icebergs from place to place; and by predicting catastrophic movements of glaciers and ice sheets, which can as time goes by do more and more harm to the economies of polar and mountainous lands. Such movements can in some cases be so great that they cause the level of the World Ocean to rise drastically.

In a previous work [32], using the Antarctic Ice Sheet as an example, we pointed out the importance of studying the thermal regime of glaciers, since this regime governs all the fundamental englacial processes, and we drew attention to the enormous potential of the 'thermophysical' approach as a means of obtaining new data (as yet unknown information) about glaciers. In that work we suggested that the branch of glaciology dealing with the application of the thermophysical method, or, conversely, the branch of thermophysics in which its methods are applied to glaciers, should be called 'thermophysical glaciology', by analogy with the 'stable isotope glaciology' of Dansgaard.

In the present work the fundamental problems of the thermophysics of glaciers are attacked with the aid of the integral approaches and the approximate and numerical methods typical of thermophysics. We hope that this study will be a useful aid in working out the methodology proper of thermophysical research in glaciology and related disciplines.

This book is based on the following source material: (1) information obtained during glacier studies in the Caucasus, the Pamirs, and the Polar Urals; (2) data gathered on Soviet and American expeditions to the Antarctic; (3) information obtained at the U.S. Army Cold Regions Research and Engineering Laboratory (CRREL). It is my pleasant duty to thank the coworkers at the institutions involved and the teams of the expeditions for their help, support, and valuable discussions.

Chapter 1

OBSERVATIONS OF THE THERMAL REGIME OF GLACIERS

Data on the thermal regime of glaciers are obtained by measuring the temperature and heat fluxes in the glacier interior. However, published data concerning such measurements do not as a rule contain enough information about the measurement techniques used and the possible errors involved. The measurement errors cited in the literature generally range from ± 0.1 to ± 0.2 °C. At the same time, the differences between temperature values obtained from repeated measurements carried out by different investigators at several points can be great. Such differences are even observed in cases where an actual temperature variation during the time elapsing between measurements is very unlikely.

These discrepancies occur, in our opinion, because certain problems connected with the observational methods have not been properly taken into account. Moreover, in the literature on glaciology there are almost no works on methodological subjects associated with measuring the parameters determining the thermal regime of a glacier.

It will be the goal of this chapter to fill this gap to some extent. We will analyze the diverse factors giving rise to errors in temperature measurements in boreholes, and we will work out criteria for choosing correctly the method of borehole observations. In addition, the instrumental errors will be analyzed and means of reducing these will be sought. Finally, we will consider techniques for measuring the heat fluxes emanating from the interior of the Earth (the geothermal flux).

Despite the development of remote methods, temperature measurements in the interior of continuous opaque media are still carried out solely by direct means, that is, by (1) inserting a thermomoter at the point of interest in the medium; (2) waiting until a certain degree of equilibrium is attained between the thermometer and its surroundings and (3) determining the temperature of the sensor. Accordingly, the number and locations of temperature measurements in glaciers are limited by the possibilities for the direct insertion of instruments into the interior of the ice. Since such insertions require drilling into the ice, a considerable portion of the methodology of these studies involves techniques for making boreholes in ice, special attention being given to thermal drilling methods.

1.1. Measurements of Temperatures inside Glaciers

1.1.1. General Description

Measurements of englacial temperatures are usually carried out in boreholes. The borehole is drilled to the end, and only then is the temperature sensor, or string of sensors, lowered into the borehole. Finally, the borehole with the sensors in it is left to 'age', or 'stabilize', until the initial temperature conditions disturbed by the drilling are re-established.

The temperature sensors hang freely in the borehole, the top of the hole being closed with a special plug while the borehole and its sensors are stabilizing. The temperature is measured remotely at the glacier surface, via lines connecting all the sensors to the measuring instrument.

It should be noted that the localities where such measurements have to be made are usually in regions that are almost inaccessible, where conditions are very inhospitable to humans. Therefore, temperature measurements in boreholes are usually made as soon as possible after drilling.

Another method is used as well. A borehole is drilled to a certain depth, after which drilling is stopped and a sensor is lowered to the bottom of the hole until it makes contact with the lower surface. Then, after a time necessary to restore the thermal regime disturbed by the drilling and the lowering of the sensor, the temperature is measured. After this, the drilling is continued down to some lower level at which a temperature measurement is desired.

However, there is as yet no standard apparatus for such measurements, and each investigator uses his own (often not standard) sensors, measuring instruments, and connecting lines. When describing the results of the observations, he usually neglects to specify the fundamental parameters of these components of the measuring system. Thus there is no real possibility for a comparative analysis of the errors of temperature measurements carried out by different persons.

Still, in view of the importance of such an analysis, we have elected to carry one out anyway, if only for the sensors, connecting lines, and measuring instruments employed during studies in Antarctica and on glaciers.

1.1.2. Temperature Sensors

The main temperature sensors used at present are resistance thermometers and thermistors. For work at temperatures which are not excessively low, we employed, for example, copper resistance thermometers and ETP-310 platinum resistance thermometers. The copper thermometers were rods 4 mm in diameter doubly wound with copper wire. The wire was subjected to a special heat treatment, in order to ensure stability of its characteristics over a long period. The rod was enclosed in a thin-walled brass case 7 mm in diameter, the space between the rod and the case being filled with electrical insulation.

Before and after use each thermometer was given a calibration check. The calibrations indicated that, for a range of temperature variation from 0 to -30 °C, the 0 °C resistances of all the thermometers, as well as their variations with temperature accurate to within 0.1 °C, satisfied the standard calibration tables for copper thermometers both before and after the experiments. Using correction tables, the errors in measuring temperatures with these thermometers can be estimated to within ± 0.03 °C.

Platinum resistance thermometers are being used more and more, since they can be used to measure temperatures down to -100 °C. Although the resistance coefficients of these thermometers are more stable, the resistance of a platinum wire is less temperature dependent than the resistance of a copper wire.

The ETP-310 platinum thermometers used by us consisted of a profiled ceramic rod 5 mm in diameter with platinum wire wound onto it. The sensitive element was enclosed in a thin-walled aluminum case 7 mm in diameter. This case was in turn placed inside a protective steel housing with walls 0.3 mm thick. The 0 °C resistance of the thermometer R_0 was 100 Ω, corresponding to one of the standard calibrations for platinum thermometers.

A calibration check of the platinum thermometers indicated that the error due to the deviation of R_0 from the nominal value and due to differences in the temperature coefficient can, taking the corrections into account, be reduced to ± 0.03 °C.

Thermistors have recently begun to be used widely for temperature measurements, because of their high temperature coefficients and high resistances, and also because of the small size of the sensor, in comparison with resistance thermometers. Their disadvantages include a nonlinear relation between temperature and resistance, widely differing characteristics from thermistor to thermistor, and instability in time of these characteristics.

On the basis of studies over many years, Tsykina [71] pointed out that sensors with aged MMT-4 thermistors retain their calibration characteristics for a long time at subzero (freezing) temperatures, and that errors greater than 0.03 °C arise only after they have been used for a year and a half.

Like resistance thermometers, thermistors have to be given a calibration check before and after an experiment.

A temperature measurement with a thermometer requires the use of connecting lines (wires), since the measuring apparatus is located at the surface of the glacier, while the sensor is in the interior, a considerable distance away. Thus the instrument measures not just the resistance of the sensor, but also the resistance of all the lines (which is variable), the contact resistances, and so forth. This constitutes a serious source of possible errors. However, there is a new kind of temperature sensor, the frequency sensor, which does not have these disadvantages. Since the natural oscillation frequency of such a sensor is temperature-dependent, the line resistance, intermediate resistances, etc., cannot influence the readings. Thus this method promises to have certain advantages, especially when the sensor is far away from the measuring instrument.

So far the only commercially available thermometer of this type is the quartz-crystal thermometer. The sensitive part of the temperature sensor of this thermometer is a specially shaped quartz crystal, which has been so cut as to make its natural frequency very precisely and linearly dependent on the crystal temperature. Consequently, by comparing the frequency of the crystal sensor with that of a standard oscillator at the glacier surface, it is easy to discern the frequency difference, which unambiguously determines the temperature. The use of the frequency method eliminates the effect of the cable on the measurement errors, and, if a high-stability frequency reference is employed, temperature variations can be measured to within 0.0001 °C. Two hermetic quartz thermometers, produced by Hewlett-Packard, were frozen into the Ross Ice Shelf by us in 1979, at a depth below the surface of about 400 m. These operated reliably for more than a year, and this was apparently the first successful use of quartz thermometers in glaciology.

The sensitive elements of these thermometers were disks cut from a quartz crystal, the disk diameter being about 10 mm and the thickness about 1 mm. The disks were placed in protective stainless-steel tubes and hermetically sealed there in a helium atmosphere. The disks were located about 0.2 mm from the ends of the protective tubes, so as to prevent deformations at the tube wall and at the same time to ensure a satisfactory thermal contact with this wall.

To conclude this discussion of temperature sensors, we should mention liquid thermometers. Mercury and alcohol glass thermometers are now scarcely used at all to measure temperatures inside of glaciers. Although, as compared with "electrical" thermometers, they have the advantage of measurement reliability within their range of rated accuracy, this accuracy is low, and the action of the thermometer has to be 'slowed down', that is, the thermometer must be placed in an envelope sufficiently massive to keep its temperature from varying while the unit is raised up out of the borehole to the glacier surface to be read. However, using such a 'slow-acting' thermometer, the time needed for the temperature measurement will be very long indeed.

1.1.3. Connecting Lines and Measuring Circuits

Because of the low temperatures and difficult working conditions on glaciers, the choice of the wires connecting the thermometers with the measuring instruments is of particular importance. For instance, PVR [PVR is the name of a type of wire used in Soviet Antarctic Expeditions - **Transl.**] korotage wire proved to be the only type of wire with an insulation suitable for measurements in Central Antarctica at temperatures of about -60 °C, which involved repeated bending of the wire as it was unwound and lowered into the borehole. This is a fine (0.35 mm^2 section) wire with a high specific resistance (about 0.2 Ω m).

We connected all the thermometers, designed for use in boreholes under these conditions, to the measuring instruments with this type of wire.

For borehole depths up to 60 m, it is advisable to use thermometer 'chains', that is, strings of thermometers spaced 5 m apart. Such a chain can be used to measure temperatures throughout the entire borehole, right down to the part passing through the active layer. The maximum distance between the thermometer and the measuring instrument (along the wire) usually does not exceed 80 m, making the maximum line resistance 32 Ω.

For temperature measurements at average borehole depths (100 to 400 m) we generally used an individual platinum resistance thermometer. The distances between the sensor and the measuring instrument in this case were more than 400 m, making the line resistance about 160 Ω.

Temperature measurements with resistance thermometers may involve bridge or potentiometer circuits. Comparative evaluations of the operation of different circuits, together with a determination of their errors, are given in [57].

The choice of the measurement circuit is dictated by the need to use wires with a high specific resistance, as well as by the fact that under glacier conditions the temperature of the connecting wires can fluctuate over a wide range.

A two-wire measuring circuit with copper and platinum resistance thermometers clearly does not provide the required accuracy of the observations, since the line resistance has a marked effect. It is sufficient to note that, for a line length of 100 m and even using a copper wire 0.75 mm^2 in section, a 30 °C variation in line temperature will change the total line resistance by 0.25 Ω. If copper resistance thermometers are used, this leads to an error of 2 °C.

A two-conductor system with PVR wire is suitable only for work with thermistors; for instance, using an MMT-4 thermistor, even a 10-Ω variation in the line resistance of such a wire corresponds to an error of less than 0.01 °C in the temperature determination.

An analysis has shown that a three-conductor circuit is feasible with copper and platinum thermometers only if the thermometer is connected to the measuring instrument by a copper wire with a quite large section.

If copper leads 1 mm^2 in section are used, for a cable length of 100 m and temperature fluctuations of ± 15 °C, then the error for the calibration of copper thermometers of 50 Ω resistance at 0 °C, which fits the so-called 2A calibration table, does not exceed 0.01 °C. With PVR-2 wire, on the other hand, the measurement error under similar conditions reaches 1 °C.

Therefore, in all our measurements with copper and platinum thermometers, only four-wire measuring circuits were employed, thereby completely eliminating any effects of the line resistance on the measurement accuracy (Figure 1.1). As seen from the figure and as the name of the circuit implies, a disadvantage of this circuit is that each thermometer terminal has to be connected to the measuring instrument via two wires.

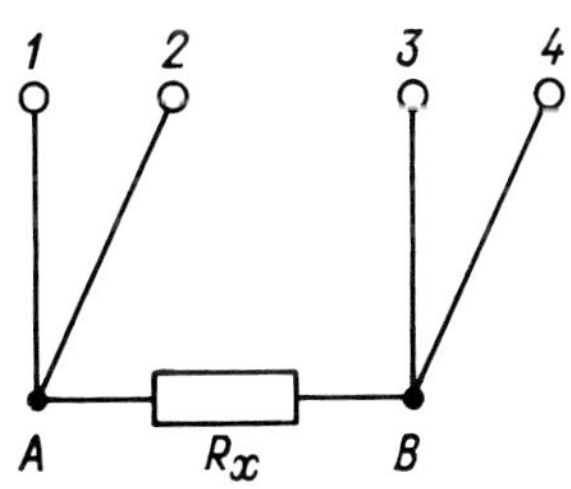

Fig. 1.1. Four-wire circuit for measuring resistance of electrical thermometer.

If the resistance of each segment of the line is denoted by the letter R, with subscripts indicating the beginning and end of the segment, then we can describe as follows the results of four measurements R_I, R_{II}, R_{III}, R_{IV}, comprising two measurements of the line resistances (1-A-2 and 3-B-4) and two measurements including the thermometer (1-A-B-3 and 2-A-B-4):

$$R_{1A} + R_{A2} = R_I, \; R_{3B} + R_{B4} = R_{II};$$
$$R_{1A} + R_{AB} + R_{B3} = R_{III}; \; R_{2A} + R_{AB} + R_{B4} = R_{IV},$$

from which we get

$$R_{AB} = [(R_{III} + R_{IV}) + (R_I + R_{II})]/2. \tag{1.1}$$

1.1.4. Measuring Equipment. Increasing the Accuracy. Errors.

Because of the specifics of work under complex field conditions, bridges with accuracies no higher than 0.1% are usually employed to measure thermometer resistances. Our main measurements on glaciers in the Caucasus, the Pamirs, the Polar Urals, and Antarctica were made, for example, with MO-62 bridges [MO-62 is the name of a particular Russian-made bridge - Translator].

When evaluating the accuracy of temperature measurements with electrical thermometers, using the described apparatus, we have to distinguish between two kinds of observations, each of which involves a different measurement accuracy:

(1) measuring the resistance of the sensor via a connecting link, the sensor being calibrated regardless of the specific resistance-measuring instrument used in the experiment and regardless of the type of link between the sensor and the measuring instrument;

(2) measuring the resistance of the sensor, the calibration being carried out with the same measuring instrument and the same connecting line as in the field experiment.

The errors in measurements of the first type with a four-wire circuit are determined by the half-sum of the errors in two values of the thermometer resistances with two of the four connecting wires and the errors in two values of the resistances of the connecting wires.

For a platinum resistance thermometer with the above-described connecting line these resistances have maximum values of about 132 and 32 Ω, respectively. The errors in measuring them with an MO-62 bridge are ± 0.132 and ± 0.032 Ω, and the total error in measuring the resistance of the thermometer proper, without the connecting wires, is ± 0.164 Ω. Thus the possible error in determining the temperature of a platinum thermometer is close to ± 0.4 °C.

The foregoing evaluation, which pertains to conditions of thermometer use with a connecting line 80 m long, clearly shows the inadequacy of resistance bridges of class 0.1 or lower for measuring absolute values of the temperature with platinum thermometers that have been calibrated with other measuring instruments.

The error of a temperature measurement can be reduced appreciably if the resistance thermometer is calibrated with the same resistance bridge and the same connecting line as are used for observations in the field. Such a calibration yields graphs on which the readings of specific resistance bridges and thermometers are plotted. The errors in these calibration curves for a given combination of instrument and thermometer at each thermometer temperature are no longer determined by the error in measuring the resistance with the instrument, but only by the sensitivity of the latter.

Calculations show that a bridge of class 0.1 used with an external galvanometer makes it possible to find the resistance of a platinum thermometer via the specified connecting line to within ± 0.03 Ω. In this case the error in a temperature measurement does not exceed ± 0.075 °C.

Rounding off these figures, we can assume a maximum instrumental error of ± 0.1 °C for temperature measurements carried out in this manner. The accuracy of the temperature measurements can be increased still more if bridges of class 0.01 are used. But, as pointed out in [57], the precision of measuring the resistance of a thermometer can be raised to ± 0.01 °C only if standard resistance coils are available for comparison.

With regard to the connecting lines of frequency thermometers, let us just mention that these are high-frequency coaxial cables, the parameters of which have to be matched to the parameters of the sensor and the measuring instrument, in this case a precision frequency meter. For instance, for the Hewlett-Packard frequency thermometer, the sensitivity of the frequency meter must be so selected that the slope of the line showing the natural frequency of the quartz crystal as a function of temperature is close to 0.98 - 0.99 kHz per degree.

The frequency meter of a quartz thermometer is an integrating device which compares the frequency being measured with the frequency of a standard source, generally located in the frequency meter. Thus the measurement accuracy depends on how accurately the standard of frequency maintains it, and also on the comparison time. For slowly varying processes, such as those taking place in glaciers far below the surface, a comparison time of 10 sec can already provide temperature measurements with a resolution as high as 0.0001 °C. For a comparison time of 100 sec, resolutions of up to 10^{-5} °C are possible. Here the true measurement accuracy will be determined by such parameters as the frequency stability of the crystals of the sensor and standard, the zero drift, the possible hysteresis of the crystal characteristics, and the variation of the temperature of the frequency standard.

1.1.5. Heating of Temperature Sensors at the Time of Measurement

Thermometers. The possibility of a distortion of the thermometer readings, caused by the current in the circuit at the time of measurement, is a very important factor. On the one hand, the resistance bridge must have a quite high current, and thus a high supply voltage as well, in order to provide bridge sensitivity without the use of very sensitive or ultrasensitive galvanometers, the latter being unsuitable for work under field conditions.

On the other hand, it is risky to increase the supply voltage, since at the time of measurement the thermometer may well be heated by the current passing through it by an amount ΔT_{cur} which is unacceptable, given the specified error.

According to [19], the temperature increase of a thermometer in a borehole, due to the current passing through it, will not exceed ΔT_{cur} if the ratio between the voltage v_T across the thermometer terminals and the thermometer resistance R_x satisfies the condition

$$\frac{V_T^2}{R_x} < \Delta T_{cur}. \tag{1.2}$$

To ensure the specified measurement accuracy (0.1 °C), the rise in thermometer temperature due to current-caused heating (the self-heating error) should not exceed 0.01 °C. Thus, if a platinum resistance thermometer has a 0 °C resistance of 100 Ω, the voltage across its terminals cannot exceed 1 V, and the current through the thermometer cannot be more than 0.01 A.

Calculations show that the maximum allowable supply voltage of the bridge, for the overheating of a platinum thermometer with a line resistance of 32 Ω to stay less than 0.01 °C, is 2.0 V. This, then, is the condition that has to be satisfied in all temperature measurements with platinum thermometers.

Thermistors. Thermistors differ from resistance thermometers in that the heat-emitting surface of a thermistor is very small compared with the volume. Therefore, the use of expression (1.2) to evaluate the allowable voltage across the thermistor terminals may give a heating of the thermistor which reduces its accuracy too much.

The quality and thickness of the layer insulating the sensor may affect the heating of the sensor considerably. At the same time, if the heat capacity of the insulation is high enough, the heating of the sensor will be practically independent of the medium around the sensor.

Kutasov [51] made an experimental study of the temperature rise of a thermistor sensor as a function of the current power P_T. Using his results, it can be shown that a temperature increase of 0.01 °C corresponds to a power of about 0.2 mW.

Consequently, condition (1.2) for a thermistor sensor with an allowable overheating of 0.01 °C becomes

$$P_T \frac{V_T^2}{R_x} < 0.2 \text{ mW}. \tag{1.3}$$

For the MMT-4 thermistors being considered as an example, which have a resistance of about 10 kΩ, this condition corresponds to a voltage across the terminals of not more than 1.41 V, so that the supply voltage of the bridge cannot exceed 2 V. For high supply voltages of the instrument, a correction for the heating of the sensors must be introduced. The idea of taking such corrections into account was proposed in [51].

1.2. Sources of Errors During Temperature Measurements in Glacier Boreholes

In order to obtain reliable measurements of ice temperatures in boreholes, several conditions have to be satisfied. Otherwise, the measurements will not reflect the true temperature in the interior of the glacier.

The main sources of error, causing the temperature measured in a borehole to differ from the true value, are: (1) the temperature of the borehole wall differs from that of the glacier interior, due to the heat produced during drilling; (2) free natural convection of the air or liquid in the borehole takes place; (3) the effect of drive pipes, if they are inserted into the borehole.

Lyubimova et al. [56, 58] analyzed these factors for boreholes which were drilled to obtain data on the geothermics of the Earth's crust. However, since there are no data in the current literature suitable for an evaluation of these factors vis-à-vis boreholes in ice, let us consider this problem in some detail.

1.2.1. The Temperature Variation Due to the Heat Generated During Drilling. The Stabilization Time.

At the end of the drilling, the disturbances of the temperature field caused by the thermal effects of the drilling gradually level off, and after some time they become negligible. Let us call the time required for this the 'stabilization time' of the borehole.

The temperature regime of the glacier interior near a borehole wall varies because of the conversion of mechanical energy into heat for rotary drilling or via the direct transfer of heat to the glacier interior from the meltwater at the hole end for thermal drilling.

Other conditions being the same, the longer the drilling time, the more severe will be the disturbance of the thermal equilibrium and the longer will be the stabilization time.

When drilling deep boreholes, it is necessary to fill the hole with liquid, to prevent closure of the borehole due to plastic deformations of the ice. This in turn increases the disturbance of the thermal regime and lengthens the stabilization time of the borehole [50, 57].

This problem is evaluated theoretically by considering the disturbed temperature field close to a borehole being drilled out to be the field of a linear heat source of constant strength. The acting time of this source is equal to the time interval between the moment the end of the borehole reaches the depth at which the disturbed temperature field is being considered and the end of the drilling. The space inside the borehole is arbitrarily assumed to be filled with a substance whose temperature coefficient of thermal conductivity is the same as that of the ice being drilled [87].

The problem so formulated is similar to the problem of the heating of a borehole wall by an auger under conditions where all the space inside the hole is filled with the mud created by the auger. Such drilling was carried out, for instance, in central Antarctica in order to measure temperatures down to depths of 40 or 50 m.

Table 1.1 gives some calculated stabilization times for this case.

This model was used to evaluate the drilling of quite deep holes in ice with a drill turned by a tube column imparting a rotational moment (torque) from above, that is, from the drill assembly (a flushing liquid was used). Here the drilling time stretches over tens of days or even months, and the temperature-recovery time of a borehole will be very long indeed.

Table 1.2 gives recovery times for the temperature of a borehole wall, accurate to 1% of the temperture disturbance at the end of the borehole.

Table 1.1. Stabilization times of shallow boreholes for temperature measurements.

Stabilization time of borehole, h	Deviation of wall temperature from true ice temperature, °C
4	0.23
12	0.08
24	0.03
48	0.01

Table 1.2. Stabilization times (in days) of deep boreholes in ice for temperature measurements.

Borehole diameter, cm	Drilling time, days		
	10	30	100
2.5	116	305	900
5.0	142	360	1040
10.0	178	450	1230

The table was compiled on the basis of a calculation using the method of [58] for the three most common borehole diameters (2.5, 5, and 10 cm) and for drilling times of 10, 30, and 100 days. It is seen that the borehole stabilization time needed to restore a temperature regime differing from that of the undisturbed ice by no more than 1% of the disturbance due to drilling will, for a long drilling time, be 10 or 12 times the time spent on drilling the hole.

For instance, the stabilization time for a borehole drilled near Mirny Station in Antarctica was estimated to be 500 days, since the net drilling time in this case was about 40 days.

If boreholes are drilled in the indicated manner in the central region of the Antarctic Ice Sheet to depths of around 3000 m, then the drilling time may be of the order of 100 days. In this case the stabilization time of a borehole, long enough to ensure reliable temperature observations, will be 1000 to 1200 days.

It should be noted once again that this stabilization time pertains to the upper parts of the borehole, where the thermal disturbance due to drilling lasts longest. For parts closer to the bottom of the hole, this time is shorter.

The picture is somewhat different if the hole is drilled with a tool which disturbs the thermal regime of the glacier interior at the hole end or close to it. Examples of this kind of drilling are thermal and mechanical drilling whereby the energy consumed in drilling is liberated only close to the end of the borehole. In this case the stabilization time of the borehole does not depend on the total drilling time, being as a first approximation determined by the time during which the drill is close to the point in question, both as it cuts and as it is lowered and raised.

In this case for a borehole about 3000 m deep around 10 days are needed for recovery of the temperature regime of the borehole to conditions differing from the regime of the undisturbed ice by no more than 1% of the temperature disturbance caused by the drilling (for a borehole 20 cm in diameter). If the borehole diameter is smaller, this time will be shorter.

It should be noted that this evaluation apparently yields borehole stabilization times which are somewhat too short, since we have not taken into account thermal disturbances caused by forced convection of the liquid in the borehole as the drill is lowered and raised.

1.2.2. The Effect of Free Natural Convection

Beginning at some distance from the glacier surface, the ice temperature increases with depth. This causes thermal expansion of the air (or liquid) in a borehole. The more heated part of the air (liquid) moves upward, and the cooler part descends, giving rise to mixing currents. Thus the temperature in the lower part of the borehole drops slightly, while that in the upper part rises somewhat.

This effect becomes more marked with an increase in the positive temperature gradient in the ice interior. If there is an inverse gradient (temperature drop with depth) all along the length of the borehole, then this mixing effect disappears.

It should be easy to eliminate the effect of free convection by plugging the borehole, that is, by placing plugs in it at various depths, to prevent the development of convection. Experiments along these lines have been carried out, but no definite conclusions on the effectiveness of this method have ensued [58].

A theoretical study of the mechanism of free convection in boreholes indicates that convection can originate in a vertical borehole in either of two situations [52]: (1) an axisymmetric flow, whereby the air or liquid ascends along the axis of the cylinder and descends along the walls; (2) an 'antisymmetric flow', whereby liquid ascends in one half of the borehole section and descends in the other half. The former type of convection is more prevalent.

To verify the presence of convective flows in boreholes, Cridge proposes using the concept of a critical gradient. If the actual temperature gradient in a borehole is greater than critical, then appreciable convection occurs; if it is less than critical, then it can be assumed that the air or liquid in the borehole does not become mixed along the vertical.

The critical gradient for the first type of motion is defined by the expression

$$\beta_{cr} = \beta_* + 216 \frac{\nu_{bh} a_{bh}}{g \alpha_{bh} R_0^4} . \qquad (1.4)$$

The first term on the right-hand side represents the critical gradient in an ideal fluid (adiabatic gradient). For air this gradient is close to 1 °C per 100 m. The second term takes into account the effect of viscous forces, which tend to keep the gas or liquid from moving because of friction at tbe borehole wall. In (1.4) ν_{bh} is the kinematic viscosity, a_{bh} is the temperature conductivity of the gas or liquid filling the borehole, R_0 is the borehole radius, and α_{bh} is the coefficient of thermal expansion of the borehole filler.

The value of β_{cr} depends mainly on the second term in (1.4), which is determined by the viscous resistance to mixing in the borehole.

Table 1.3 gives some critical gradients in boreholes 14 and 10 cm in diameter, calculated by us neglecting β_*. The calculations were carried out for unfilled boreholes (for air) as well as for holes filled with nonfreezing liquids that should be suitable for use in superdeep boreholes. The physical parameters of these fillers were taken from Eckert and Drake [78].

Table 1.3. Critical gradient β_{cr} for boreholes in ice

Filler	T °C	ν_{bh} m^2/s	a_{bh} m^2/h	α_{bh} 1/°C	β_{cr} °C/100 m
Air	- 23	9.49×10^{-6}	47.4×10^{-3}	3.3×10^{-3}	$\frac{3.46^*}{14.8}$
Glycerine	0	0.083×10^{-6}	3.54×10^{-4}	0.504×10^{-3}	$\frac{1.5\times10^{-3}}{6\times10^{-3}}$
Machine oil (fresh)	0	0.0428×10^{-6}	3.28×10^{-4}	0.7×10^{-3}	$\frac{0.5\times10^{-3}}{2\times10^{-3}}$

* The numerator pertains to a borehole 14 cm in diameter, and the denominator to a borehole 10 cm in diameter.

Remarkably enough, as the table shows, the filler which eliminates convection in a borehole most effectively is just air. In this case even for a borehole diameter of 15 cm a temperature increase of 3.5 °C/100 m with increasing distance from the glacier surface, that is, a gradient close to the geothermic step in ice, will not cause natural convection. A reduction of the borehole diameter sharply increases the viscous friction at the wall, preventing convection.

It should be noted that the temperature gradients in real glacier boreholes are almost always less than critical. The foregoing calculation suggests that observational data from open (unplugged) boreholes may be useful for evaluating the temperature field in the glacier. This is an important result, since measurements in open boreholes at present provide the main observational data on temperatures in glaciers.

Filling a borehole in ice with a liquid greatly increases the risk that convection will appear. With a liquid filler, convection will be absent in boreholes 5 cm in diameter only if the temperature gradient in the borehole does not exceed 0.1 °C/100 m, that is, if it is tens of times less than the geothermal gradient.

Finally, calculations for boreholes filled with water (the usual case for drilling in temperate mountain glaciers) indicate that natural convection in a 5 cm borehole will be absent only if the gradient is less than 1.5 °C/100 m.

1.2.3. Effects of Other Factors

Effect of forced convection. Moving a thermometer up and down in a borehole inevitably causes some of the air or liquid in the hole to move with the thermometer and become mixed. The time needed to establish a disturbed thermal equilibrium in the borehole, as well as the time needed to arrive at equilibrium between the thermometer and its surroundings, will depend on the kind of material filling the borehole. According to our data, for an ETP-310 thermometer in a steel protective case, this time for measurements in an unfilled (except with air) borehole is 3 to 3.5 hr. According to the data of Lyubimova [58], for water-filled boreholes this time is several minutes.

Effect of drive pipes. The presence of a drive pipe between the drill and the ice causes some return flow of heat along the pipe, which usually has a thermal conductivity higher than that of the ice. Studies have shown, however, that this effect cannot distort the temperature field appreciably, except at points close to the upper and lower ends of the pipe [58].

Effect of an air flow entering a borehole through snow or firn. Some interesting studies by Dubrovin and Petrov [22] have revealed disturbances of the temperature in boreholes drilled through snow or firn. These observations showed that the disturbances were caused by air flows entering the borehole from the surface via the porous snow or firn. Such disturbances are observed, however, only at distances of the order of ten meters below the surface. Because of the shallowness of this phenomenon and because it has been little studied, we have not considered it theoretically, and we only note here that temperature data from the upper part of a snow-firn mass may be distorted if the data are obtained from borehole measurements.

1.2.4. Temperature Measurements at the End (Bottom) of a Borehole

Earlier we mentioned that one method of temperature measurement in a borehole is to measure the temperature at the bottom of the hole. Let us assume that the bottom part of the borehole is filled with liquid (for deep drilling) or chips (for drilling with an auger). The temperatur of this liquid or chips is assumed to be as high as possible, close to the melting point of ice. In view of the short time of action of the heat source (drill) close to the bottom of the hole, the corresponding disturbance of the temperature of the surroundings is assumed to be negligible. Then, for a little while after the drilling stops, heat will flow out radially from the liquid or mud into the surrounding ice.

Cooper and Jones [89] stated the problem in a similar manner. Using their calculations, we constructed curves of the ratio of the liquid or chips temperature close to the borehole end after stabilization to the true temperature of the surrounding ice, as a function of the borehole stabilization time (Figure 1.2). The curves pertain to boreholes 12 and 24 cm in diameter, drilled through pure ice.

As the figure shows, stabilization (aging) of a borehole 12 cm in diameter for 5 hr reduces the difference between the temperature of the liquid (or mud) at the bottom of the hole and the true ice temperature to less than 2.5% (provided that when the drilling stops the temperature of the liquid or chips is around 0 °C). After 10 hr, the temperature difference is further reduced, to less than 1% of the true ice temperature.

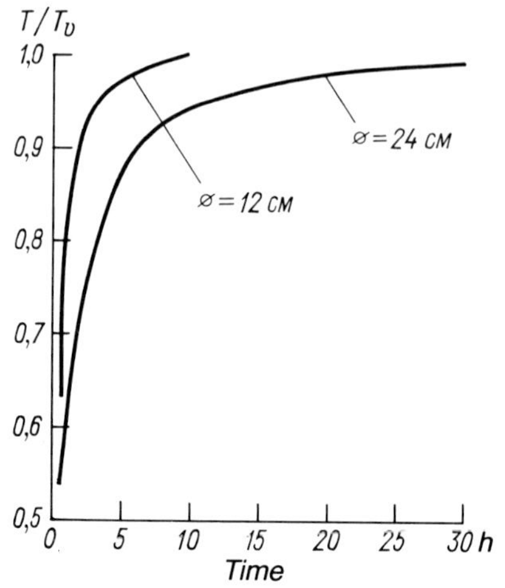

Fig. 1.2. Dimensionless temperature near end wall as function of borehole stabilization time, for ice drilling.

The figure also indicates that an increase in borehole diameter leads to a sharp rise in the time needed to stabilize the borehole.

These calculations were made for boreholes whose lower portions were filled with liquid or deposits of chips. If the borehole can be kept practically 'clean' right down to the bottom, so that a chips layer several borehole diameters in height does not form, then the stabilization time of the borehole before temperature measurements can be made at the bottom is reduced almost to zero.

Thus, these data show that both methods are applicable to shallow boreholes (depth of about 50 m) if the hole bottom is clean. With the use of the first method about 2 days must pass before measurements can be made.

Using the method of end measurements at various levels, the stabilization time at each level, 3 or 4 hr, is determined by the time needed for the thermometer to reach the ambient ice temperature. In this case the total time of a measurement in the borehole (at intervals of 5 m) is about 1.5 days, that is, comparable to the time required when measurements are made simultaneously in a hole drilled to the very end.

Since long breaks in drilling are inconvenient, for reasons related to organizing the work of the drilling crew, it is advisable to make measurements in shallow boreholes by drilling out the entire hole, and then inserting a string of sensors for simultaneous temperature measurements at several points in the borehole.

For deep boreholes, on the other hand, when the stabilization time of a completely drilled out hole is measured in months, it is apparently preferable to measure the temperature at the end of the hole at various levels. For boreholes filled with liquid, this method also makes it possible to eliminate errors related to free convection in the borehole. The instruments devised by Vostretsov [13] to measure the temperature at the end of a deep borehole in a glacier are useful for this.

Chapter 2

THERMAL DRILLING OF GLACIERS

Important information about the structure of the deep regions of a glacier, and about the processes taking place there, is obtained by means of direct measurements in boreholes and by studying cores. However, so far the difficulty of drilling sufficiently deep holes in ice has been a major obstacle to such studies.

Thermal drilling is a method of ice drilling which is being used more and more. The idea of drilling holes in glaciers by thermal means has attracted the attention of scientists for a long time, in view of the surprisingly low temperature of the ice making up a glacier. The first successful attempts at the thermal drilling of glaciers date back to the beginning of this century [11].

More recently, thermal drilling has been employed to make boreholes in mountain glaciers, as well as in the ice sheets in Greenland and Antarctica. But, despite its comparatively long history, the method of thermal drilling of glaciers is still in the experimental stage of trying out different ideas and types of drills. As yet no standard portable thermal drill is available for making comparatively shallow holes in glaciers (holes hundreds of meters deep), and there is no standard drill for reliable deep drilling in ice (to depths of thousands of meters).

One reason for the slow progress in devising equipment for the thermal drilling of glaciers is that an overly simplified view of thermal drilling leads to a simplified technical implementation of the method as well. Thus thermal drills are often manufactured in a slipshod manner, by nonspecialists, and they are put aside when the first trials show that the technical realization of a simple idea requires considerably more time and attention than at first thought. Another reason is that comparatively little attention has been devoted to a theoretical analysis of the operating conditions of a thermal drill in ice, especially with regard to such topics as the temperature disturbances in the ice caused by thermal drilling. Accordingly, some specialists were afraid that thermal drilling would disturb the temperature regime in a glacier close to the borehole too much, and this fear has tended to restrict the use of thermal drills.

In this chapter we will attempt to classify the main methods of thermal drilling of glaciers, to present some results obtained by thermal drilling, and also to analyze theoretically the main processes governing the regime of the thermal drill as it cuts through the ice and characterizing the conditions in the ice interior both ahead of the moving thermal drill and at the side walls of the borehole.

2.1. The Theory of Drilling by Contact with a Heated Solid Surface

2.1.1. The Drill Efficiency

Let us consider the uniform motion through solid ice of a thermal drill with a solid heater. We distinguish two zones in the part of the ice affected by the heater: zone I, located directly under the heater, having the form of a cylinder with a diameter equal to that of the heater, and zone II, at the side of the borehole (see Figure 2.1).

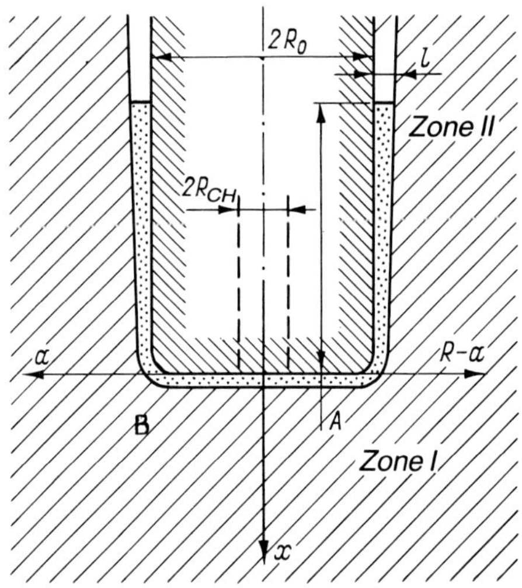

Fig. 2.1. Contact drilling with a solid heater.

Because of the weight of the heater, its heated lower surface transfers heat through the water space A to the ice surface B below.

This heat is expended: (1) to warm up the ice in zone I from its natural temperature to the melting point ($q_{\lambda x}$); (2) to melt the ice in zone I at a rate equal to the rate of the heater (q_{rx}); (3) to overheat the water in the space between the heater and the ice surface below ($q_{\delta x}$). Some of the heat ($q_{\delta R}$) goes to melt the side walls of the borehole, thereby making the borehole diameter greater than the diameter of the heater. Heat is also expended to warm up the side walls of the borehole ($q_{\lambda R}$). Amounts of heat $q_{\delta R}$ and $q_{\lambda R}$ are expended in zone II. If P_H is the heat output of the source, then for steady motion of the heater through the ice we have

$$P_H = q_{\lambda x} + q_{rx} + q_{\delta x} + q_{\lambda R} + q_{\delta R}, \tag{2.1}$$

where $q_{\lambda x} + q_{rx}$ is the useful heat expenditure, that is, the heat supply needed for drilling. The efficiency of the thermal drill η_d is defined as the ratio of the useful heat expenditure to the total amount of heat supplied by the heater:

$$\eta_d = \frac{q_{\lambda x} + q_{rx}}{P_H}. \tag{2.2}$$

2.1.2. The Ice Temperature Ahead of a Moving Thermal Drill

For a process which is steady in time, it is advisable to describe the motion of a thermal drill through ice with the aid of a coordinate system that moves through the ice with the drill. In order to analyze the heat transfer in zone I ahead of the moving drill, let us consider the one-dimensional problem of a melting front advancing in an unbounded medium at a constant velocity u_d.

The x axis, lying along the heater axis, points in the direction of motion of the front, and the origin moves with the melting front. Then, under steady-state conditions, the temperature field in the ice ahead of the melting front will be described by the equation

$$a \frac{\partial^2 T}{\partial x^2} + u_d \frac{\partial T}{\partial x} = 0 \tag{2.3}$$

for boundary conditions $x = 0$, $T = T_m$ and $x = \infty$, $T = T_i$. Here a is the thermal diffusivity of the ice, T is the temperature at a point located a distance x from the melting front, T_m is the melting point of the ice, and T_i is the temperature of the ice which is undisturbed by the melting front [74].

The solution of this equation gives unambiguously the temperature field in a glacier ahead of a melting surface, in the form

$$\frac{T - T_m}{T_i - T_m} = 1 - \exp\left(- \frac{u_d}{a} x\right). \tag{2.4}$$

Figure 2.2 shows the temperature distribution ahead of the melting front for various velocities of a thermal drill through ice with an initial temperature of -50 °C. The melting point of the ice is taken to be 0 °C. It is seen that for drill velocities of about 0.1 cm/s, a typical value for present-day thermal drills, the temperature at a distance of about 1 cm from the front is already essentially equal to the natural initial ice temperature. This implies that a thermal drill does not disturb the thermal state of the ice much, a fact which is important with regard to evaluating the use of such drills for glaciological studies.

Taking formula (2.4) into account, we can write the heat flow $q_{\lambda x}$ from the melting front to the ice as

$$q_{\lambda x} = -\pi R_0^2 \lambda \frac{d}{dx} \left\{ (T_i - T_m) \left[1 - \exp\left(- \frac{u_d}{a} x \right) \right] \right\} =$$

$$= u_d c \rho g (T_m - T_i) \pi R_0^2, \tag{2.5}$$

where πR_0^2 is the cross-sectional area of the borehole, and c, ρ are the specific heat and density of the substance being drilled.

Accordingly, the drilling efficiency is found by substituting (2.5) into (2.1), which gives

$$\eta_d = u_d \rho g[r + c(T_m - T_i)]\pi R_0^2/P_H.$$

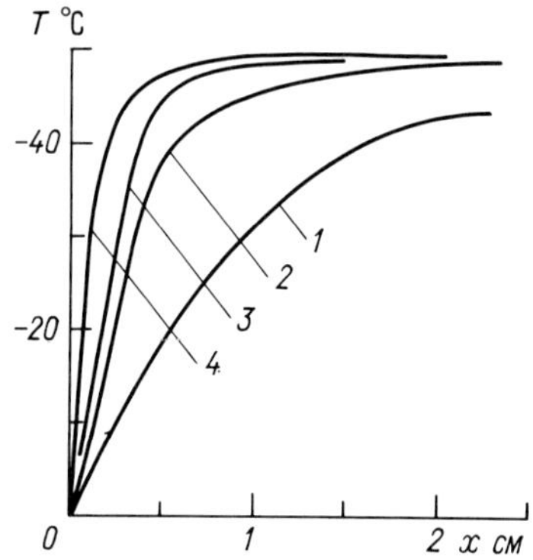

Fig. 2.2. Temperature distribution in ice interior ahead of thermal drill, for various cutting speeds.

(1) u_d = 0.01 cm/s; (2) u_d = 0.03 cm/s; (3) u_d = 0.06 cm/s; (4) u_d = 0.1 cm/s.

As was to be expected, the heat flow from the melting front to the ice via heat conduction, in the one-dimensional problem, goes entirely to raise the enthalpy of the ice at the melting front to the phase-transition temperature. For an ice-coring drill, which has a central core-sampling hole, expression (2.5) must be multiplied by $1 - R_{CH}^2/R_0^2$ (where $R_{CH}/R_0 = \bar{R}_{CH}$ is the dimensionless radius of the central hole).

2.1.3. The Ice Temperature at the Borehole Walls and Bottom

The ice temperature at the side of the borehole (in zone II of Figure 2.1) is higher than the temperature of the undisturbed ice. An amount of heat $q_{\lambda R}$ goes to heat the side walls of the borehole. To determine the temperature field in these walls and the amount of heat entering zone II at the sides from zone I, bounded by the surface of the borehole of radius $R = R_0$, let us consider the advance of a point source of heat located on the heater axis and moving downward at a constant velocity u_d.

It will be assumed that the strength of this source is sufficient to keep the temperature in plane a - a (see Figure 2.1), located a distance R_0 from the heater axis, equal to the melting temperature of the ice. The problem of the temperature distribution in the interior of a medium through which a point source moves with a constant velocity was solved by Rosenthal [119]. Taking into account his solution and the foregoing observation, we can represent the temperature field in a plane a - a perpendicular to the direction of motion of a thermal drill as

$$\frac{T - T_i}{T_m - T_i} = \exp\left[- \frac{u_d R_0}{2a} (\bar{R} - 1) \frac{1}{\bar{R}}\right], \qquad (2.6)$$

where $\bar{R} = R/R_0$ is the dimensionless distance from the borehole wall.

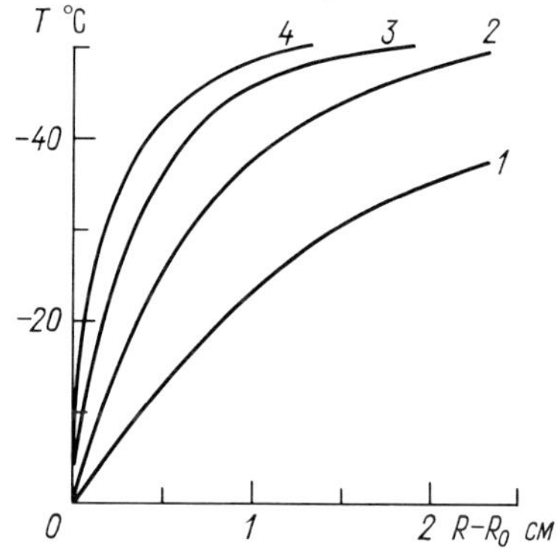

Fig. 2.3. Temperature distribution in ice interior at side walls of borehole, for various drilling rates.

(1) u_d = 0.01 cm/s; (2) u_d = 0.03 cm/s; (3) u_d = 0.06 cm/s; (4) u_d = 0.1 cm/s.

Figure 2.3 shows how the temperature varies as a function of the distance from the borehole wall, for various speeds of a thermal drill moving through the ice. The curves were plotted for ice with an initial temperature of -50 °C and a melting point of 0 °C. The borehole diameter 2 R_0 is 100 mm. The heating of the borehole wall caused by the drilling is seen to affect only a comparatively thin ice layer. The effect of varying the borehole diameter from 2.5 to 10 cm, moreover, is negligible; if the distance from the wall is 5 mm, it amounts to about 1 °C.

The amount of heat $q_{\lambda R}$ supplied to the ice in zone II can readily be found, since it goes wholly to vary the enthalpy of the ice, which passed through plane a - a with a velocity u_d, as its temperature varies from T_1 to T. This heat is

$$q_{\lambda R} = \int_{R_0}^{\infty} \rho g c u_d \, (T - T_i) \cdot 2\pi R \, dR.$$

After substituting the value of T into this expression and carrying out some transformations, we get

$$q_{\lambda R} = (T_m - T_i) \rho g c u_d \pi R_0^2 \frac{4a}{u_d R_0} . \tag{2.7}$$

If this expression is now compared with formula (2.5), for the amount of heat passing into zone I to melt the ice there, it is seen that formulas (2.7) and (2.5) differ only by the dimensionless coefficient $\bar{K}_R$:

$$\bar{K}_R = \frac{4a}{u_d R_0} \frac{1}{1 + \bar{R}_{CH}} . \tag{2.8}$$

This coefficient, which represents the amount of heat supplied to the ice in zone II at the sides from zone I, can be called the coefficient of heat dissipation in the walls. Figure 2.4 shows $\bar{K}_R$ as a function of the drill velocity and the drill diameter for $\bar{R}_{CH} = 0$ (drill without central core-sampling hole). This graph shows that for low velocities of the thermal drill and small heater diameters the heat supplied to the side walls of the borehole becomes quite considerable.

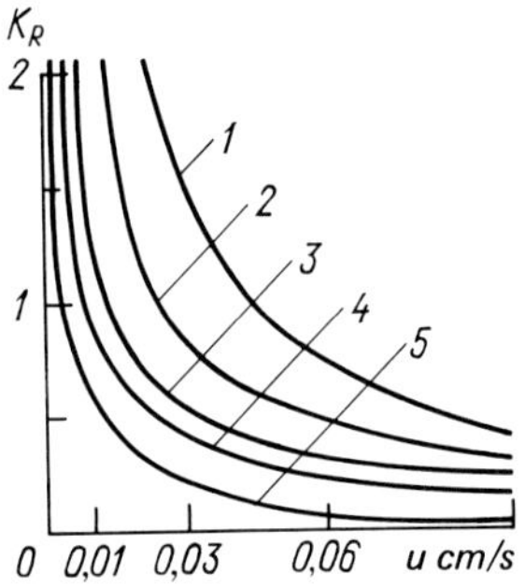

Fig. 2.4. Coefficient of sidewall dissipation of heat from borehole bottom, as function of drilling rate and radius R_0 of thermal drill. (1) $R_0 = 1$ cm; (2) $R_0 = 2$ cm; (3) $R_0 = 4$ cm; (4) $R_0 = 5$ cm; (5) $R_0 = 8$ cm.

2.1.4. The Temperature of the Drill Heating Surface

Although the ice at the bottom of a borehole is at the melting point, the temperature of the heater of the thermal drill and the water temperature in the space between the heater and the borehole end may differ appreciably from the melting temperature. It is necessary to know this temperature, in order to specify the safe operating regimes of the heater and also in order to determine the amount of heat expended in heating the water in the space to some temperature above the melting point T_m.

The temperature of the heater surface is found from the equation of heat balance at the melting surface of the ice under the heater:

$$-\lambda_w \left.\frac{\partial T}{\partial x}\right|_{x=H_w} = u_d \rho g[r + c(T_m - T_i)], \tag{2.9}$$

where λ_w is the thermal conductivity of water, $dt/dx|_{x=H_w}$ is the temperature gradient in the water at the ice surface (here and below we consider a heater with a flat forward heating surface perpendicular to the axis), r is the latent heat of fusion of the ice, and H_w is the thickness of the water layer between the drill and the ice.

If the temperature distribution over the thickness of the water space between the heater and the ice is linear, then condition (2.9) readily gives the temperature of the heater surface T_d:

$$T_d = \frac{u_d \rho g H_w}{\lambda_w}[r + c(T_m - T_i)] + T_m. \tag{2.10}$$

However, the water in the space has a vertical velocity, because of the outflow of water from the melting surface. Thus, the temperature distribution in the water layer cannot be assumed to be linear, even if the thermal conductivity is taken to be constant throughout the water space.

Let us consider the transfer of heat from the heater to the ice through the water space, for the indicated motion of the water in the space. The temperature of the forward surface of the heater is assumed to be the same everywhere, and the thickness of the water layer between the heater and the ice is taken to be constant. We consider an elementary column of water oriented along the heater axis. The coordinate origin is at the heater surface, the x axis pointing downward as before. In this coordinate system the velocity of the water in the x direction will vary from zero at the heater surface to a value equal to the velocity of water outflow w from the melting surface of the ice in the x direction; the latter is equal to the drilling rate u_d times the ratio of the densities of the water ρ_w and the ice ρ: $w_w = u_d\rho/\rho_w$.

In order to simplify the problem, we assume that the distribution of the water velocities in the space in the x direction is linear and satisfies only the following conditions at the boundaries: $x = 0$, $w_w = 0$; and $x = H_w$, $w_w = u_d\, \rho/\rho_w$. In this case the heat transfer in the x direction is given by the following equation:

$$\frac{d^2T}{dx^2} + \frac{\rho}{\rho_w}\frac{u_d x}{a_w H_w}\frac{dT}{dx} = 0, \tag{2.11}$$

with the following boundary conditions: $x = 0$, $T = T_d$; and $x = H_w$, $T = T_m$. Here a_w is the thermal diffusivity of water.

The solution of this equation, which gives the distribution of temperature gradients in the water layer, has the form [27]

$$\frac{dT}{dx} = \frac{T_d - T_m}{H_d}\frac{2}{\sqrt{\pi}}\sqrt{\frac{u_d H_w \rho}{2a_w \rho_w}}\;\frac{\exp - \left(\sqrt{\frac{u_d H_w \rho}{2a_w \rho_w}}\frac{x}{H_w}\right)^2}{\operatorname{erf}\sqrt{\frac{u_d H_w}{2a_w}\frac{\rho}{\rho_w}}}, \tag{2.12}$$

where

$$\operatorname{erf} z = \int_0^z \exp(-f^2)\, df$$

is the error integral [27].

If we take the gradient of the water temperature at the boundary with the melting ice $dt/dx\big|_{x = H_w}$ from expression (2.12), then after an appropriate transformation we get the following expression for the temperature of the heater surface, taking into account the motion of the water in the space:

$$T_w = \frac{u_d\, g[r + c(T_m - T_i)]}{\lambda_w} H_w \times$$

$$\times \left[\frac{\sqrt{\pi}\operatorname{erf}\sqrt{\frac{u_d H_w}{2a_w}\frac{\rho}{\rho_w}}}{2\sqrt{\frac{u_d H_w}{2a_w}\frac{\rho}{\rho_w}}\exp\frac{u_d H_w \rho}{2a_w \rho_w}}\right] + T_m. \tag{2.13}$$

Since the multiplier in square brackets on the right-hand side of (2.13) is greater than unity, it follows that the surface temperature of a heater moving with a speed u_d through ice at a temperature T_i will be higher than indicated by expression (2.10). Formula (2.10) does not take into account the effect of the water outflow from the surface of the melting ice, since it was derived assuming that the entire heat flux leaving the heater surface arrives at the surface of the melting ice.

It follows from expression (2.12) that the temperature gradient in the water layer at the heater surface (for $x = 0$) is larger than the temperature gradient at the ice surface ($x = H_w$). Consequently, the amount of heat leaving the heater surface will be greater than the amount of heat needed for the melting proper. Here some of the heat obviously goes to heat the water formed by the melting. Therefore, the heat transfer from the heater to the melting surface can be characterized by some efficiency value η_1 less than unity. This value will be equal to the ratio of the amount of heat arriving at the ice to the amount of heat leaving the heater surface of the drill:

$$\eta_1 = \frac{q_{rx} + q_{\lambda x}}{dT/dx\,|_{x=0}} = \exp\left(-\frac{u_d H_w}{2a_w}\frac{\rho}{\rho_w}\right). \tag{2.14}$$

Expressions (2.13) and (2.14) show that the temperature of the heater surface and the efficiency of the heat transfer through the water space both depend markedly on the thickness of this space.

Figure 2.5 shows the temperature of the heater surface and the efficiency of the drill heater η_1 as functions of the product $u_d H_w$. It is seen that for low values of product $u_d H_w$ the overheating of the water layer can be neglected, so that the simple formula (2.10) can be used to find the temperature of the heater surface.

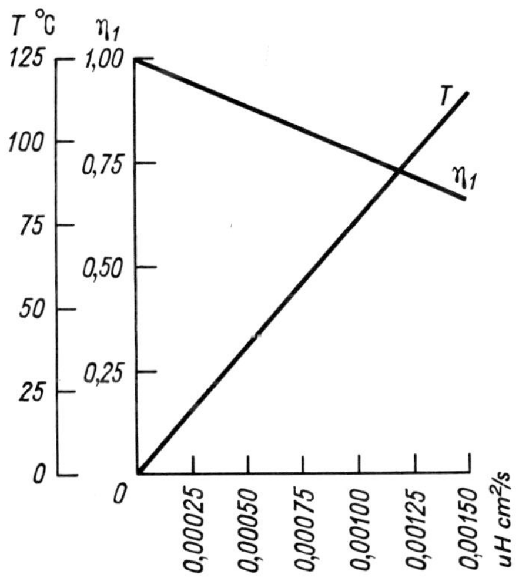

Fig. 2.5. Temperature of heater surface of thermal drill as function of thickness of water layer under it and drilling speed.

2.1.5. The Thickness of the Water Layer Between the Drill Heater and the Ice.

If as a first approximation we ignore the slight effect of friction in the water layer between the drill heater and the ice, then the thickness of the water layer can be found from the following condition: the gravity work of the heater, moving downward in the gravitational field, is equal to the kinetic energy of the water flowing out from under the heater. For simplicity, we assume that the thermal drill does not have a central hole for taking core samples or for sampling the meltwater. Then the meltwater formed in the borehole during drilling can be assumed to emerge from under the heater only along the borehole walls. For a drilling rate u_d and a drill mass G_d, this equality of the energies can be represented as

$$u_d G_d = \rho_w - \frac{W_{R_0}^2}{2} \cdot 2\pi R_0 H_w W_{R_0} \tag{2.15}$$

where H_w is the thickness of the water layer under the heater, and W_{R_0} is the mean (over the layer thickness) rate of radial outflow of the water at a distance R_0 from the center, that is, at the place where the water leaves the space between the drill heater and the melting ice and enters the region between the heater sides and the borehole walls (the space between the borehole walls and the side of the heater is assumed to be quite large). The rate W_{R_0} in expression (2.15) is found from the continuity equation, which in integral form can be written as

$$\pi u_d R_0^2 = W_{R_0} \cdot 2\pi R_0 H_w.$$

If this expression is now substituted into (2.15), then after some transformations we get the following formula for the thickness of the water layer between the drill heater and the ice:

$$H_w = \sqrt{\frac{\pi}{8} \frac{R_0^4}{G_d} u_d \rho_w} \cdot \tag{2.16}$$

For a more precise evaluation of the thickness of the water layer, the effect of the friction of the water at the walls of the ice and heater has to be taken into account.

The thickness of the water layer between the forward part of the heater and the ice, assuming steady motion of the heater, is defined as that thickness for which the pressure of the heater on the ice is exactly compensated by the pressure in the layer of meltwater squeezed out from under the heater. Let us assume, for generality, that a hole of radius R_{CH} is provided in the middle of the thermal drill, for taking core samples or for drawing off water.

Let $p(R)$ be the pressure in the water in the space over the ice, which is a function of the distance R from the heater axis, and let G be the pressure of the heater on the ice. Then, the condition stipulating invariance in time of the thickness of the water layer is

$$G = 2\pi \int_{R_{CH}}^{R_0} p(R)R\, dR. \qquad (2.17)$$

The pressure distribution in the water layer depends markedly on the layer thickness, being found in general form from the equations of motion, energy, and continuity in the water layer. This problem was considered in detail by Shreve [121], whose results are presented in the form of tables, obtained with a computer. It seems to us, however, that the problem should be reconsidered, so as to introduce certain limitations, or assumptions related to these, which would make it possible to derive working formulas relating the thickness of the water layer to the parameters of the thermal drill.

Following Shreve, we will neglect the variation of the pressure in the water layer in the x direction, in comparison with the radial pressure variations. Moreover, here, in contrast to the case considered by us previously, inertial terms will be neglected. Then, for the axisymmetric problem, the equation of motion in cylindrical coordinates has the form

$$H_w^2 \frac{\partial p(R)}{\partial R} = \frac{\partial}{\partial \bar{x}} \left[\mu(T) \frac{\partial v_w}{\partial \bar{x}}\right]. \qquad (2.18)$$

Here $\bar{x} = x/H_w$ is the dimensionless distance from the heater surface; $\mu(T)$ is the viscosity of the water, which is temperature dependent and thus varies over the thickness of the water layer; and v_w is the rate of radial outflow of the meltwater in the space.

The boundary conditions at the surfaces of the heater and ice are the adhesion conditions: $\bar{x} = 0$, $v_w = 0$; and $\bar{x} = 1$, $v_w = 0$.

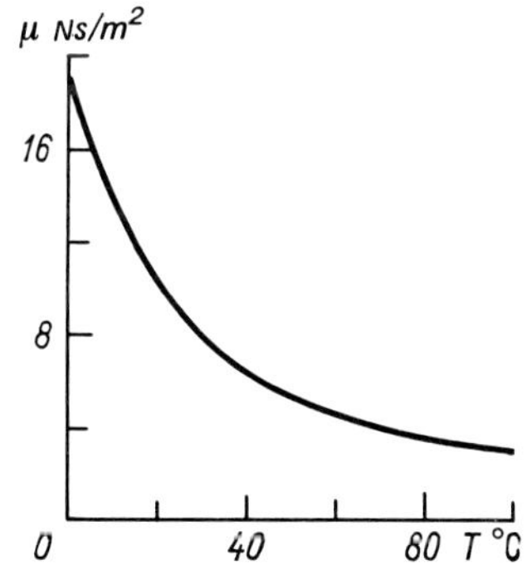

Fig. 2.6. Viscosity of water as function of temperature.

Figure 2.6 shows the dynamic viscosity of water as a function of temperature, according to the data of [121]. The viscosity is seen to vary considerably with temperature, so that Equation (2.18) has to be solved simultaneously with the heat-transfer **Equation** (2.11). This is what Shreve concluded, too. However, our evaluations have indicated that in the given problem the viscosity distribution over the thickness of the water layer can be assumed to be close to linear:

$$\mu(\bar{x}) = \mu_H + (\mu_0 - \mu_H)\bar{x}, \tag{2.19}$$

where μ_0 is the viscosity of water at its freezing point, and μ_H is the viscosity of water at the temperature of the heater surface. This assumption allows us to proceed without resorting to Equation (2.11).

We obtained the following solution of Equation (2.18), assuming the viscosity to vary with the coordinates only according to formula (2.19):

$$v_w = \frac{\partial p(R)}{\partial R} H_w^2 \times \frac{\bar{x} \ln \frac{\mu_0}{\mu_H} - \ln\left(1 + \frac{\mu_0 - \mu_H}{\mu_H}\bar{x}\right)}{\mu_0 - \mu_H \ln \frac{\mu_0}{\mu_H}} . \tag{2.20}$$

If the temperature differential in the layer is small, so that the effect of the temperature on the viscosity of the water is negligible ($\mu_H \to \mu_0$), then l'Hôpital's rule can be applied twice to expression (2.20) and it can be simplified to Poiseuille's formula

$$v_w = \frac{\partial p(R)}{\partial R} H_w^2 \frac{1}{2\mu_0} (\bar{x}^2 - \bar{x}) . \tag{2.21}$$

The continuity equation for the given approximation can conveniently be represented as

$$2\pi R H_w \int_0^1 v_w \, d\bar{x} = \pi(R^2 - R_m^2) u_d \frac{\rho}{\rho_w}. \tag{2.22}$$

Here, as in Shreve's paper, R_m is the radius at which the meltwater divides into a stream moving toward the outer edge of the drill head and a stream moving toward the central hole of the drill (if there is a central hole).

The pressure distribution in the R direction in the water layer can be found by simultaneously considering the equations of motion and continuity. The pressure in the water layer at the points of transition from the forward surface of the drill to its inner or outer sides is assumed to be atmospheric. If we arbitrarily take it to be zero, then

$$R = R_0, \; p(R_0); \quad R = R_{CH}, \; p(R_{CH}) = 0. \tag{2.23}$$

Now, by substituting (2.20) into (2.22) and integrating over x and R taking (2.23) into account, we arrive at an expression for the distribution of the excess (above atmospheric) pressure of the water in the layer between the melting ice and the flat forward surface of the heater:

$$p(R) = \frac{u_d \mu_0 \rho g \left(\dfrac{R_0^2 - R_{CH}^2}{2} - R_m^2 \ln \dfrac{R}{R_{CH}} \right)}{2H_w^3 \rho_w g \dfrac{\ln \dfrac{\mu_0}{\mu_H} - \dfrac{\mu_H}{\mu_0 - \mu_H} \left(1 + \dfrac{\mu_0 - \mu_H}{\mu_H}\right) \ln \left(1 + \dfrac{\mu_0 - \mu_H}{\mu_H}\right) + 1}{\left(1 - \dfrac{\mu_H}{\mu_0}\right) \ln \dfrac{\mu_0}{\mu_H}}},$$

where (2.24)

$$R_m = \sqrt{\frac{R_0 - R_{CH}}{2 \ln \dfrac{R_0}{R_{CH}}}} .$$

Next, by integrating formula (2.17), in which p(R) is given by expression (2.24), we get the following relation for the thickness of the meltwater layer, taking the variable viscosity into account:

$$H_w = \bar{K}_T \bar{K}_0 \sqrt[3]{\frac{u_d \mu_0 \rho R_0^4 \cdot 3\pi}{16 G \rho_w}} \tag{2.25}$$

where $\bar{K}_T$ is a dimensionless coefficient allowing for the effect of the heater temperature on the thickness of the water layer:

$$\bar{K}_T = \sqrt[3]{\frac{1 - \dfrac{\mu_H}{\mu_0} \ln \dfrac{\mu_0}{\mu_H}}{12\left[\ln \dfrac{\mu_0}{\mu_H} - \dfrac{\mu_H}{\mu_0 - \mu_H}\left(1 + \dfrac{\mu_0 - \mu_H}{\mu_H}\right) \ln\left(1 + \dfrac{\mu_0 - \mu_H}{\mu_H}\right) + 1\right]}}, \tag{2.26}$$

and $\bar{K}_0$ is a coefficient representing the effect of a central hole on the thickness of the water layer:

$$\bar{K}_0 = \sqrt[3]{\bar{R}_{CH}^4 - 1 + \frac{1 + \bar{R}_{CH} - 2R_{CH}^2}{\ln \dfrac{1}{R_{CH}}}} . \tag{2.27}$$

Here $\bar{R}_{CH} = R_{CH}/R_0$ is the dimensionless radius of the central hole in the end surface of the heater.

Figure 2.7 shows a plot of coefficient $\bar{K}_T$ as a function of heater temperature, obtained on the basis of expression (2.26) and the curve in Figure 2.6 for the water viscosity as a function of temperature.

Figure 2.8 shows coefficient $\bar{K}_0$, representing the effect of a central hole on the thickness of the water layer, as a function of the hole size. In practice, it is convenient to start calculating the thickness of the water layer under the heater by assuming that the temperature of the water in the space does not vary over the layer thickness and equals zero. Then $\bar{K}_T$ is unity. After this, the temperature of the heater surface is found using formula (2.13), and the thickness of the water layer is corrected by introducing coefficient $\bar{K}_T$ with the aid of the curve in Figure 2.7. Then formula (2.10) is used again to determine the temperature of the heater surface corresponding to the new value of the water-layer thickness. Estimates show that two repetitions of these operations are usually sufficient to arrive at water-layer thicknesses and heater-surface temperatures that are close to the true values.

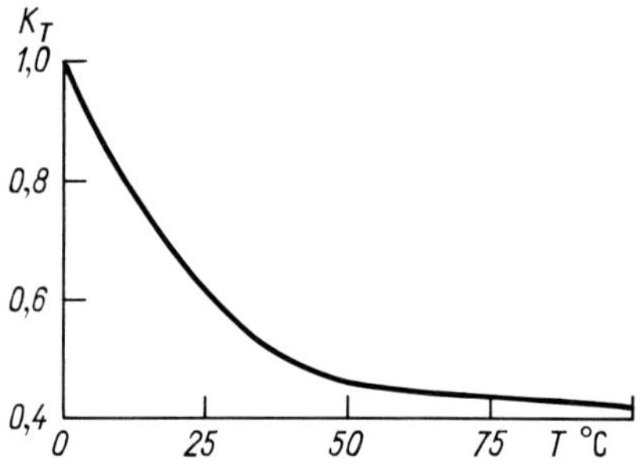

Fig. 2.7. Dimensionless coefficient $\bar{K}_T$ representing effect of variable viscosity of water in space between ice and heater, as function of temperature of drill heater.

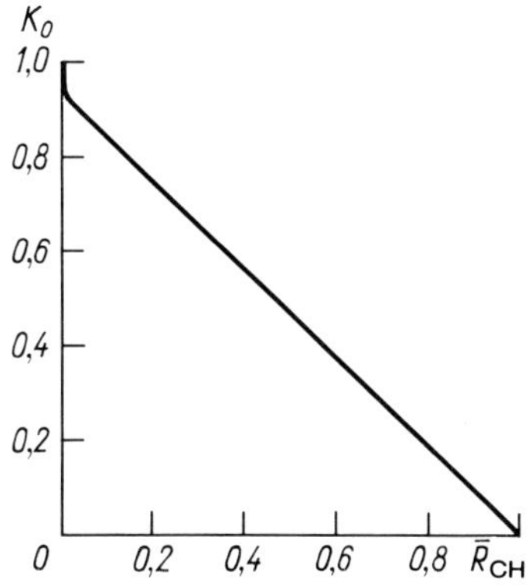

Fig. 2.8. Dimensionless coefficient $\bar{K}_0$ representing effect of central hole on thickness of meltwater layer under heater, as function of dimensionless hole radius $\bar{R}_{CH}$.

2.1.6. Melting or Freezing of the Borehole Side Walls

An appreciable part of the energy of the drill heater may be used to melt the side walls of the borehole. If the heater is submerged in water, then heat will be transferred to melt these walls by convection as well as by molecular thermal conduction. Since the sides of the heater are at a temperature close to the boiling point of water, liquid should move upward along the heater walls and downward along the cold walls of the borehole. In connection with this motion, additional heat exchange will take place due to natural free convection. The heat flux through the liquid layer between these walls can be calculated using methods of steady-state heat exchange through a medium in which there is no heat transfer via free convection. In this case, instead of the molecular thermal conductivity λ_w, we have to use an equivalent thermal conductivity λ_{we} allowing for the effect of free convection [78].

To determine the thickness of the ice layer l melted at the sides of the heater, at a distance h from its front end, let us return to Figure 2.1. The heat transfer in direction R from the side wall of the heater, which has a temperature T_s, to the borehole wall, which has a temperature T_m, is through a water layer with an effective thermal conductivity λ_{we}. Since the layer thickness l is small, the heat transfer in the x direction, along the side wall, is negligible in comparison with the heat transfer through this layer. The thinness of the layer compared to the drill diameter also allows us to disregard the curvature of the side walls of the drill and borehole. Then the increase in the thickness of the water layer Δl during the time $\Delta\tau_w$ required to melt the side wall of the borehole by the transfer of heat through a layer of thickness l is determined by the balance of the heat fluxes passing through the layer and melting the ice:

$$\lambda_{we} \frac{T_s - T_m}{l} \Delta\tau_w = \rho_w g[r + c(T_m - T_i)]\Delta l .$$

By decreasing the time $\Delta\tau_w$, passing to the limit, and integrating, we obtain the following relation between the thickness of the water layer and the time during which melting takes place:

$$\int_0^l \frac{1}{\lambda_{we}} dl = \frac{(T_s - T_m)\tau_w}{[r + c(T_m - T_i)]\rho_w g} . \tag{2.28}$$

Note that here the quantity λ_{we} is under the integral sign, since in general it depends on many parameters, including the layer thickness. The value of λ_{we} and its relation to the molecular thermal conductivity under different conditions were obtained by Kraussold [78] on the basis of a generalization of diverse experimental data on free convection in a liquid layer between flat or concentric vertical walls at a different temperature. This relation was represented as

$$\frac{\lambda_w}{\lambda_{we}} = F(Gr, Pr), \tag{2.29}$$

where Gr is the dimensionless Grashof number, giving the intensity of free convection, which in our case is

$$Gr = \frac{\beta_w l^3 (T_s - T_m)}{g\rho_w^2},$$

and Pr is the dimensionless Prandtl number: $Pr = c_w \mu/\lambda_w$. Here c_w is the specific heat, μ is the dynamic viscosity, λ_w is the molecular thermal conductivity, and β_w is the coefficient of thermal expansion of water.

Our analysis of Kraussold's relation (2.29) indicated that, if the temperature differential in the water-filled space between the heater and the borehole wall is about 50 °C, then free convection already has an appreciable effect for a layer thickness of about 0.5 mm. In this case, for practical purposes, for $T_d - T_m$ values from 50 to 100 °C, relation (2.29) can be approximated by a linear expression of the form

$$\frac{\lambda_{we}}{\lambda_w} = 1 + \alpha^* l. \tag{2.30}$$

For water, coefficient α^* is close to 6.5 cm^{-1}.

By substituting (2.30) into (2.28) and integrating, we now obtain the following expression relating the thickness l of the meltwater layer at the side wall of the borehole to the time τ_w during which the borehole wall is affected by the drill heater (which has a temperature T_d):

$$\tau_w = \left[\frac{1}{\alpha^*} - \frac{\ln(1 + \alpha^* l)}{\alpha^{*2}}\right] \frac{\rho g[r + c(T_m - T_i)]}{(T_i - T_m)\lambda_w}. \tag{2.31}$$

The time during which the drill heater at a temperature T_d affects the side wall of the borehole depends on the velocity of motion of the heater through the ice, that is, the rate of drilling of the borehole u_d, and also on the distance h of the point on the borehole from the end of the hole (see Figure 2.1): $\tau_w = h/u_d$. Substitution of this expression into (2.31) gives a formula indicating the nature of the increase in the thickness of the lateral space between the heater and the borehole:

$$h \frac{(T_i - T_m)\lambda_w}{u_d \rho g[r + c(T_m - T_i)]} = \frac{1}{\alpha^*} - \frac{\ln(1 + \alpha^* l)}{\alpha^{*2}}. \tag{2.32}$$

Let us call the second multiplier on the left-hand side of (2.32) the coefficient of intensity of melting of the borehole walls ξ. If the thermal conductivity of the liquid is constant, and independent of the layer thickness (free natural convection is absent), then formula (2.28) leads to the familiar formula of Stefan:

$$\tau_w = \frac{l^2}{2} \frac{\rho g[r + c(T_m - T_i)]}{(T_i - T_m)\lambda_w} \qquad (2.33)$$

or

$$h \frac{(T_i - T_m)\lambda_w}{u_d \rho g[r + c(T_m - T_i)]} = \frac{l^2}{2} . \qquad (2.34)$$

Figure 2.9 shows the size of the water space produced by melting of the borehole side walls as a function of the length h of the heated lateral surface of the drill. The curve was plotted on the basis of expression (2.32) for $\alpha^* = 6.5\ \text{cm}^{-1}$. For comparison, the curve calculated using Stefan's formula (2.34) is also given. The abscissa in Figure 2.9 is the true distance, that is, the length h of the lateral surface of the drill below the water level in the space, multiplied by the coefficient of melting intensity ξ. Such a scale gives a clear picture of the influence of various factors on the melting of the side walls of the borehole. Once the thickness of the ice layer melted from the side walls is known, we can readily find the heat flux $q_{\delta R}$ supplied by the heater to melt the side walls. As a first approximation, this flux is

$$q_{\delta R} = 2\pi R l \rho g[r + c(T_m - T_i)]. \qquad (2.35)$$

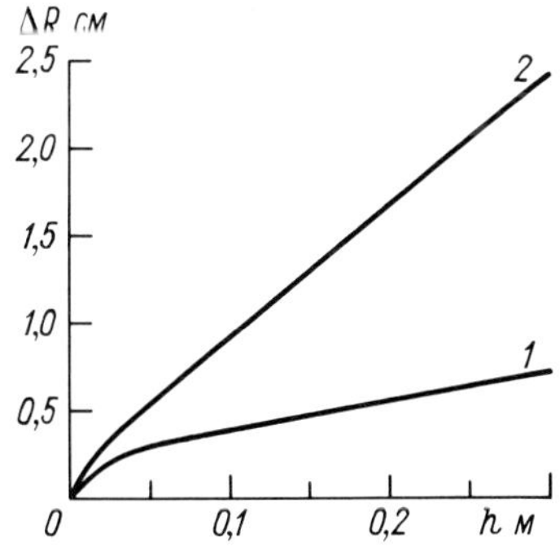

Fig. 2.9. Melting of borehole side walls as function of length of heated lateral surface of thermal drill.

(1) Stefan's formula; (2) formula (2.32).

2.2. Methods of Thermal Drilling of Ice

2.2.1. Drilling with a High-Speed Jet of Gas or Liquid Without Melting the Ice

If the temperature of a jet of gas or liquid differs from the ice temperature, then the rapidity of the motion of the jet along the ice surface will create high heat-transfer coefficients there. This in turn produces such appreciable temperature gradients at the end surface of the borehole that the ice may be broken by the resulting thermal stresses and removed from the borehole without reaching the melting point. The temperature of the jet can be either higher or lower than the ice temperature at the end of the borehole. This method of thermal drilling is now very widely used to drill holes in rocks [79, 80], but it has not been used to drill ice. The comparatively small difference under natural conditions between the ice temperature and the melting temperature is one of the obstacles to the application of this method for drilling ice, since the temperature range containing the gradient determining the breaking thermal stresses is limited.

2.2.2. Drilling with a Gas or Liquid Jet that Melts or Vaporizes the Ice

The temperature of the gas or liquid jet must exceed the melting point or evaporation temperature of the ice or water. This case is also characterized by very high coefficients of heat transfer to the ice surface, but the gradients created are not large enough to break the ice at the end of the borehole before it melts or vaporizes.

One of the methods of this type used to drill glaciers is "flame-jet" drilling. This is drilling with the aid of hot combustion products produced in a combustion chamber in the immediate vicinity of the end of the borehole. The first attempt to use this method known to us was that of S. V. Mikheev [60] in 1971 on the Central Tuyuksu Glacier. In a combustion chamber (1 in Figure 2.10) at the bottom of a borehole, gasoline was burned with compressed air, the fuel being supplied from the surface through hoses. Boreholes as deep as 53 m were drilled using this method. The second attempt to use this method was that of Browning in 1978 in the vicinity of Camp J-9 on the Ross Ice Shelf. This time the experiment was performed on a much larger scale [85]. A series of holes was drilled through a glacier 420 m thick. The borehole diameters ranged from 60 to 80 cm, the drilling time for each hole being about 8 hr. The working element of the drill in this case was a water-cooled combustion chamber, which burned kerosene in compressed air. The air and fuel were piped in from the surface of the glacier. A simple calculation shows that to obtain each of these boreholes about 900 kg of kerosene had to be burned in the chamber alone. During the air-kerosene flame-jet drilling of boreholes to such depths, however, it is very difficult to supply the combustion chamber with a large amount of air compressed to more than 4.9 MPa. Thus this method will probably not be applied to drill deeper holes, since this difficulty would then become even more serious.

This method also includes drilling with the aid of hot water or steam. Portable units for such drilling were used until 1978 just to drill shallow boreholes in mountain glaciers. In 1979 Browning used a hot-water drill to make three holes about 80 cm in diameter and about

416 m deep through the Ross Ice Shelf. He also demonstrated the feasibility of using a special annular setup to draw off the hot water at the end of the hole, so as to obtain cylindrical ice samples (cores) while drilling the ice with hot water.

To conclude our discussion of equipment for drilling ice with the aid of a jet of fluid, let us note that charges of solid rocket fuel could also be used for such drilling. With regard to this, the proposal and attempts of Tsyferov [72] to carry out effective drilling of sedimentary rock by this means are of particular interest. He used for such drilling standard aviation boosters, that is, solid-fuel rockets designed to be mounted on aircraft to assist their takeoff. For drilling, the nozzle part of the booster has to be replaced by a special adapter, which turns some of the ejected combustion products around, so that the rocket thrust is reduced to zero, or so that the rocket is even pushed against the end of the borehole. Figure 2.11 shows schematically a thermal drill utilizing a rocket converted in this manner.

As an alternative to aviation boosters, small inexpensive rockets used to protect against hail are also suitable for the thermal drilling of glaciers.

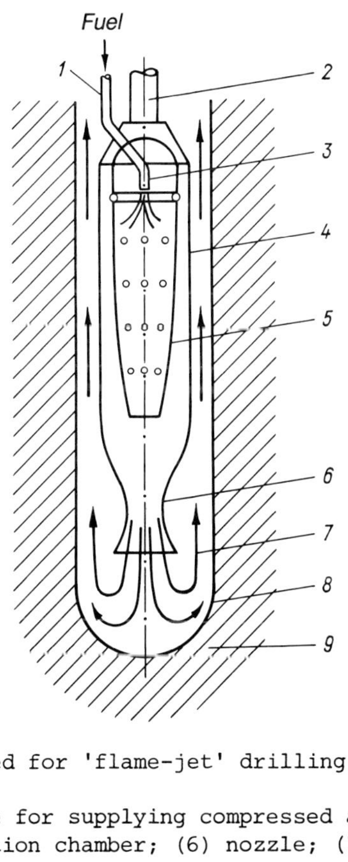

Fig. 2.10. Assembly used for 'flame-jet' drilling.

(1) fuel pipe; (2) pipe for supplying compressed air; (3) fuel injector; (4) casing; (5) combustion chamber; (6) nozzle; (7) direction of motion of combustion products in borehole; (8) borehole walls; (9) interior of glacier.

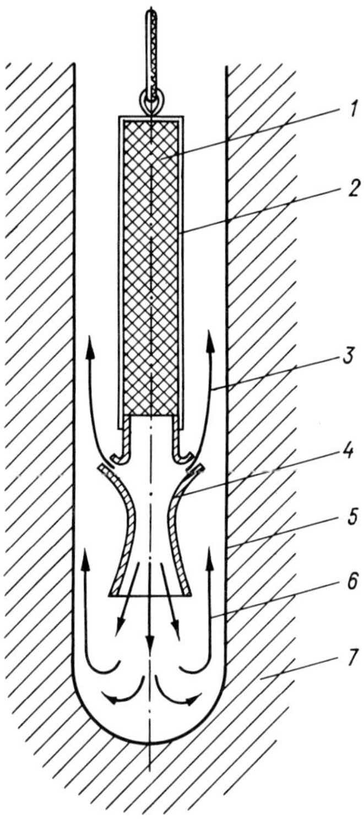

Fig. 2.11. Thermal drilling with solid-fuel rockets.

(1) charge of solid fuel; (2) walls of fuel compartment of rocket (drill body); (3) flows for reducing rocket thrust; (4) modified nozzle; (5) borehole walls; (6) direction of motion of combustion products in borehole; (7) interior of glacier.

Now let us evaluate the efficiency of the use of rockets to drill ice. Assuming the heating value of any fuel to be around 29.3 MJ/kg and the fuel density to be 2, and assuming that half of the combustion products participate in the heat exchange with the ice at the end of the borehole, and also that the borehole cross-sectional area is four times the cross-sectional area of the fuel in the fuel compartment of the rocket, it follows that each unit length of the part of the rocket charged with fuel can melt and remove to the surface twenty length units of the borehole. Thus, even a light antihail rocket with a fuel charge length of 0.5 m can, after its nozzle has been adapted as described above, drill a hole up to 10 m deep in ice during the calculated number of seconds of fuel combustion. Then, of course, the rocket can be changed and the drilling continued. In spite of the arbitrariness and approximate nature of this evaluation, it should still demonstrate the possibility of using a solid-fuel rocket to drill glacier ice.

2.2.3. Ice Drilling by Contact with a Heated Solid Surface

Heat is transferred from the heater surface to the melting ice surface at the end of the borehole by direct contact between the solid heater and the solid ice or else through a meltwater layer produced between the heater and the end of the hole. The temperature of the heater surface is kept high by a continual supply of energy to the heater. This may be electrical energy, chemical energy, or energy of radioactive decay. This is the method most often used for the thermal drilling of ice.

The heater head of the drill is usually lowered into the borehole on a rope or cable, a reserve supply of which is provided at the glacier surface. The borehole has to be preserved as drilled for the duration of the experiments carried out in it. The simplest way to carry out such drilling is through the layer of snow and firn on the glacier. The drill used is generally very simple, comprising a heavy body with electrical heating elements inside it. The energy needed to maintain the required temperature of the drill body is supplied via cable from the surface of the glacier. This method was used in 1959 by Ignatov [39] to drill a borehole at Vostok Station in Antarctica, by Kruchinin at Lazarev Station, by Kapitsa and Krasnushkin on Drygalski Island, and in 1960 by Dubrovin at Lazarev Station. This technique was also used successfully by the author in Antarctica in the vicinity of Mirny and Komsomol'skaya Stations.

Such thermal drills have been used quite successfully to penetrate the Antarctic Ice Sheet to depths of about 40 m. At these depths the glacier interior is still made up of snow and firn. All the meltwater produced by the drilling leaves the borehole from the bottom through pores in the glacier ice, leaving the borehole dry and its walls case-hardened. At great depths, however, the structure of the glacier is different, and this water cannot exit through pores, so that as the heater penetrates the glacier it becomes submerged in meltwater. The rate of drilling then drops sharply, and difficulties arise in connection with extracting the heater, since the water freezes and the top of the heater becomes frozen into the ice. Thus the use of the simplest thermal drills to make holes in cold ice is very restricted.

The need to obtain deeper boreholes in cold ice led to the development of more complicated thermal drills, including a system for pumping meltwater out of the borehole. Thermal drills of this type can also usually be adapted to take core samples. The first attempts to study and develop such drills, with provision for meltwater pumping and core sampling, were undertaken during the Soviet Antarctic Expedition in 1959 by the author and in 1960 by Barkov [7, 61]; these were continued in 1961 by Morev [62]. In 1966 Kudryavtsev [35] tested a drill with provision for water pumping and core sampling near Mirny Station. The maximum borehole depth attained was 77 m (the drill reached bedrock). In 1968 an improved version of such a drill was used by Morev 25 km from Mirny Station to make a borehole 212 m deep. In that same year Kudryashov, Fisenkov, and their colleagues designed a new thermal drill with provision for pumping water and taking core samples, and in 1969 Fisenkov managed to drill a borehole 250 m deep 50 km from Mirny Station. In 1970 Barkov started to drill a deep borehole with a tool of this type in central Antarctica near Vostok Station. By the end of 1970 a depth of more than 500 m had been reached, and by 1980 the hole was 1200 m deep.

A similar drill was constructed in America at the U.S. Army Cold Regions Laboratory, and it was used to make a test borehole 537 m deep in Greenland [35].

Another interesting tool is a thermal drill which descends in the ice only because of its own mass, without leaving a borehole behind it. In this case all the necessary cable is stored in the upper part of the drill itself, being paid out steadily as the drill descends. No borehole remains, since the meltwater freezes. This drill resembles a spider descending along its web. Naturally, such a drill assembly must have inside it all the instruments needed to make measurements in the ice interior during drilling and after the drill stops.

Such a self-contained drill was tested successfully in Greenland, where in 1968 a borehole 1000 m deep was made [82].

These experimental drillings of cold glacier ice showed that there are two possible means of thermal drilling in cold ice via contact with a hot solid drill:

(a) by constantly pumping all the meltwater out of the borehole, in which case the quantity H_w in expressions (2.10) and (2.13) is equal to zero;

(b) by so selecting the shape of the drill head and the distribution of heat sources in the drill that over the entire drill surface the water layer between the heater drill and the ice remains thin enough to preclude the appearance of intensive natural convection in this layer, which would lead to large heat losses in overheating the water.

A drill assembly employing the first method (pumping all the water out to leave a dry borehole bottom) was designed by Kudryashov and his colleagues [49], and it was used successfully to drill a hole more than 900 m deep through the Antarctic Ice Sheet near Vostok Station.

A disadvantage of this type of drilling is that it cannot be used to make very deep boreholes. Very deep boreholes in the ice cannot be left dry, since the pressure of the overlying ice layers would cause the borehole to 'pull in' or constrict, if there were no liquid to compensate for the pressure of the ice. For instance, it was impossible to make the above-mentioned borehole at Vostok Station much more than 900 m deep. The reason was that, while the apparatus was being raised to the surface and lowered again, the borehole diameter at the bottom so decreased that further drilling was prevented. The Antarctic Ice Sheet in the vicinity of Vostok Station remains very cold down to 900 m (surface temperature of -55 °C, rising to -40 °C at the bottom of the hole), so that the constriction of a dry borehole is appreciable only at great depths. In the case of a temperate glacier, on the other hand, this effect manifests itself at much shallower depths.

Thermal-drill assemblies employing the second method (water layer at bottom of borehole) were constructed in the USSR by Morev [62] and Sukhanov [68]. This method is especially recommended for drilling holes in mountain glaciers whose entire interior is at a temperature close to the melting point of ice ('temperate' glaciers). It was found that the drilling of such glaciers did not require the pumping of meltwater out of the borehole, since the water in the hole did not freeze for quite a while, and further drilling and other work in the borehole could be carried out unhindered.

The heater head of each drill assembly comprises an electrical heating element (Nichrome wire) enclosed in a hermetic case. The size and shape of the drill case are chosen so as to ensure an effective supply of heat from the heater to the drill surface and then to the ice. This makes the specific power output of the drill cross section quite high: about 2 MW/m^2. The efficiency of such a heater may be as high as 85%. A drill rod, that is, a thick-walled tube 1.5 m long, has to be provided to stabilize the vertical position of the drill during operation, and also to produce the required specific pressure at the working surface of the drill. A device consisting of flat springs is attached to the rod, in order to center the upper end of the drill. Energy is supplied to the heater via a cable, a reserve supply of which is provided at the glacier surface, cable being unreeled into the borehole as the drill advances into the ice. Power is supplied by light portable gasoline-electric units with outputs of 1 or 2 kW at 220 volts.

Using such drill assemblies, we and our teams have succeeded in making boreholes hundreds of meters deep in glaciers of the Caucasus, the Pamirs, and the Polar Urals, and on Spitsbergen [24, 62].

Experience in working with thermal needles indicates the following recommended maximum powers for a drill 4 cm in diameter [24]: 2000 W for ice; 1500 W for firn; 1000 W for snow. The corresponding drilling rates are as follows:

P W	Pure ice	Contaminated ice	Firn	Snow
2000	12-14 m/h	-	-	-
1500	9-10	4-5 m/h	10 m/h	-
1000	3-6	1-3	6-7	14-15 m/h

Similar equipment is also widely used to drill holes in temperate glaciers in other countries.

One of the drill assemblies designed by Sukhanov [60] utilizes electronic heating of the solid heater wall. This device, which is actually a kind of 'electron gun', makes it possible to increase considerably the density of the heat flux arriving at the heating surface of the drill, and also to vary slightly the distribution of the energy fluxes arriving at different parts of the heater. Consequently, drilling rates of 50 or 60 m/h can be attained.

Now let us return to expression (2.7), which involves the rate of advance of a thermal drill into snow, firn, or ice. It is clear from this formula that the rate of drilling is a function of the density of the medium. Thus, if the drilling power, borehole diameter, and heat dissipation at the sides do not vary, then the rate of thermal drilling can be used to get an idea of the ice density, to discover inhomogeneities of the ice layers, etc. In addition, rate variations may give us information about interfaces between snow and firn, firn and ice, etc. Sukhanov [68] was the first to point out and apply this feature of thermal drilling.

Figure 2.12 shows a typical record of the thermal drilling of a glacier (Dzhankuat Glacier in the Central Caucasus). The drilling time (which is inversely proportional to the drilling rate) is seen to vary appreciably even over short distances. This reflects the complicated structure of a glacier, something which can be revealed by thermal drilling [68].

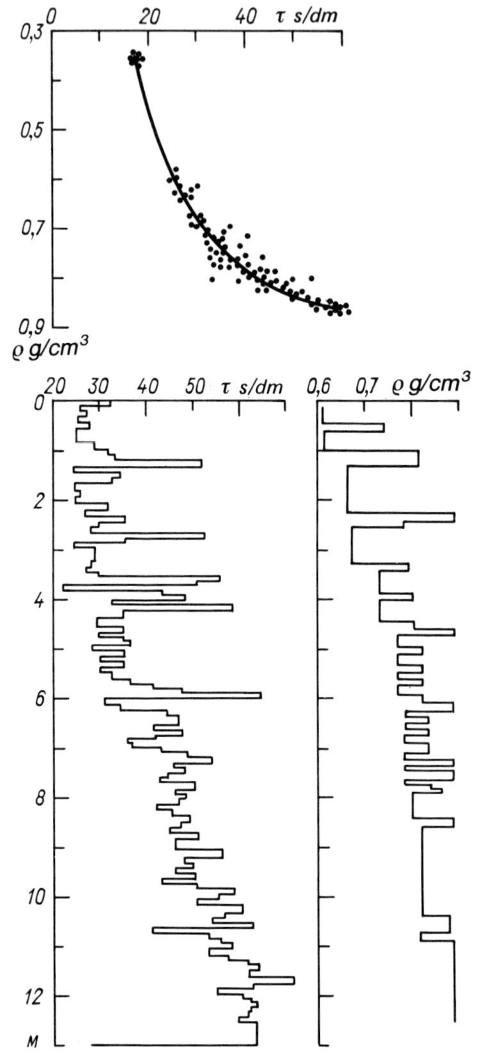

Fig. 2.12. Typical record of drilling time (τ) for short sections of glacier. Marked fluctuations of drilling rate are evident, revealing variations in density (ρ) of glacier interior [68].

2.3. Antifreeze Thermal Drilling by Direct Contact with a Solid Heater

There is a great difference between a hot point drilling through a temperate glacier, the meltwater being left in the borehole, and a complicated drill assembly which pumps the water out of the borehole into special containers that constitute part of the assembly. Obviously, in the latter case the drill assembly will have to be longer, at least by the length of the meltwater container plus the length of the section containing the pumping and pump-monitoring equipment. The drill assembly also has to be heavier, and the cable thicker, since the latter must contain additional wires to power and monitor the pumping, making a larger winch necessary as well; thus the entire drill complex will be less reliable and much more massive.

2.3.1. Drilling Cold Ice with Water Retention in the Borehole

It might be asked whether a thermal drill with an annular heater, designed to be used in cold ice, could not be simplified by completely eliminating the pumping of meltwater from the borehole. The author, together with V. A. Morev [134], hypothesized that this would indeed be possible if the conditions in the borehole were similar to those obtaining during the drilling of a temperate glacier, that is, if throughout the boring process the liquid in the borehole had a temperature close to that of the borehole walls but at the same time did not freeze. This would be the case if a special substance inhibiting freezing (antifreeze) was added to the meltwater in the borehole, in such quantity that the freezing point of the meltwater-antifreeze mixture would equal the temperature of the borehole walls when no heat flows through the wall from the borehole. However, then the temperature of the borehole wall would equal that of the undisturbed ice at the corresponding level, a temperature which is known to vary greatly over the thickness of a cold glacier. This means that the antifreeze concentration at each level would have to be so adjusted as to make the freezing point of the meltwater-antifreeze mixture equal to the ice temperature there.

But can this be done? Theoretically, yes, provided that the antifreeze is supplied to the bottom of the borehole and the temperature varies monotonically over the glacier thickness. For instance, if, as often happens, the glacier temperature increases with depth, then the antifreeze density has to be lower than the density of water. In this case near the glacier surface layers of mixture will have a maximum antifreeze content, and thus a minimum density. Near the bottom of the borehole, where the temperature is higher, the antifreeze concentration will be lower and the mixture density higher. The column of liquid in the borehole will then be resistant to natural convection and mixing. It is not clear, however, whether such a concentration distribution in the liquid column would persist if, after every 2 or 3 m of drilling, the assembly were lowered to the bottom and then raised up again with a core sample. Nor is it clear what kind of antifreeze should be used. On the one hand, it might be advisable to select an additive providing the Cridge-Hels maximum critical gradient (expression (1.4), Chapter 1). On the other hand, an antifreeze with a low viscosity might be preferable. This would reduce the danger of mixing due to the upward and downward movement of liquid in the borehole, in the wake eddies behind the moving assembly.

Tests and experiments carried out by Morev have verified the above hypothesis. It was also found that ethyl alcohol is the best antifreeze for such purposes.

2.3.2. Equipment for Antifreeze Thermal Drilling

The research of Morev in 1972 [62], described briefly above, led to the development of special equipment for the antifreeze thermal drilling of cold glaciers with core sampling. Using this drill assembly, ethyl alcohol is supplied to the end of the borehole each time the drill is lowered for the next drilling session. With the aid of such a rig, many boreholes were successfully drilled in mountain glaciers and in arctic and antarctic glaciers, for purposes of thermophysical and other studies.

In view of the simplicity of this equipment and the importance of the results obtained with it, let us consider it in more detail. Figure 2.13 shows a cross-sectional view of this drill assembly. Tube 3, the main part of the assembly, is made of stainless steel. The bit 1, which is attached to the bottom of the tube, is a ring made out of two thin-walled copper cylinders hermetically welded together at the top and bottom. A heating filament with a high electrical resistance is sandwiched between these cylinders. Good electrical insulation is provided between the heating filament and the copper cylinders, which serve as a shield for the drill bit. The bit is bolted to the tube in such a way that it can be replaced easily in the event of breakage.

The electrical circuit of the drill assembly is simple. One line is an insulated wire which becomes a strand of the sheathed cable, and the other line is the 'massive' part, that is, the cylinders of the drill bit, the drill casing 3, the steel braiding of the cable, and the winch housing.

Inside the drill casing is a tube 2 with its lower end open. This tube can be used either to hold an ice core or as a storage tank for the antifreeze (alcohol). It is possible to use one and the same space for two different purposes because a plunger 10 is so mounted inside the tube that it can be moved freely up and down along the entire length. When the drill is lowered into the borehole to start drilling, the plunger is in its lowest position, at the bottom of tube 2. The tube is at this time filled with an alcohol solution via inlet 4. To keep the solution from running out of the tube through the space between the tube walls and the plunger, the latter is provided with felt seals. After the drill assembly reaches the bottom of the hole, the bit is energized and the drill begins to advance more deeply into the ice.

As the bit moves downward due to melting of the ice beneath it, the central part of the ice inside the bit, which has not melted, enters inner tube 2, pushes against plunger 10, and begins to move the latter upward. As it ascends, the plunger forces the alcohol-water mixture out of the inner tube through opening 9 into the space between this tube and the casing, and then down to the bottom of the borehole via the lower part of the drill assembly. When the plunger reaches its highest point, the entire volume of tube 2 will be occupied by the core. Thus in this case the tube serves as a core holder. At this point the drilling stops, the drill assembly with the core inside it is raised to the surface, and more solution is pumped into tube 2 through inlet 4, until the tube is full and the plunger is again pushed down to its lowest position. By the time the core has been removed, and at the end of this stage of the drilling cycle the core is out of the drill assembly and tube 2 is refilled with the alcohol-water mixture. Now tube 2 is once again an antifreeze container, and the drill assembly is ready for lowering into the borehole, at the start of the new drilling cycle.

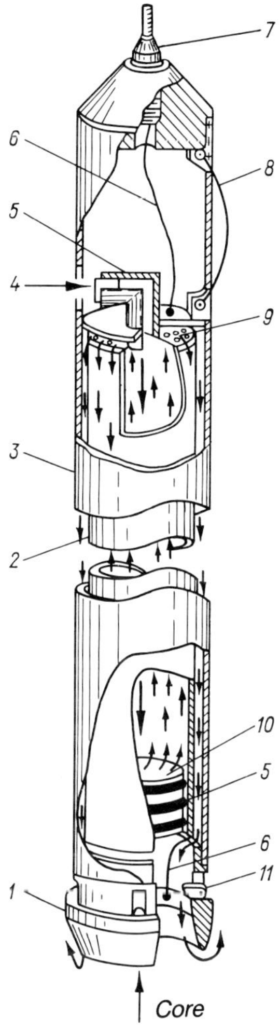

Fig. 2.13. Assembly for antifreeze thermal drilling.

(1) drill bit, electrically heated; (2) inner tube (holds ice core or alcohol solution); (3) casing; (4) inlet for filling tube 2 with alcohol solution; (5) seals of plunger; (6) voltage-supply wires of bit; (7) device for connecting cable; (8) springs for centering drill; (9) holes for outflow of solution from tube 2; (10) plunger; (11) teeth for extracting core (corer).

In order to detach the core inside the tube from the bottom of the borehole, and to keep it from falling as the drill assembly is raised, a rotating corer 11 is provided at the bottom. Laminated springs 8, mounted in the upper part of the assembly, serve to keep the drill centered in the borehole; linking device 7 connects the cable to the

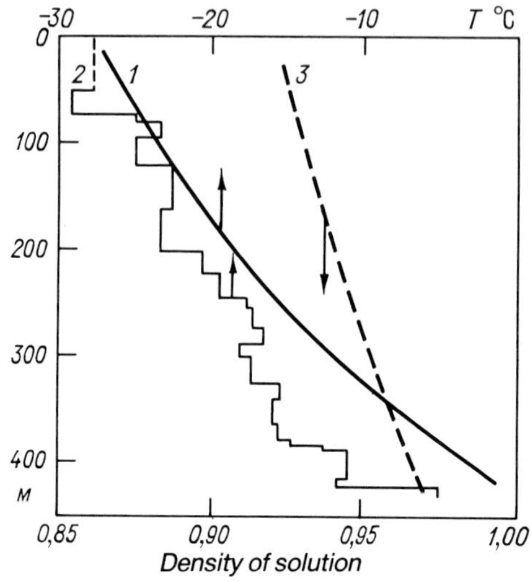

Fig. 2.14. Freezing point of alcohol solution over depth of borehole, during antifreeze thermal drilling of Ross Ice Shelf, compared with temperature in glacier interior.

(1) temperature of ice interior [88]; (2) calculated freezing point of alcohol solution in borehole at various levels; (3) mean variation of density of alcohol solution over depth of borehole.

drill column.

Because it is so simple, this drill assembly is very small. For instance, the equipment used to drill through the 416-m-deep Ross Ice Shelf in Antarctica had an outside casing diameter of 10.3 cm, a bit diameter of 108 mm, and an inside container-tube 8 cm in diameter and 2.6 m long; cores of this length (2.6 m) could thus be taken. The total length of the drill assembly was 3.16 m.

2.3.3. Thermal-Drilling Methods

In order to complete our description of the methods and procedure of thermal drilling with antifreeze, let us consider the aforementioned drilling operation through the Ross Ice Shelf, carried out in 1979 by Zagarodnov, Raikovskii, and the author [33, 134].

The drilling equipment, aside from the drill assembly and a single-strand sheathed cable 600 m long, included a mast and an electric winch. The total weight of the equipment was less than 600 kg (as compared with a weight of more than 10 tons for ordinary drilling with meltwater pumping).

Each drilling cycle started with a selection of the appropriate concentration of the alcohol-water mixture, so as to ensure that the right amount of antifreeze was added in the borehole. This choice depended on the temperature in the glacier interior at the given level. Figure 2.14 shows the temperature distribution over the depth of this glacier (curve 1). The calculated freezing temperatures of the alcohol mixture filling the borehole are also shown (curve 2). It is seen that the actual alcohol concentrations in the borehole were higher than the calculated values.

For the final choice of the alcohol concentrations, the behavior of the solution in the borehole had to be taken into account. What happens is — and this occurred in our case as well, for the theoretical alcohol concentrations in the borehole — ice crystals begin to form, and these sometimes increase the ice viscosity so much that it is difficult for the drill to descend in the borehole. Only by increasing the alcohol concentration can this undesirable effect be avoided.

To conclude this discussion, it should be mentioned that curve 3 gives the average density of the solution over the borehole depth.

2.3.4. Extracting a 'Bottom' Core

Recently, increasing interest has been shown in bringing to the glacier surface the undisturbed lowermost portion of a 'bottom' core, that is, a core whose lower end just reaches the lower boundary of the glacier. The shape and structure of this lower boundary may well provide information about whether melting or freezing occurs at the bottom of the glacier, and in addition they may even indicate the intensities of these processes, which would be very important for all thermophysical calculations.

The extraction of an undisturbed lower portion of a glacier is by no means easy, because the near approach of the drill assembly to the undersurface of the glacier places large stresses on the bridging section Δ (Figure 2.15) between the bottom of the borehole and the

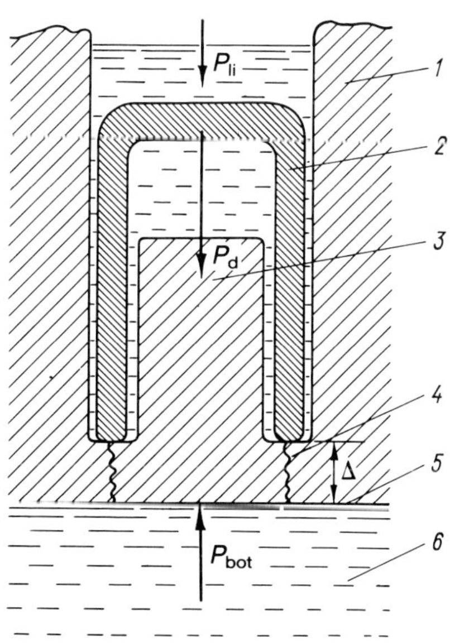

Fig. 2.15. Advance of thermal drill through lower boundary of glacier.

(1) glacier; (2) drill; (3) core; (4) bridging section; (5) undersurface; (6) water under glacier.

glacier undersurface. These stresses are caused by the downward pressure of the drill G and the force arising from the difference between the hydrostatic pressure G_1 of the liquid (if any is present) in the borehole and the hydrostatic pressure G_2 of the water under the glacier. The total force $G_\Sigma = G + G_1 + G_2$ may be directed either downward, if the sum of the hydrostatic pressure and the drill pressure exceeds the hydrostatic pressure under the glacier, or upward, as is almost always the case when drilling a 'dry' borehole or a borehole with a 'little' liquid in it, so as to obtain an undisturbed bottom core. If $G_\Sigma \neq 0$, then the bridging section will start to break not when the drill itself melts this bridge and the core is completely inside the core holder, but rather when the shearing stress τ_{sh} caused by the pressure sum G_Σ becomes sufficient to force the drill and core downward ($G_\Sigma > 0$) or else to break the bridge and push the drill and core upward ($G_\Sigma < 0$). This is evidently what happened when the drill reached the lower surface of the Antarctic Ice Sheet near Byrd Station [124]. The drill assembly was thrown several meters upward and broken. Beneath the glacier there was found to be fresh water at a pressure corresponding to $G_\Sigma < 0$. The bottom core and its lower surface were damaged.

If $G_\Sigma > 0$ and this pressure is low, then after the bridge is broken and the drill penetrates below the glacier boundary, some of the core will protrude from the drill assembly. This part of the core will have a length equal to that of the bridge at the moment when the drill broke through the lower surface of the glacier, at which time the tangential shearing stress on the bridge was equal to the ultimate shearing strength of the ice τ^*_{sh}. Since $\tau^*_{sh} = G^*_\Sigma/F_\Delta$, where G^*_Σ is the value of G_Σ for which the drill broke through the undersurface and F_Δ is the shear area, equal to $2\pi\Delta R_0$ (here $2R_0$ is the borehole diameter), therefore

$$\tau^*_{sh} = \frac{G^*_\Sigma}{2\pi\Delta R_0} \,. \tag{2.36}$$

The bridge of the core end protruding from the drill should not be too long. Obviously, it should not be longer than the core radius R_c (that is, $\Delta \leqq R_c$). Thus, if we wish to preserve the last piece of the core, the expression for the maximum, breaking, difference between the forces acting on the upper and lower sides of the bridge at the bottom of the glacier prior to the breaking of the bridge will be

$$G^*_\Sigma < 2\pi R_0 R_c \tau^*_{sh}. \tag{2.37}$$

For the drill considered above and for $\tau^*_{sh} \approx 49$ kN/m^2, we get $G^*_\Sigma = 343$ N. Then, passing from G_Σ to the pressure difference P^*_Σ

equivalent to this force and taking into account that $P^*_\Sigma = G_\Sigma/(\pi R_c^2)$, we obtain

$$P^*_\Sigma < \frac{2R_c}{R_0}\tau^*_{sh} \tag{2.38}$$

or, for the parameter values assumed above, $P^*_\Sigma = 98\ \mathrm{kN/m^2}$.

It is clear from the foregoing evaluation that a bottom core can be obtained from a thick glacier only with great difficulty, if the pressure of the liquid on both sides of the bridging section holding the last, bottom, core is tens or hundreds of times the maximum allowed

Fig. 2.16. Lower surface of Ross Ice Shelf and portion of bottom core, brought to glacier surface from a depth of 416 m by thermal drilling. The core is seen to protrude from the drill by an amount Δ, equal to the thickness of the bridge (Figure 2.15) which was broken through due to the difference in pressures on the bridge from above (from the bore-hole) and below (from under the glacier) (author's photo).

pressure P^*_Σ for successful core extraction.

To conclude this section, we note that, during the above-mentioned drilling of the Ross Ice Shelf, we succeeded in bringing to the surface an undisturbed bottom part of the core whose lower end reached the bottom of the ice shelf (Figure 2.16).

As this figure shows, the core protrudes from the drill, so that the pressure in the borehole exceeded the opposing force from below. Since the protrusion Δ of the core was 3 cm, and since the drill broke through the glacier at a moment when the hydrostatic pressures in the borehole and under the glacier were equalized due to hydraulic contact between the borehole and the sea under the glacier through the cracks, therefore in our case G^*_Σ equals the drill pressure on the borehole bottom in the water G, which is close to 245 N. The true value of the ultimate shearing strength of the ice can be estimated from the values of G^*_Σ and Δ. According to (2.36), it was about $0.3 N/cm^2$; this was the first such value determined in situ. The extraction of a bottom core from such depths was a piece of rare good fortune, and we do not know of another such instance. However, the evaluations carried out should help create conditions favorable for the extraction of such cores during future drilling.

Chapter 3

THE THEORY OF THE THERMAL REGIME OF GLACIERS

The fundamental processes taking place in a glacier can be described briefly as follows. In the nourishment region of a glacier, or in the central part of an ice sheet, snow will accumulate, gradually settle, become compacted, and turn into ice. Because different parts of the ice will be at different heights, ice located at higher elevations and possessing a higher energy of position will attempt to reach a more stable location, corresponding to the lowest possible energy of position, and it will flow outward. Thus, the velocity of ice motion will be higher, the greater the distance from the glacier margin or from the center of the ice cap. As it moves from a higher region to a lower, warmer region, the ice may become heated as it descends. Considerable heat will also be supplied by internal friction during the ice motion and from geothermal sources, if the ice is over land, or from the subglacial water, if the ice is floating.

Consequently, the heat regime of a glacier depends on many factors, all acting together. The motion of the glacier influences its thermal regime substantially. In turn, the temperature state of the glacier, which affects the rheological properties of the ice, determines to a large extent the regime of glacier motion. The problem is complicated, however, by the fact that these processes may not be steady-state, because of variations in climate or due to the growth or deterioration of the glacier.

Studies of glacier motion and the heat exchange in a glacier should be carried out assuming the glacier to be a continuum. Then the corresponding equations describing the heat and mass transfer of a continuous medium can be used, taking into account the specific features of glaciers. These equations are the equations of continuity, motion, energy, and state, to which must be added an experimental law relating the deformations and stresses in the ice, plus the appropriate uniqueness conditions (initial and boundary conditions).

3.1. The Fundamental Equations and Conditions Describing the State of a Glacier

The continuity equation is determined by the law of conservation of mass, and it embodies the following statement: the difference between the mass entering a certain volume and the mass leaving this volume during the same time interval is equal to the mass variation occurring per unit time due to the density variation. As applied to a moving continuous medium of variable density in an unsteady state, the continuity equation can be written in rectangular coordinates as:

$$\frac{\partial \rho}{\partial t} + \frac{\partial (\rho u)}{\partial x} + \frac{\partial (\rho v)}{\partial y} + \frac{\partial (\rho w)}{\partial z} = 0, \tag{3.1}$$

where u, v, w are the components of velocity U in directions x, y, z, ρ is the density, and t is time.

Since the density of the ice almost does not vary at all, let us assume the ice to be incompressible, so that the continuity equation simplifies to

$$\frac{\partial u}{\partial x} + \frac{\partial v}{\partial y} + \frac{\partial w}{\partial z} = 0. \tag{3.2}$$

Equation (3.1) and (3.2) can be expressed in vector form as

$$\frac{\partial \rho}{\partial t} + \mathrm{div}(\rho \vec{U}) = 0.$$

<u>Equations of motion</u>. In order to construct the equations of motion, let us start from the fundamental law of mechanics which states that mass times acceleration equals the sum of all the forces acting on the mass. Our discussion will follow that of [73].

Let us consider a volume element of size dx, dy, dz inside a glacier and the forces acting on this element. First we will find an expression for the surface forces acting on a parallelepiped of volume V = dx dy dz (see Figure 3.1).

The stresses $\vec{P}$ on the left-hand and right-hand faces, perpendicular to axis X and having an area dy dz, are, respectively,

$$\vec{P}_x \, dx \, dz; \; \vec{P}_x \, dx \, dz + \frac{\partial P_x}{\partial x} dx \, dy \, dz. \tag{3.3}$$

These stresses are vector quantities, and the subscript x indicates that these vectors act on areas perpendicular to the x axis. Similarly, we can obtain expressions for vectors acting on area elements perpendicular to axes Y and Z. The components of the resultant force for these axes will be: for the x direction $(\partial \vec{P}_x/\partial x)$ dx dy dz; for the y direction $(\partial \vec{P}_y/\partial y)$ dx dy dz; for the z direction $(\partial \vec{P}_z/\partial z)$ dx dy dz. Now, by adding these components and dividing by the volume of the parallelepiped, we obtain the surface force P per unit volume produced by the state of stress:

$$P = \frac{\partial \vec{P}_x}{\partial x} + \frac{\partial \vec{P}_y}{\partial y} + \frac{\partial \vec{P}_z}{\partial z} . \tag{3.4}$$

Vectors $\vec{P}_x$, $\vec{P}_y$, $\vec{P}_z$ can be resolved into components parallel to axes X, Y, Z (Figure 3.1). To do this, let us introduce the following notation: components perpendicular to the area elements, that is, normal stresses, will be designated by the symbol σ, with a subscript indicating the axis along which the stress acts. Tangential stresses directed along area elements will be designated by the symbol τ, with two subscripts, the first indicating to which axis the area element is perpendicular and the second indicating the axis to which the tan-

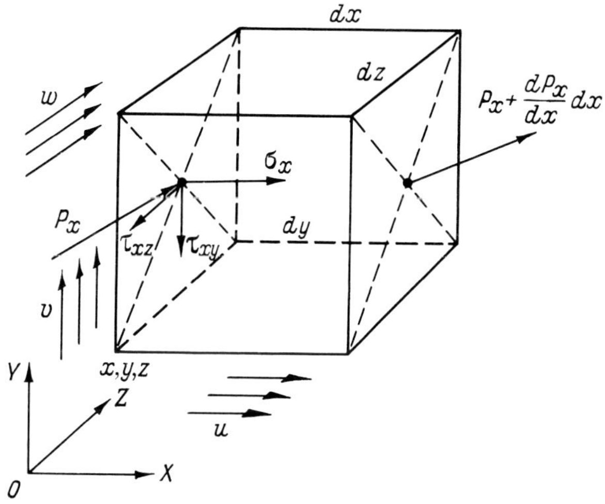

Fig. 3.1. Derivation of equations of motion.

gential stress is parallel. Thus we have

$$\begin{aligned}\vec{P}_x &= i\sigma_x + j\tau_{xy} + k\tau_{xz};\\ \vec{P}_y &= i\tau_{yz} + j\sigma_y + k\sigma_{yz};\\ \vec{P}_z &= i\tau_{zx} + j\tau_{zy} + k\sigma_z.\end{aligned} \tag{3.5}$$

where i, j, k are the unit vectors (vectors with a length equal to unity) along axes X, Y, Z.

Consequently, the stressed state is described by nine scalar quantities, the aggregate of which forms the stress tensor.

Note that, because of the equality of moments relative to an arbitrary axis, tangential stresses with the same, but opposite in order, subscripts will be equal. For example, since the moments relative to axis Z must be equal, we have τ_{xy} dy dz dx = τ_{yx} dx dz dy, so that $\tau_{xy} = \tau_{yz}$; similarly, $\tau_{xz} = \tau_{zx}$ and $\tau_{yz} = \tau_{zy}$.

Now, after substituting quantities (3.5) into (3.4) and taking into account the indicated pairing of tangential stresses, we get the following expression for the surface force:

$$\vec{P}_x = i\left(\frac{\partial\sigma_x}{\partial x} + \frac{\partial\tau_{xy}}{\partial y} + \frac{\partial\tau_{xz}}{\partial z}\right) + j\left(\frac{\partial\tau_{xy}}{\partial x} + \frac{\partial\sigma_y}{\partial y} + \frac{\partial\tau_{yz}}{\partial z}\right) +$$

$$+ k\left(\frac{\partial\tau_{xz}}{\partial x} + \frac{\partial\tau_{yz}}{\partial y} + \frac{\partial\sigma_z}{\partial z}\right). \tag{3.6}$$

Next let us introduce the vector $\vec{K} = \rho g$, the mass force per unit volume in the form $\vec{K} = iK_x + jK_y + kK_z$, where K_x, K_y, K_z are the components of the mass force along axes X, Y, Z, and g is the free-fall acceleration.

Then the equation of motion in vector form can be written as

$$\rho\frac{D\vec{U}}{Dt} = \vec{K} + \vec{P}, \tag{3.7}$$

where $D\vec{U}/Dt$ is the substantial acceleration, for instance,

$$\frac{D\vec{U}}{Dt} = \frac{\partial\vec{U}}{\partial t} + u\frac{\partial u}{\partial x} + v\frac{\partial u}{\partial y} + w\frac{\partial u}{\partial z}.$$

By substituting (3.6) into (3.7), expanding the substantial derivative, and projecting onto the coordinate axes, we obtain the equations of motion along axes X, Y, Z in the form

$$\begin{aligned}
\rho\left(\frac{\partial u}{\partial t} + u\frac{\partial u}{\partial x} + v\frac{\partial u}{\partial y} + w\frac{\partial u}{\partial z}\right) &= K_x + \frac{\partial\sigma_x}{\partial x} + \frac{\partial\tau_{xy}}{\partial y} + \frac{\partial\tau_{xz}}{\partial z};\\
\rho\left(\frac{\partial v}{\partial t} + u\frac{\partial v}{\partial x} + v\frac{\partial v}{\partial y} + w\frac{\partial v}{\partial z}\right) &= K_y + \frac{\partial\tau_{xy}}{\partial x} + \frac{\partial\sigma_y}{\partial y} + \frac{\partial\tau_{yz}}{\partial z};\\
\rho\left(\frac{\partial w}{\partial t} + u\frac{\partial w}{\partial x} + v\frac{\partial w}{\partial y} + w\frac{\partial w}{\partial z}\right) &= K_z + \frac{\partial\tau_{xz}}{\partial x} + \frac{\partial\tau_{yz}}{\partial y} + \frac{\partial\sigma_z}{\partial z};
\end{aligned} \tag{3.8}$$

This is a system of three equations, containing, in addition to the velocities and their derivatives, the six components σ_x, σ_y, σ_z, τ_{xy}, τ_{xz}, τ_{yz} of the stress tensor.

For a process which is steady-state in time, the equations of motion can be written in tensor form more briefly as

$$\rho\frac{\partial v_i}{\partial t} + v_i\frac{\partial v_i}{\partial x_i} = \frac{\partial\sigma_{ij}}{\partial x_i} + \rho F_i, \quad i = 1, 2, 3; \tag{3.9}$$

where v_i is the velocity component in the x_i direction, F_i is the component of the mass forces in the i direction, and σ_{ij} is the component of the stress tensor.

The energy equation is derived on the basis of the law of conservation and transformation of energy. For heat processes this law takes the form of the first law of thermodynamics, which states that the amount of energy dQ supplied to a volume dV goes to increase the internal energy of this volume by an amount dQ_u and to perform an amount of mechanical work dM by the system:

$$dQ = dQ_u + dM. \tag{3.10}$$

From the outset it should be noted that, when considering the components of the energy equation of a glacier, the ice should be assumed to be incompressible.

The amount of heat dQ supplied to a volume dV is made up of three components: the heat dQ_T supplied by thermal conduction; the heat dQ_z appearing because of internal friction; and the heat dQ_q related to internal sources or sinks of heat, created by phase transitions, penetrating radiation, etc.,

$$dQ = dQ_T + dQ_z + dQ_q. \tag{3.11}$$

The heat due to thermal conduction is given by Fourier's law. If T is the temperature and λ is the thermal conductivity of the ice, then through the left-hand face of a volume element dV = dxdydz (Figure 3.2) perpendicular to the x direction the following amount of heat per unit time will reach dV:

$$-\lambda \frac{\partial T}{\partial x} dy\, dz.$$

Correspondingly, through the right-hand face, also perpendicular to the x direction, the following amount of heat will leave:

$$-\left[\lambda \frac{\partial T}{\partial x} + \frac{\partial}{\partial x}\left(\lambda \frac{\partial T}{\partial x}\right) dx\right] dy\, dz.$$

Consequently, in time dt the following amount of heat will arrive in volume dV in the x direction:

$$dQ_{Tx} = dt\, dx\, dy\, dz \frac{\partial}{\partial x}\left(\lambda \frac{\partial T}{\partial x}\right).$$

The amount of heat supplied to element dV in time dt through all faces is, accordingly,

$$dQ_T = dt\, dV\left[\frac{\partial}{\partial x}\left(\lambda \frac{\partial T}{\partial x}\right) + \frac{\partial}{\partial y}\left(\lambda \frac{\partial T}{\partial y}\right) + \frac{\partial}{\partial z}\left(\lambda \frac{\partial T}{\partial x}\right)\right]. \tag{3.12}$$

The heat of internal friction is defined as the difference between all the work due to the normal and tangential components of the friction forces and the portion of the work converted into mechanical energy (pressure energy and kinetic energy).

The stress σ_x applied to volume element dV performs the following work:

$$dy\, dz\left[-\sigma_x u + \left(\sigma_x + \frac{\partial \sigma_x}{\partial x} dx\right)\left(u + \frac{\partial u}{\partial x} dx\right)\right] = dV \frac{\partial(\sigma_x u)}{\partial x}.$$

Consequently, the total work done per unit time by the normal and tangential force components at all faces of volume dV will be

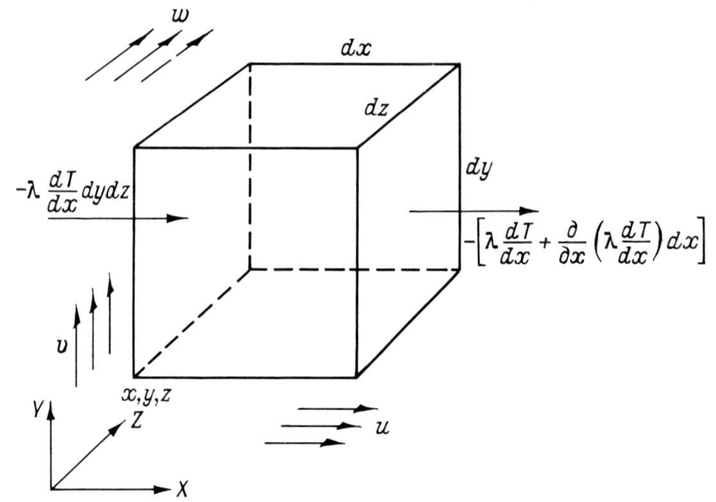

Fig. 3.2. Derivation of energy equation.

$$dV\left[\frac{\partial}{\partial x}(\sigma_x u + \tau_{xy} v + \tau_{xz} w) + \frac{\partial}{\partial y}(\tau_{yx} u + \sigma_y v + \tau_{yz} w) + \right.$$

$$\left. + \frac{\partial}{\partial z}(\tau_{zx} u + \tau_{zy} v + \sigma_z w)\right]. \qquad (3.13)$$

The part of this work converted into mechanical energy per unit time is, accordingly,

$$u\left(\frac{\partial \sigma_x}{\partial x} + \frac{\partial \tau_{yx}}{\partial y} + \frac{\partial \tau_{zx}}{\partial z}\right) + v\left(\frac{\partial \tau_{xy}}{\partial x} + \frac{\partial \sigma_y}{\partial y} + \frac{\partial \tau_{xy}}{\partial z}\right) +$$

$$+ w\left(\frac{\partial \tau_{xz}}{\partial x} + \frac{\partial \tau_{yz}}{\partial y} + \frac{\partial \sigma_z}{\partial z}\right). \qquad (3.14)$$

Now, by subtracting from the total work (3.13) the part of the work (3.14) converted into mechanical energy, we get the heat supply to element dV during time dt because of the internal friction. This amount of heat is given in mechanical units by the expression

$$dQ_z = dt\; dV\left[\left(\sigma_x \frac{\partial u}{\partial x} + \tau_{xy} \frac{\partial v}{\partial x} + \tau_{xz} \frac{\partial w}{\partial x}\right) + \left(\tau_{yx} \frac{\partial u}{\partial y} + \right.\right.$$

$$\left.\left. + \sigma_y \frac{\partial v}{\partial y} + \tau_{yz} \frac{\partial w}{\partial y}\right) + \left(\tau_{zx} \frac{\partial u}{\partial z} + \tau_{zy} \frac{\partial v}{\partial z} + \sigma_z \frac{\partial w}{\partial z}\right)\right]. \qquad (3.15)$$

The heat supplied by phase transitions and penetrating radiation is

$$dQ_q = dt\ dV\ r\kappa + dt\ dV\ R_Q, \tag{3.16}$$

where κ is the relative specific rate of the phase transitions, R_Q is the volume power of the absorbed radiant energy, and r is the latent heat of the phase transitions.

After introducing quantities (3.12), (3.15), and (3.16) into Equation (3.11) and dividing both sides of the equation by dt dV, we get

$$\frac{1}{dV}\frac{dQ}{dt} = \left[\frac{\partial}{\partial x}\left(\lambda \frac{\partial T}{\partial x}\right) + \frac{\partial}{\partial y}\left(\lambda \frac{\partial T}{\partial y}\right) + \frac{\partial}{\partial z}\left(\lambda \frac{\partial T}{\partial z}\right)\right] +$$
$$+ N\left[\left(\sigma_x \frac{\partial u}{\partial x} + \tau_{xy}\frac{\partial v}{\partial x} + \tau_{xz}\frac{\partial w}{\partial x}\right) + \left(\tau_{yx}\frac{\partial u}{\partial y} + \right.\right.$$
$$\left.\left. + \sigma_y \frac{\partial v}{\partial y} + \tau_{yz}\frac{\partial w}{\partial y}\right) + \left(\tau_{xz}\frac{\partial u}{\partial z} + \tau_{zy}\frac{\partial v}{\partial z} + \sigma_z \frac{\partial w}{\partial z}\right)\right] +$$
$$+ r\kappa + R_Q, \tag{3.17}$$

where N is the thermal equivalent of work.

Let us consider the components on the right-hand side of Equation (3.17) which are defined by expression (3.10). If c is the specific heat of ice, then the increase in the internal energy of volume dV for incompressible ice will be $dQ_u = \rho g\ dV\ (cT)$. Thus, after dividing both sides of the equation by dV dt, we have

$$\frac{1}{dV}\frac{dQ_u}{dt} = \rho g \frac{d(cT)}{dt}. \tag{3.18}$$

Note that the derivative on the right-hand side of this equation is the substantial derivative, since during time dt the element moved in space. Therefore

$$\frac{1}{dV}\frac{dQ_u}{dt} = \rho g\left[u\frac{\partial(cT)}{\partial x} + v\frac{\partial(cT)}{\partial y} + w\frac{\partial(cT)}{\partial z} + \frac{\partial(cT)}{\partial t}\right].$$

With regard to the mechanical work dM done by the system, note that it is proportional to the square of the velocity of the medium, and for the velocities observed in glaciers this work will be negligible in comparison with the variation in internal energy.

Then the energy equation can be written in the form

$$\rho g\left[\frac{\partial(cT)}{\partial t} + u\frac{\partial(cT)}{\partial x} + v\frac{\partial(cT)}{\partial y} + w\frac{\partial(cT)}{\partial z}\right] =$$
$$= \left[\frac{\partial}{\partial x}\left(\lambda \frac{\partial T}{\partial x}\right) + \frac{\partial}{\partial y}\left(\lambda \frac{\partial T}{\partial y}\right) + \frac{\partial}{\partial z}\left(\lambda \frac{\partial T}{\partial x}\right)\right] +$$

$$+ N\left[\left(\sigma_x \frac{\partial u}{\partial x} + \tau_{xy} \frac{\partial v}{\partial x} + \tau_{xz} \frac{\partial w}{\partial x}\right) + \left(\tau_{yx} \frac{\partial u}{\partial y} + \right.\right.$$

$$+ \sigma_y \frac{\partial v}{\partial y} + \tau_{yz} \frac{\partial w}{\partial y}\Bigg) + \Bigg(\tau_{zx} \frac{\partial u}{\partial z} + \tau_{zy} \frac{\partial v}{\partial z} +$$

$$\left.\left. + \sigma_z \frac{\partial w}{\partial z}\right)\right] + r\kappa + R_Q. \tag{3.19}$$

In concise vector form this equation becomes

$$\rho g \frac{\partial (cT)}{\partial t} + \sum_{i=1}^{3} u_i \frac{\partial (cT)}{\partial x_i} = \sum_{i=1}^{3} \frac{\partial}{\partial x_i}\left(\lambda \frac{\partial T}{\partial x_i}\right) +$$

$$+ N\sigma_{ij}\dot{\varepsilon}_{ij} + r\kappa + R_Q, \tag{3.20}$$

or

$$\rho g \frac{\partial (cT)}{\partial t} = \vec{U}\rho g \text{ grad } (cT) + \text{div } \lambda \text{ (grad } T) + q, \tag{3.21}$$

where q is the volume power of the heat sources

$$q = N\sigma_{ij}\dot{\varepsilon} + r\kappa + R_Q. \tag{3.22}$$

Relation between deformations and stresses in ice. Ice possesses properties intermediate between those of an ideally viscous body and those of an ideally plastic body. The law relating the deformations and stresses in ice (The stress-strain law) was first established experimentally by Glen [99], on the basis of studies of the compression of ice cubes. The compressive deformations ε were found to be related as follows to the stress σ:

$$\dot{\varepsilon} = \frac{d\varepsilon}{dt} = B'\sigma^n, \tag{3.23}$$

where B' and n are coefficients which are obtained experimentally, and ε is the relative contraction (elongation). For example, the relative contraction ε_x is $\varepsilon_x = d(\Delta x)/\Delta x$, where Δx is the initial length and $d(\Delta x)$ is its variation due to the stress.

Later studies by Steinemann [123], Voitkovskii [12], Vyalov [15], and others verified the correctness of this relation for ice in mountainous and polar regions. It was shown that relation (3.23) holds true for both compression and shear:

$$\dot{\gamma} = \frac{d\dot{\gamma}}{dt} = B\tau^n, \tag{3.24}$$

where B and n are coefficients which are obtained experimentally, γ is the angle of shear (for instance, γ_{xy} is the variation of the right angle between the X and Y axes caused by a tangential stress γ_{xy}).

Parameter n, by the way, is practically independent of the temperature, while B varies considerably with T.

Experiments also showed that there is a certain stress τ_0 such that for $\tau > \tau_0$ the flow velocity increases sharply (exponent n in expression (3.24) becomes 2 to 3 times as great). In this case a slight increment in stress causes a sizeable increase in the deformation rate. The values of τ_0 range from 1 to 1.65 kg/cm^2 (for temperatures close to the melting point).

Different deformation mechanisms correspond to conditions $\tau < \tau_0$ (slow flow) and $\tau > \tau_0$. For slow flow the crystals break up and become reoriented. The flow in this case is close to viscous flow. For $\tau > \tau_0$ the breakup of the crystals is accompanied by mechanical disintegration and rotation relative to one another. The shear deformation becomes a sliding deformation, which corresponds to block sliding (Blockschollen movement).

Note that for n = 1 Equation (3.24) becomes an equation describing an ideally viscous liquid with a viscosity $\mu = 1/B$, whereas for $n = \infty$ Equation (3.23) defines the conditions of ideal plasticity: $\tau = \tau_0 =$ const, where $\tau_0 = 1/B^n$. Ice occupies an intermediate position between these limiting cases.

The temperature dependences of coefficient B vary if different sets of data are used. A dependence of the type of the exponential Arrhenius law relating reaction rate and temperature has the best physical basis.

When using Glen's law, however, it must be kept in mind that it only describes the nature of the process approximately. There would appear to be considerable differences between the values of B and n obtained from the data of different investigators. But the results of these experiments are not yet complete enough to lead to suitable working formulas.

To conclude this discussion of experimental data on the relation between stresses and deformations in ice, let us just mention that even the use of expressions like (3.23) and (3.24) for simple shear and expansion-contraction involves appreciable uncertainty, because of the discrepancies between the experimental findings of different investigators (different values of B, and even of n).

Inspection of Figure 3.3 indicates that n (the exponent in Glen's law) varies greatly with temperature. The range of this variation is large, from 1.5 to 4.0. The value of n increases for greater stresses and deformation rates.

Many attempts have been made to take the temperature into account in the rheological law. For example, Voitkovskii [12] proposed a relation of the type

$$\dot{\gamma} = \frac{B}{1 + |T|} \tau^n .$$

Glen used the following expression, known from the theory of the creep of metals:

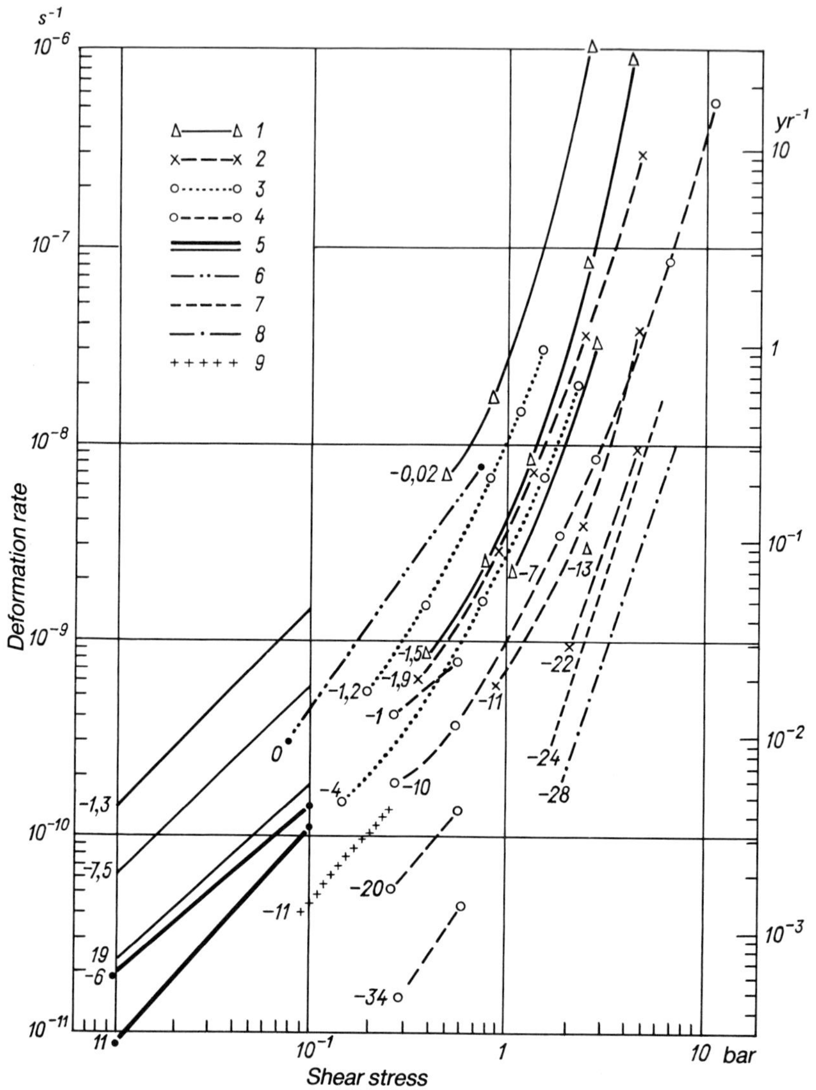

Fig. 3.3. Experimental relation between stresses and rates of deformations in ice [5].

(1) Glen; (2) Steinemann; (3) Voitkovskii; (4) Mellor and Smith; (6) Butkovich and Landauer; (6) Gerard (Jungfraufirn); (7) Hansen and Landauer (Greenland); (8) Gow (Byrd Station); (9) Wilson (Greenland).

$$\dot{\gamma} = B \exp\left(-\frac{Q^*}{R^* T^*}\right),$$

where Q^* is the activation energy, R^* is the gas constant, and T^* is the absolute temperature.

Budd [5] generalized the curves shown in Figure 3.3 and suggested, without even taking the temperature into account, a two-termed equation to describe the relation between the deformation and stress rates for a simple stressed state. Still more uncertainty is introduced by the presence in actual glaciers of a complex, rather than simple, stressed state of the ice.

In view of this, it has recently seemed more and more advisable to return to describing the flow of ice in a glacier with the aid of the Navier-Stokes equations of motion. However, then we feel it necessary to introduce the concept of the apparent viscosity of the glacier ice. Taking into account expression (3.26), for simple one-dimensional shear in a glacier, in the y direction, say, we can write Glen's experimental law (3.24) in the form

$$\tau = \frac{1}{B^{1/n}}\left(\frac{\partial u}{\partial y}\right)^{1/n} = \left[\frac{1}{B^{1/n}}\left(\frac{\partial u}{\partial y}\right)^{(1/n)-1}\right]\frac{\partial u}{\partial y}.$$

At the same time, Newton's law of friction for an ideally viscous liquid has the form

$$\tau = \mu \frac{\partial u}{\partial y}.$$

A comparison of these two expressions shows readily that the law of motion for ice can be put into the form of Newton's law if the quantity representing the viscosity is defined as follows:

$$\mu_a = \frac{1}{B^{1/n}}\left(\frac{\partial u}{\partial y}\right)^{(1/n)-1}. \qquad (3.25)$$

In the theory of motion of a non-Newtonian or pseudoplastic liquid, this quantity is known as the apparent viscosity. Since this approach has been very fruitful in engineering studies of the flow of non-Newtonian liquids, for which $(1/n) - 1$ is low, it should be useful for ice as well. It should be noted that tables giving data on the viscosity of ice generally do not contain any indication of the value of $\partial u/\partial y$ for which the experiments were conducted. Therefore, these data are to be used for calculations only with extreme caution.

Cauchy conditions. The physical law of ice motion in a glacier was defined above as a relation between the deformation rates $\dot{\varepsilon}$ and $\dot{\gamma}$ and the applied stresses. At the same time, in the equations of continuity, motion, and energy, it is the velocities of motion of the glacier ice, and the derivatives of these velocities, which are involved.

The deformation rates and the velocities of ice motion are related by the Cauchy equations, which in rectangular coordinates can be written as

$$\dot{\varepsilon}_x = \frac{\partial u}{\partial x}; \quad \dot{\varepsilon}_y = \frac{\partial v}{\partial y}; \quad \dot{\varepsilon}_z = \frac{\partial w}{\partial z};$$

$$\dot{\gamma}_{xy} = \frac{\partial u}{\partial y} + \frac{\partial v}{\partial y}; \quad \dot{\gamma}_{yz} = \frac{\partial v}{\partial z} + \frac{\partial w}{\partial y}; \quad \dot{\gamma}_{zx} = \frac{\partial w}{\partial x} + \frac{\partial u}{\partial z} .$$

3.2. Equations of Motion of Viscoplastic Ice

Substitution of the stress-strain relations for a Newtonian (ideally viscous) fluid into the equations of motion in form (3.8) led to the Navier-Stokes equations, making possible a direct analysis of the motion of a viscous liquid. It would be extremely tempting to try to derive equations of this type to describe the flow of ice.

3.2.1. Relation Between Stresses and Velocity Gradients

According to the law of ice motion proposed by Glen (3.23), (3.24), the deformations and stresses along axes X, Y, and Z are related as follows:

$$\frac{\partial \varepsilon_x}{\partial t} = B'\sigma_x^n; \quad \frac{\partial \varepsilon_y}{\partial t} = B'\sigma_y^n; \quad \frac{\partial \varepsilon_z}{\partial t} = B'\sigma_z^n;$$

$$\frac{\partial \gamma_{xy}}{dt} = B\tau_{xy}; \quad \frac{\partial \gamma_{zx}}{\partial t} = B\tau_{xz}; \quad \frac{\partial \gamma_{yz}}{\partial t} = B\tau_{yz}. \tag{3.26}$$

Here ε_x, ε_y, ε_z are the contraction-expansion deformations along axes X, Y, Z; and γ_{xy}, γ_{zx}, γ_{yz} are the angular deformations in planes XY, XZ, YZ. These six components of the deformation of a cube are related to one another. First of all, let us note that the relative variation of volume ΔV is

$$\frac{d(\Delta V)}{dV} = e = \varepsilon_x + \varepsilon_y + \varepsilon_z.$$

In order to find out more about this relation, let us introduce for each point of the glacier the concept of the displacement. Let the initial coordiantes of our point be x, y, z; after deformation these coordinates will be $x + \zeta$; $y + \eta$; $z + \xi$, where ζ, η, ξ are the components of dispacement along axes X, Y, Z. Deformation components ε_x, ε_y, ε_z, γ_{xy}, γ_{yz}, γ_{zx} are related to displacement components ζ, η, ξ. This relation, which is easily discerned from Figure 3.4, has the form:

$$\varepsilon_x = \frac{\partial \zeta}{\partial x}; \quad \varepsilon_y = \frac{\partial \eta}{\partial y}; \quad \varepsilon_z = \frac{\partial \xi}{\partial z};$$

$$\gamma_{xy} = \frac{\partial \zeta}{\partial y} + \frac{\partial \eta}{\partial x}; \quad \gamma_{yz} = \frac{\partial \eta}{\partial z} + \frac{\partial \xi}{\partial y}; \quad \gamma_{zx} = \frac{\partial \xi}{\partial x} + \frac{\partial \zeta}{\partial z} . \tag{3.27}$$

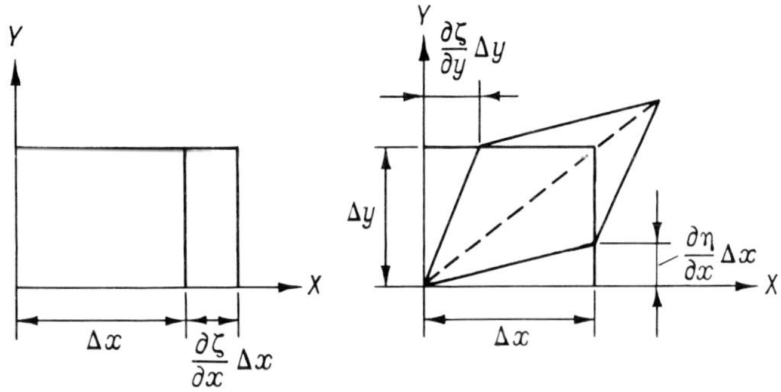

Fig. 3.4. Derivation of relation between displacements in ice and velocities of ice motion [73].

At the same time, the velocities of motion of points in the medium are defined as time variations of the displacement coordinates:

$$\frac{\partial\zeta}{\partial t} = u; \quad \frac{\partial\eta}{\partial t} = v; \quad \frac{\partial\xi}{\partial t} = w. \tag{3.28}$$

Now, changing to the time rates of variation of the deformations and taking (3.27) and (3.28) into account, we obtain:

$$\frac{\partial\varepsilon_x}{\partial t} = \frac{\partial\zeta}{\partial x\,\partial t} = \frac{\partial u}{\partial x}; \quad \frac{\partial\varepsilon_y}{\partial t} = \frac{\partial v}{\partial y}; \quad \frac{\partial\varepsilon_z}{\partial t} = \frac{\partial w}{\partial z};$$

$$\frac{\partial\gamma_{xy}}{\partial t} = \frac{\partial\zeta}{\partial y\,\partial t} + \frac{\partial\eta}{\partial x\,\partial t} = \frac{\partial u}{\partial y} + \frac{\partial v}{\partial x}; \quad \frac{\partial\gamma_{yz}}{\partial t} = \frac{\partial v}{\partial z} + \frac{\partial w}{\partial y};$$

$$\frac{\partial\gamma_{zx}}{\partial t} = \frac{\partial w}{\partial x} + \frac{\partial u}{\partial z}\,. \tag{3.29}$$

Next, by considering expressions (3.26) and (3.29) together, we are able to exclude the deformations not entering into the equations of motion. This gives us an expression for Glen's law, that is, the experimental law of ice flow, in the form:

$$\sigma_x^n = \frac{1}{B'}\frac{\partial u}{\partial x}; \quad \sigma_y^n = \frac{1}{B'}\frac{\partial v}{\partial y}; \quad \sigma_z^n = \frac{1}{B'}\frac{\partial w}{\partial z};$$

$$\tau_{xy}^n = \frac{1}{B}\left(\frac{\partial u}{\partial y} + \frac{\partial v}{\partial x}\right); \quad \tau_{yz}^n = \frac{1}{B}\left(\frac{\partial v}{\partial z} + \frac{\partial w}{\partial y}\right); \tag{3.30}$$

$$\tau_{zx}^n = \frac{1}{B}\left(\frac{\partial w}{\partial x} + \frac{\partial u}{\partial z}\right).$$

Here B, B', and n are experimental constants of the rheological law. Expression (3.30) is thus an experimental law of ice flow, expressed in terms of the velocities. It should be restated at this point that, like Glen's original law, these expressions are, strictly speaking, valid only for a simple stressed state, due to the action of only one of the six components. Consequently, the use of these expressions describing a complex stressed state is not wholly justified, a shortcoming which is, however, now common to all models of ice deformations. Accordingly, if we consider a complex stressed state, when the deformation is caused by the action of all the normal and tangential stress components, then we can assume that the relation between the tangential stresses τ and shears γ remains the same as indicated earlier in (3.26). At the same time, the relation of the relative elongations ε and their rates of variation $\partial\varepsilon/\partial t$ with the normal stresses ε cannot in general be portrayed by Equations (3.26).

In this case the normal stress σ_x, for example, which acts in the x direction, causes not only a certain rate of relative elongation in direction X, but also a relative contraction in the transverse directions. Let m be the ratio of the rate of longitudinal deformation in direction X under the influence of stress σ_x to the associated rate of transverse deformation:

$$m = -\frac{\partial\varepsilon_x/\partial t}{\partial\varepsilon_y/\partial t} = -\frac{\varepsilon_x}{\varepsilon_y}\ .$$

The quantity m defined here is none other than the coefficient of transverse contraction, the reciprocal of Poisson's ratio. Since for solids this coefficient is usually close to 10/3, we will assume this value for ice as well. In this case the total rate of relative elongation in direction X due to stresses σ_x, σ_y, σ_z will be equal to the rate of relative elongation due just to stress σ_x, that is, $B'\ \sigma_x^n$, minus the rates of transverse contraction caused by stresses σ_y and σ_z, that is, $(B'/m)\ \sigma_y^n$ and $(B'/m)\ \sigma_z^n$. Consequently,

$$\frac{\partial\varepsilon_x}{\partial t} = B'\sigma_x^n - \frac{B'}{m}(\sigma_y^n + \sigma_z^n).$$

If we apply these same considerations to the deformation rates relative to axes Y and Z and add the relations for τ and γ (3.30), which are unchanged, then we obtain the following relations:

$$\frac{1}{B'}\ \frac{\partial\varepsilon_x}{\partial t} = \sigma_x^n - \frac{1}{m}\ (\sigma_y^n + \sigma_z^n);$$

$$\frac{1}{B'}\ \frac{\partial\varepsilon_y}{\partial t} = \sigma_y^n - \frac{1}{m}\ (\sigma_z^n + \sigma_x^n);$$

$$\frac{1}{B'}\frac{\partial \varepsilon_z}{\partial t} = \sigma_z^n - \frac{1}{m}(\sigma_x^n + \sigma_y^n);$$

$$\frac{1}{B}\frac{\partial \gamma_{xy}}{\partial t} = \tau_{xy}^n; \quad \frac{1}{B}\frac{\partial \gamma_{yz}}{\partial t} = \tau_{yz}^n; \quad \frac{1}{B}\frac{\partial \gamma_{zx}}{\partial t} = \tau_{zx}^n. \tag{3.31}$$

Accordingly, the rate of variation of the relative volume is found by substituting (3.31) into the expression for the variation of this volume:

$$\frac{\partial e}{\partial t} = B' \frac{m-2}{m}(\sigma_x^n + \sigma_y^n + \sigma_z^n). \tag{3.32}$$

Thus, on the one hand, we have six equations (3.29) relating the six components of the deformation rate $\partial\varepsilon_x/\partial t, \ldots, \partial\gamma_{ZX}/\partial t$ to the six derivatives of the ice velocity (u, v, w) with respect to the coordinates; on the other hand, we now have six equations (3.31) relating the six components of this deformation rate to the six stress components: $\sigma_x, \sigma_y, \sigma_z, \tau_{xy}, \tau_{yz}, \tau_{zx}$.

3.2.2. Glen's Rheological Law in a Form Analogous to Stokes Law

In order to use the foregoing material in the equations of motion in form (3.8) and to transform these into equations similar to the Navier-Stokes equations, we have to express the indicated six stress components directly in terms of the velocity components, that is, we have to put them into a form similar to Stokes law for a viscous fluid. To do this, we have to eliminate from the twelve equations (3.31) and (3.29) the six components

$$\partial\varepsilon_x/\partial t, \ \partial c_y/\partial t, \ \partial\varepsilon_z/\partial t, \ \partial\gamma_{xy}/\partial t, \ \partial\gamma_{yz}/\partial t, \ \partial\gamma_{zx}/\partial t.$$

Actually, the rates of the angular deformations were already eliminated in (3.30). In order to eliminate the rates of relative elongation, we make use of the relation between (1) the parameters determining the resistance to shear and expansion-contraction and (2) the coefficient of transverse contraction [73]:

$$\frac{1}{B} = \frac{m}{B' \cdot 2(m+1)}. \tag{3.33}$$

Then, expressing B' in terms of B in Equation (3.32), we get

$$\sigma_x^n + \sigma_y^n + \sigma_z^n = 2\,\frac{m+1}{m-2}\,\frac{1}{B}\,\frac{\partial e}{\partial t}. \tag{3.34}$$

Next we replace the sum $\sigma_y^n + \sigma_z^n$ in the first of equations (3.31) by its value from (3.32) to obtain

$$\frac{1}{B'}\frac{\partial \varepsilon_x}{\partial t} = \frac{m+1}{m}\sigma_x^n - \frac{1}{B'}\frac{1}{m-2}\frac{\partial e}{\partial t},$$

or, using (3.33) to go from B' to B,

$$\sigma_x^n = \frac{1}{B'}\left(\frac{\partial \varepsilon_x}{\partial t} + \frac{1}{m-2}\frac{\partial e}{\partial t}\right) = \frac{2}{B}\left(\frac{\partial \varepsilon_x}{\partial t} + \frac{1}{m-2}\frac{\partial e}{\partial t}\right). \tag{3.35}$$

If we introduce the arithmetic mean of the three n-th powers of the normal stress $\bar{\sigma}$:

$$\bar{\sigma}^n = \frac{1}{3}(\sigma_x^n + \sigma_y^n + \sigma_z^n)$$

and substitute this quantity into (3.35), then, taking (3.34) into account, we obtain

$$\sigma_x^n = \bar{\sigma}^n + 2\frac{1}{B}\frac{\partial u}{\partial x} - \frac{2}{3}\frac{1}{B}\,\text{div U}. \tag{3.36}$$

Analogous relations are found for σ_y^n and σ_z^n. For viscoplastic ice, with an exponent n close to unity ($n \approx 1.5 - 2$), under conditions when the values of σ_x, σ_y, σ_z differ little from one another, we suppose $\sigma_x^n + \sigma_y^n + \sigma_z^n = \frac{1}{3}(\sigma_x + \sigma_y + \sigma_z)^n$. Then $\sigma^n = 3^{n-1} \times \left(\frac{\sigma_x + \sigma_y + \sigma_z}{3}\right)^n = -3^{n-1}\,\bar{P}_h^n$, where $\bar{P}_h$ is the hydrostatic pressure in the ice.

Taking this relation and (3.36) into account, we find that the six components of the stress tensor are related only to the velocities. These relations have the form

$$\sigma_x = \left(-3^{n-1}\,\bar{P}_h^n + 2\frac{1}{B}\frac{\partial u}{\partial x} - \frac{2}{3}\frac{1}{B}\,\text{div}\,\vec{U}\right)^{1/n};$$

$$\sigma_y = \left(-3^{n-1}\,\bar{P}_h^n + 2\frac{1}{B}\frac{\partial v}{\partial y} - \frac{2}{3}\frac{1}{B}\,\text{div}\,\vec{U}\right)^{1/n};$$

$$\sigma_z = \left(-3^{n-1}\,\bar{P}_h^n + 2\frac{1}{B}\frac{\partial w}{\partial z} - \frac{2}{3}\frac{1}{B}\,\text{div}\,\vec{U}\right)^{1/n};$$

$$\tau_{xy} = \left[\frac{1}{B}\left(\frac{\partial u}{\partial y} + \frac{\partial v}{\partial x}\right)\right]^{1/n}; \tag{3.37}$$

$$\tau_{yz} = \left[\frac{1}{B}\left(\frac{\partial v}{\partial z} + \frac{\partial w}{\partial y}\right)\right]^{1/n};$$

$$\tau_{zx} = \left[\frac{1}{B}\left(\frac{\partial w}{\partial x} + \frac{\partial u}{\partial z}\right)\right]^{1/n};$$

Relations (3.37) can be represented simply in matrix form:

$$\begin{pmatrix} \sigma_x & \tau_{xy} & \tau_{xz} \\ \tau_{xy} & \sigma_y & \tau_{yz} \\ \tau_{xz} & \tau_{yz} & \sigma_z \end{pmatrix}^n = \begin{pmatrix} -3^{n-1}\,\bar{P}_h^n & 0 & 0 \\ 0 & -3^{n-1}\,\bar{P}_h^n & 0 \\ 0 & 0 & -3^{n-1}\,\bar{P}_h^n \end{pmatrix} +$$

$$+ \frac{1}{B}\begin{pmatrix} \frac{\partial u}{\partial x} & \frac{\partial u}{\partial y} & \frac{\partial u}{\partial z} \\ \frac{\partial v}{\partial x} & \frac{\partial v}{\partial y} & \frac{\partial v}{\partial z} \\ \frac{\partial w}{\partial x} & \frac{\partial w}{\partial y} & \frac{\partial w}{\partial z} \end{pmatrix} + \begin{pmatrix} \frac{\partial u}{\partial x} & \frac{\partial v}{\partial x} & \frac{\partial w}{\partial x} \\ \frac{\partial u}{\partial y} & \frac{\partial v}{\partial y} & \frac{\partial w}{\partial y} \\ \frac{\partial u}{\partial z} & \frac{\partial v}{\partial z} & \frac{\partial w}{\partial z} \end{pmatrix} -$$

$$- \frac{2}{3}\frac{1}{B}\begin{pmatrix} \text{div } U & 0 & 0 \\ 0 & \text{div } U & 0 \\ 0 & 0 & \text{div } U \end{pmatrix}. \tag{3.38}$$

Expressions (3.37) and (3.38) obtained by us represent a general expression of the law of resistance for the motion of viscoplastic ice, which obeys the experimental rheological power law of the type of Glen's law. It is the first such expression which as been derived.

Now let us assume that n equals unity, which corresponds to ideal viscous motion with a viscosity μ equal to $1/B$. In this case expression (3.38) becomes

$$\begin{pmatrix} \sigma_x & \tau_{xy} & \tau_{xz} \\ \tau_{xy} & \sigma_y & \tau_{yz} \\ \tau_{xz} & \tau_{yz} & \sigma_z \end{pmatrix} = \begin{pmatrix} -\bar{P}_h & 0 & 0 \\ 0 & -\bar{P}_h & 0 \\ 0 & 0 & -\bar{P}_h \end{pmatrix} +$$

$$+ \mu\begin{pmatrix} \frac{\partial u}{\partial x} & \frac{\partial u}{\partial y} & \frac{\partial u}{\partial z} \\ \frac{\partial v}{\partial x} & \frac{\partial v}{\partial y} & \frac{\partial v}{\partial z} \\ \frac{\partial w}{\partial x} & \frac{\partial w}{\partial y} & \frac{\partial w}{\partial z} \end{pmatrix} + \begin{pmatrix} \frac{\partial u}{\partial x} & \frac{\partial v}{\partial x} & \frac{\partial w}{\partial x} \\ \frac{\partial u}{\partial y} & \frac{\partial v}{\partial y} & \frac{\partial w}{\partial y} \\ \frac{\partial u}{\partial z} & \frac{\partial v}{\partial z} & \frac{\partial w}{\partial z} \end{pmatrix} -$$

$$- \frac{2}{3}\mu\begin{pmatrix} \text{div } \vec{U} & 0 & 0 \\ 0 & \text{div } \vec{U} & 0 \\ 0 & 0 & \text{div } \vec{U} \end{pmatrix}. \tag{3.39}$$

Relation (3.39) is known as the mathematical expression of the Stokes law of friction for an ideally viscous fluid.

Consequently, expressions (3.37) for n = 1 also take the form of the following well-known relations between the stresses and velocities for a Newtonian fluid:

$$\sigma_x = -\bar{P}_h + \mu\left(2\frac{\partial u}{\partial x} - \frac{2}{3}\operatorname{div} U\right); \quad \tau_{xy} = \mu\left(\frac{\partial u}{\partial y} + \frac{\partial v}{\partial x}\right);$$

$$\sigma_y = -\bar{P}_h + \mu\left(2\frac{\partial v}{\partial y} - \frac{2}{3}\operatorname{div} U\right); \quad \tau_{yz} = \mu\left(\frac{\partial v}{\partial z} + \frac{\partial w}{\partial y}\right);$$

$$\sigma_z = -\bar{P}_h + \mu\left(2\frac{\partial w}{\partial z} - \frac{2}{3}\operatorname{div} U\right); \quad \tau_{zx} = \mu\left(\frac{\partial w}{\partial x} + \frac{\partial u}{\partial z}\right). \qquad (3.40)$$

If the compressibility of the ice is neglected, then these expressions become even simpler. In this case div U = 0 and the last term in (3.38) and (3.39) disappears.

3.2.3. The Equation of Motion of Ice in a Form Analogous to the Navier-Stokes Equations

The foregoing mathematical description of the law of resistance to ice flow (3.37) and (3.38) makes it possible to obtain the equations of motion for ice not considered to be a Newtonian liquid. In order to do this, we replace the stress components in the system of equations of motion (3.8) by the corresponding equivalents (3.37). As a result we obtain

$$\rho\left(\frac{\partial u}{\partial t} + u\frac{\partial u}{\partial x} + v\frac{\partial u}{\partial y} + w\frac{\partial u}{\partial z}\right) = K_x +$$

$$+ \frac{\partial}{\partial x}\sqrt[n]{-3^{n-1}\bar{P}_h^n + \frac{2}{B}\frac{\partial u}{\partial x} - \frac{2}{3B}\operatorname{div}\vec{U}} +$$

$$+ \frac{\partial}{\partial y}\sqrt[n]{\frac{1}{B}\left(\frac{\partial u}{\partial y} + \frac{\partial v}{\partial x}\right)} + \frac{\partial}{\partial z}\sqrt[n]{\frac{1}{B}\left(\frac{\partial u}{\partial z} + \frac{\partial w}{\partial x}\right)};$$

$$\rho\left(\frac{\partial v}{\partial t} + u\frac{\partial v}{\partial x} + v\frac{\partial v}{\partial y} + w\frac{\partial v}{\partial z}\right) = K_y +$$

$$+ \frac{\partial}{\partial y}\sqrt[n]{-3^{n-1}\bar{P}_h^n + \frac{2}{B}\frac{\partial v}{\partial y} - \frac{2}{3B}\operatorname{div}\vec{U}} +$$

$$+ \frac{\partial}{\partial x}\sqrt[n]{\frac{1}{B}\left(\frac{\partial u}{\partial y} + \frac{\partial v}{\partial x}\right)} + \frac{\partial}{\partial z}\sqrt[n]{\frac{1}{B}\left(\frac{\partial v}{\partial z} + \frac{\partial w}{\partial y}\right)};$$

$$\rho\left(\frac{\partial w}{\partial t} + u\frac{\partial w}{\partial x} + v\frac{\partial w}{\partial y} + w\frac{\partial w}{\partial z}\right) =$$

$$= K_z + \frac{\partial}{\partial z}\sqrt[n]{-3^{n-1}\bar{P}_h^n + \frac{2}{B}\frac{\partial w}{\partial z} - \frac{2}{3B}\operatorname{div}\vec{U}} +$$

$$+ \frac{\partial}{\partial x}\sqrt[n]{\frac{1}{B}\left(\frac{\partial w}{\partial x} + \frac{\partial u}{\partial z}\right)} + \frac{\partial}{\partial y}\sqrt[n]{\frac{1}{B}\left(\frac{\partial w}{\partial y} + \frac{\partial v}{\partial z}\right)}. \quad (3.41)$$

Expressions (3.41) are the equations of motion of viscoplastic ice, which obeys a rheological power law with an experimental coefficient B and an exponent n. This system is analogous to the Navier-Stokes equations for a Newtonian fluid. If we set $n = 1$ and $1/B = \mu$ in Equations (3.41), which corresponds to a viscous liquid, then we obtain:

$$\rho\left(\frac{\partial u}{\partial t} + u\frac{\partial u}{\partial x} + v\frac{\partial u}{\partial y} + w\frac{\partial u}{\partial z}\right) = K_x + \frac{\partial \bar{P}_h}{\partial x} +$$

$$+ \frac{\partial}{\partial x}\mu\left(2\frac{\partial u}{\partial x} + \frac{2}{3}\operatorname{div}\vec{U}\right) + \frac{\partial}{\partial y}\mu\left(\frac{\partial u}{\partial y} + \frac{\partial v}{\partial x}\right) -$$

$$+ \frac{\partial}{\partial z}\mu\left(\frac{\partial u}{\partial z} + \frac{\partial w}{\partial x}\right);$$

$$\rho\left(\frac{\partial v}{\partial t} + u\frac{\partial v}{\partial x} + v\frac{\partial v}{\partial y} + w\frac{\partial v}{\partial z}\right) = K_y + \frac{\partial \bar{P}_h}{\partial y} +$$

$$+ \frac{\partial}{\partial y}\mu\left(2\frac{\partial v}{\partial y} - \frac{2}{3}\operatorname{div}\vec{U}\right) + \frac{\partial}{\partial x}\mu\left(\frac{\partial u}{\partial y} + \frac{\partial v}{\partial x}\right) +$$

$$+ \frac{\partial}{\partial z}\mu\left(\frac{\partial u}{\partial x} + \frac{\partial w}{\partial x}\right);$$

$$\rho\left(\frac{\partial w}{\partial t} + u\frac{\partial w}{\partial x} + v\frac{\partial w}{\partial y} + w\frac{\partial w}{\partial z}\right) = K_z + \frac{\partial \bar{P}_h}{\partial z} +$$

$$+ \frac{\partial}{\partial z}\mu\left(\frac{\partial w}{\partial z} - \frac{2}{3}\operatorname{div}\vec{U}\right) + \frac{\partial}{\partial x}\mu\left(\frac{\partial w}{\partial x} + \frac{\partial u}{\partial z}\right) +$$

$$+ \frac{\partial}{\partial y}\mu\left(\frac{\partial w}{\partial y} + \frac{\partial v}{\partial z}\right). \quad (3.42)$$

This system of equations is the familiar system of Navier-Stokes equations, the equations of motion of an ideally viscous fluid.

3.3. Internal Heat Sources Created by Ice Motion

3.3.1. The Dissipation Function for Viscoplastic Ice Motion

As mentioned in Section 3.1, the amount of heat released per unit volume of a moving continuous medium is specified by Equation (3.15), as a function of the velocity components and the components of the stress tensor. Equations (3.37) obtained by us make it possible to get rid of these components. Actually, by substituting (3.37) into (3.15), we get

$$\frac{dQ_z}{dt}\frac{1}{dV} = \frac{\partial u}{\partial x}\sqrt[n]{3^{n-1}\bar{P}_h^n + \frac{2}{B}\frac{\partial u}{\partial x} - \frac{2}{3B}\operatorname{div} U} +$$

$$+ \frac{\partial v}{\partial y}\sqrt[n]{3^{n-1}\bar{P}_h^n + \frac{2}{B}\frac{\partial v}{\partial y} - \frac{2}{3B}\operatorname{div} U} +$$

$$+ \frac{\partial w}{\partial z}\sqrt[n]{3^{n-1}\bar{P}_h^n + \frac{2}{B}\frac{\partial w}{\partial z} - \frac{2}{3B}\operatorname{div} U} +$$

$$+ \frac{\partial v}{\partial x}\sqrt[n]{\frac{1}{B}\left(\frac{\partial u}{\partial y} + \frac{\partial v}{\partial x}\right)} + \frac{\partial w}{\partial x}\sqrt[n]{\frac{1}{B}\left(\frac{\partial w}{\partial x} + \frac{\partial u}{\partial z}\right)} +$$

$$+ \frac{\partial u}{\partial y}\sqrt[n]{\frac{1}{B}\left(\frac{\partial u}{\partial y} + \frac{\partial v}{\partial x}\right)} + \frac{\partial w}{\partial y}\sqrt[n]{\frac{1}{B}\left(\frac{\partial v}{\partial z} + \frac{\partial w}{\partial y}\right)} +$$

$$+ \frac{\partial u}{\partial z}\sqrt[n]{\frac{1}{B}\left(\frac{\partial w}{\partial x} + \frac{\partial u}{\partial z}\right)} + \frac{\partial v}{\partial z}\sqrt[n]{\frac{1}{B}\left(\frac{\partial v}{\partial z} + \frac{\partial w}{\partial y}\right)}. \tag{3.43}$$

If we then set $n = 1$ and $1/B = \mu$ in these expressions, we obtain

$$\frac{dQ_z}{dt}\frac{1}{dV} = \bar{P}_h\left(\frac{\partial u}{\partial x} + \frac{\partial v}{\partial y} + \frac{\partial w}{\partial z}\right) + 2\mu\left[\left(\frac{\partial u}{\partial x}\right)^2 + \right.$$

$$+ \left(\frac{\partial v}{\partial y}\right)^2 + \left(\frac{\partial w}{\partial z}\right)^2 - \frac{1}{3}\left(\frac{\partial u}{\partial x} + \frac{\partial v}{\partial y} + \frac{\partial w}{\partial z}\right)^2 +$$

$$+ \frac{1}{2}\frac{\partial v}{\partial x}\left(\frac{\partial u}{\partial y} + \frac{\partial v}{\partial x}\right) + \frac{1}{2}\frac{\partial w}{\partial x}\left(\frac{\partial w}{\partial x} + \frac{\partial u}{\partial z}\right) +$$

$$+ \frac{1}{2}\frac{\partial u}{\partial y}\left(\frac{\partial u}{\partial y} + \frac{\partial v}{\partial x}\right) + \frac{1}{2}\frac{\partial w}{\partial y}\left(\frac{\partial v}{\partial z} + \frac{\partial w}{\partial y}\right) +$$

$$\left. + \frac{1}{2}\frac{\partial u}{\partial z}\left(\frac{\partial w}{\partial x} + \frac{\partial v}{\partial z}\right) + \frac{1}{2}\frac{\partial v}{\partial z}\left(\frac{\partial v}{\partial z} + \frac{\partial w}{\partial y}\right)\right].$$

Next we drop the first term on the left-hand side, which is none other than the work of expansion or contraction, which is not converted into heat, and we combine pairs in the last six terms on the right-hand side, to arrive at the simplified expression

$$\frac{\partial Q_z}{\partial t}\frac{1}{dV} = 2\mu\left[-\frac{1}{3}\left(\frac{\partial u}{\partial x} + \frac{\partial v}{\partial y} + \frac{\partial w}{\partial z}\right)^2 + \right.$$

$$+ \frac{1}{2}\left(\frac{\partial w}{\partial x} + \frac{\partial u}{\partial y}\right)^2 + \frac{1}{2}\left(\frac{\partial w}{\partial x} + \frac{\partial u}{\partial z}\right)^2 +$$

$$+ \frac{1}{2}\left(\frac{\partial w}{\partial y} + \frac{\partial v}{\partial z}\right)^2\Bigg] = \Phi \; . \tag{3.44}$$

The expression in square brackets in (3.44) is well-known in the hydrodynamics of an ideally viscous fluid as the <u>dissipation function</u>, giving the intensity of heat release during the motion of a continuous medium. In the following this function will be denoted by the symbol Φ. By analogy with it, Equation (3.43) can be called an expression for the dissipation function of viscoplastic ice Φ'.

3.3.2. The Energy Equation with the Dissipation Function of Viscoplastic Ice Φ'

The energy equation (3.19) can be transformed similarly. By substituting expressions (3.37) into it, we obtain

$$\begin{aligned}
&\rho g\left[\frac{\partial(cT)}{\partial t} + u\,\frac{\partial(cT)}{\partial x} + v\,\frac{\partial(cT)}{\partial y} + w\,\frac{\partial(cT)}{\partial z}\right] = \\
&= \left[\frac{\partial}{\partial x}\left(\lambda\,\frac{\partial T}{\partial x}\right) + \frac{\partial}{\partial y}\left(\lambda\,\frac{\partial T}{\partial y}\right) + \frac{\partial}{\partial z}\left(\lambda\,\frac{\partial T}{\partial z}\right)\right] + \\
&+ N\Bigg\{\Bigg[\sqrt[n]{3^{n-1}\,\bar{P}_h^n - 2\,\frac{1}{B}\,\frac{\partial u}{\partial x} - \frac{2}{3}\,\frac{1}{B}\,\mathrm{div}\,U}\;\frac{\partial u}{\partial x} + \\
&+ \sqrt[n]{\frac{1}{B}\left(\frac{\partial u}{\partial y} + \frac{\partial v}{\partial x}\right)}\;\frac{\partial v}{\partial x} + \sqrt[n]{\frac{1}{B}\left(\frac{\partial w}{\partial x} + \frac{\partial u}{\partial z}\right)}\;\frac{\partial w}{\partial x}\Bigg] + \\
&+ \Bigg[\sqrt[n]{\frac{1}{B}\left(\frac{\partial u}{\partial y} + \frac{\partial v}{\partial x}\right)}\;\frac{\partial u}{\partial z} + \\
&+ \sqrt[n]{3^{n-1}\,\bar{P}_h^n + 2\,\frac{1}{B}\,\frac{\partial v}{\partial y} - \frac{2}{3}\,\frac{1}{B}\,\mathrm{div}\,U}\;\frac{\partial v}{\partial y} + \\
&+ \sqrt[n]{\frac{1}{B}\left(\frac{\partial v}{\partial z} + \frac{\partial w}{\partial y}\right)}\;\frac{\partial v}{\partial y}\Bigg] + \Bigg[\sqrt[n]{\frac{1}{B}\left(\frac{\partial w}{\partial x} + \frac{\partial u}{\partial z}\right)}\;\frac{\partial u}{\partial z} + \\
&+ \sqrt[n]{\frac{1}{B}\left(\frac{\partial v}{\partial z} + \frac{\partial w}{\partial y}\right)}\;\frac{\partial v}{\partial z} + \\
&+ \sqrt[n]{3^{n-1}\,\bar{P}_h^n + 2\,\frac{1}{B}\,\frac{\partial w}{\partial z} - \frac{2}{3}\,\frac{1}{B}\,\mathrm{div}\,U}\;\frac{\partial w}{\partial z}\Bigg]\Bigg\} + \\
&+ r\kappa + R_q .
\end{aligned} \tag{3.45}$$

For n = 1 and 1/B = μ this becomes

$$\rho g\left[\frac{\partial(cT)}{\partial t} + u\,\frac{\partial(cT)}{\partial x} + v\,\frac{\partial(cT)}{\partial y} + w\,\frac{\partial(cT)}{\partial z}\right] =$$

$$= \frac{\partial}{\partial x}\left(\lambda\,\frac{\partial T}{\partial x}\right) + \frac{\partial}{\partial y}\left(\lambda\,\frac{\partial T}{\partial y}\right) + \frac{\partial}{\partial z}\left(\lambda\,\frac{\partial T}{\partial z}\right) +$$

$$+ N\Bigg[\left(P_h + 2\mu\frac{\partial u}{\partial x} - \frac{2}{3}\,\mu\,\mathrm{div}\,U\right)\frac{\partial u}{\partial x} +$$

$$+ \mu\left(\frac{\partial u}{\partial y} + \frac{\partial v}{\partial x}\right)\frac{\partial v}{\partial x} + \mu\left(\frac{\partial w}{\partial x} + \frac{\partial u}{\partial z}\right)\frac{\partial w}{\partial x} +$$

$$+ \mu\left(\frac{\partial u}{\partial y} + \frac{\partial v}{\partial x}\right)\frac{\partial u}{\partial y} + \left(P_h + 2\mu\,\frac{\partial v}{\partial y} - \frac{2}{3}\,\mu\,\mathrm{div}\,U\right)\frac{\partial v}{\partial y} +$$

$$+ \mu\left(\frac{\partial v}{\partial z} + \frac{\partial w}{\partial y}\right)\frac{\partial v}{\partial y} + \mu\left(\frac{\partial w}{\partial x} + \frac{\partial u}{\partial z}\right)\frac{\partial u}{\partial z} +$$

$$+ \mu\left(\frac{\partial v}{\partial z} + \frac{\partial w}{\partial y}\right)\frac{\partial v}{\partial z} + \Bigg(3P_h + 2\mu\,\frac{\partial w}{\partial z} -$$

$$- \frac{2}{3}\,\mu\,\mathrm{div}\,\frac{\partial w}{\partial z}\Bigg)\frac{\partial w}{\partial z}\Bigg] + r\kappa + R_Q. \tag{3.46}$$

However, since the expression in braces in Equation (3.45) is the dissipation function of viscoplastic flow Φ', while the expression in square brackets in (3.46) is none other than the dissipation function of the Navier-Stokes equation, it follows that the energy equations for viscoplastic flow and for purely viscous flow will be, respectively,

$$\rho g\left[\frac{\partial(cT)}{\partial t} + u\,\frac{\partial(cT)}{\partial x} + v\,\frac{\partial(cT)}{\partial y} + w\,\frac{\partial(cT)}{\partial z}\right] =$$

$$= \left[\frac{\partial}{\partial x}\left(\lambda\,\frac{\partial T}{\partial x}\right) + \frac{\partial}{\partial y}\left(\lambda\,\frac{\partial T}{\partial y}\right) + \frac{\partial}{\partial z}\left(\lambda\,\frac{\partial T}{\partial z}\right)\right] + N\Phi', \tag{3.47}$$

where Φ' is defined by Equation (3.43), and

$$\rho g\left[\frac{\partial(cT)}{\partial t} + u\,\frac{\partial(cT)}{\partial x} + v\,\frac{\partial(cT)}{\partial y} + w\,\frac{\partial(cT)}{\partial z}\right] =$$

$$= \rho g\left[\frac{\partial}{\partial x}\left(\lambda\,\frac{\partial T}{\partial x}\right) + \frac{\partial}{\partial y}\left(\lambda\,\frac{\partial T}{\partial y}\right) + \frac{\partial}{\partial z}\left(\lambda\,\frac{\partial T}{\partial z}\right)\right] + N\Phi, \tag{3.48}$$

where Φ is defined by **Equation (3.44)**.

3.4. Uniqueness Conditions

The above system of equations of continuity, motion, and energy is, however, not sufficient for solving problems involving specific glaciers. In order to study specific conditions, we must add uniqueness conditions to this system.

The geometrical conditions determine the configuration and specific dimensions of a glacier, the numerical values of the rheological constants of the ice, and its thermophysical parameters; these conditions contain information about internal heat sources and sinks produced by internal friction.

The temporal conditions indicate the features of the evolution of the processes in time (whether they are steady-state or not, the initial distributions of the parameters and their variations in time, etc.).

The boundary conditions give the values of the velocities, temperatures, or heat fluxes, as well as the conditions of ice sliding, at the glacier boundaries.

Note, too, that problems related to motion and heat transfer in a glacier can be solved correctly only after a strict proof of the theorems of existence and uniqueness for the solutions of the corresponding equations. Such proofs have been obtained, however, only for a limited class of problems involving the flow of an ideally viscous fluid. Strictly speaking, the correctness of the statement of a problem involving motion and heat transfer in a glacier should always be verified experimentally.

With regard to the geometrical conditions, it should be mentioned that the range of sizes, shapes, and types of glacier formations is very wide indeed. For our purposes it will be advisable to consider here only three basic types and to indicate their actual sizes.

Inland ice sheets have a shape which, seen from above, is close to axisymmetric; they flow radially outward on a comparatively level base. The largest present-day glacier formations are the inland ice caps or ice sheets. These include the Antarctic Ice Sheet and the Greenland Ice Sheet. Figure 3.5 shows a plan view of the Antarctic Ice Sheet, indicating its dimensions and the directions of its flow lines. At the bottom is shown a vertical section of the ice sheet along the line I-I [4].

The system of ice caps making up the ice sheet is seen to be close to axisymmetric in shape, with mean rates of outflow at the periphery of about 200 m per year. The maximum distance from the outflow center to the margin is 1500 km, and the ice thickness can reach 3500 m; the surface temperature at the seaward edges is around 11-15 °C, but in the central high-lying regions of the sheet it drops to -55 °C. The ice sheet stays the same size because each year 3 to 5 cm of snow falls onto its surface in the central regions and up to 1 m of snow falls at its margins.

Figure 3.6 shows, as an example, the same characteristics for an ice cap known as 'Drygalski Island'. This island ice cap is also in Antarctica, in the Davis Sea, about 100 km from the Soviet Mirny Station [4].

The rates of its outflow at the periphery range from 20 to 30 m per year, that is, they are only 1/10 of the outflow rates at the margins of the Antarctic Ice Sheet. The distance from the outflow center of Drygalski Island to the margin is about 7 km, or about 1/500 of that of the Antarctic Ice Sheet, while its maximum ice thickness is only 1/10 of the thickness of the ice sheet (about 350 m). The surface temperature of this glacier is about -10 °C.

Ice shelves are glaciers floating on the surface of the sea, which have a free ice outflow toward their seaward margins. These formations are usually located in regions of the sea close to inland ice sheets, and they give the impression of being floating extensions of the body

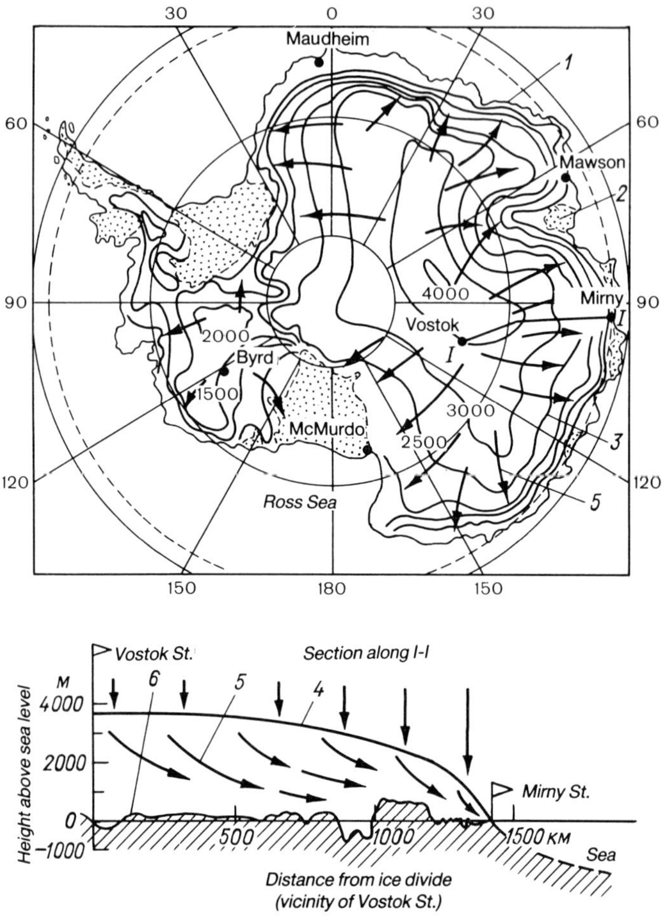

Fig. 3.5. Largest present-day inland ice sheet, the Antarctic Ice Sheet, plan view and section through line I-I.

(1) outer contours of ice sheet; (2) ice shelves; (3) isolines of height of glacier surface above sea level; (4) upper surface of glacier along line I-I [4]; (5) direction of ice motion; (6) glacier-bed boundary along line I-I [4].

of the ice sheet. As a rule, these glaciers are situated in bays, gulfs, or inland seas, where the conditions are favorable for their preservation.

The largest of the present-day ice shelves is the Ross Ice Shelf, which covers the southern part of the Ross Sea in Antarctica and is about 600 km long, with a width of this same magnitude (Figure 9.9). It is formed in its southern, rear part by flows of ice from inland Antarctica, moving at a speed of about 300 m per year. The ice thickness here can reach 800 m. Further into the sea, this thickness drops to

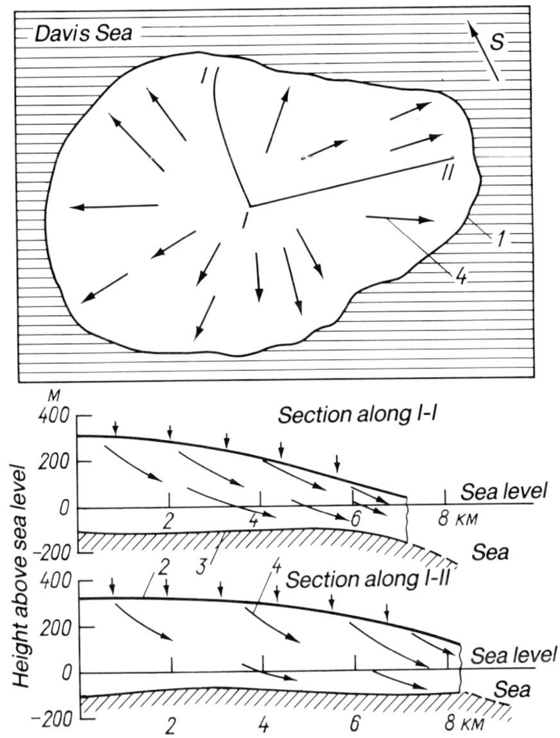

Fig. 3.6. Small ice cap 'Drygalski Island' (Antarctica, Davis Sea), plan view and sections along lines I-I and I-II [4].

(1) outer contours of glacier; (2) upper surface; (3) glacier-bed boundary; (4) direction of ice motion.

200-300 m, beyond which its margin breaks off to form icebergs. As it moves northward, seaward, from the southern regions, the glacier receives a constant supply of precipitation at its upper surface. At the lower surface of the glacier melting or freezing occurs, due to heat exchange with the seawater. A typical feature of an ice shelf is the absence of any resistance to glacier motion at its lower surface.

The sizes of ice shelves also vary over a wide range. For example, the shelf on Ellesmere Island in the Arctic Ocean is only about 30 m thick.

Mountain glaciers. These glaciers also differ greatly from one another. As an example of a mountain glacier, let us consider Fedchenko Glacier (Figure 3.7), one of the longest glaciers of this type. It is located in the basin of the Sel'dar River in the Pamirs. Fedchenko Glacier is about 77 km long and has a mean width of 2.6 km; the ice thickness in its central part reaches 1000 m. The upper portion of the glacier lies in a region **where there is an annual accumulation of** precipitation. This is thus called the nourishment zone of the glacier. Here the vertical components of the flow-line vectors are directed

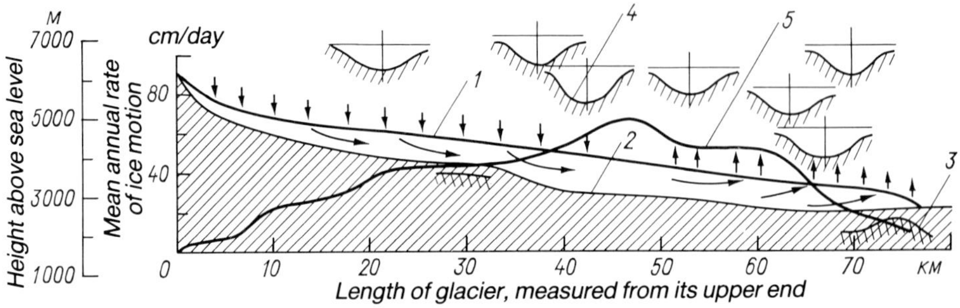

Fig. 3.7. Largest existing mountain glacier: Fedchenko Glacier (Pamirs).

(1) upper surface of glacier; (2) ice-bedrock boundary; (3) lower boundary of intermediate layer under glacier, detected seismically; this layer is apparently the moraine layer; (4) transverse profiles of glacier; (5) variation of velocity of longitudinal ice motion along glacier length.

downward. The rate of downslope motion of the glacier increases with the approach to the lower boundary of the nourishment zone, and it may be as high as 300 m per year. Further down, the melting is greater than the supply of precipitation via snow accumulation, so that this zone is called the ablation zone. Here the vertical components of the flow-line vectors are directed upward.

Obruchev Glacier, in the Polar Urals, presents a sharp contrast to Fedchenko Glacier, in that it is one of the smallest mountain glaciers: only 0.9 km long and 450 m wide (Figure 3.8). The upper part of the glacier is the nourishment region. The rate of motion of the glacier close to the lower boundary of the nourishment zone is about 10 m per year, and the maximum thickness of the ice is about 130 m.

The brief descriptions of even these few glaciers show them to be diverse and dissimilar.

The boundary conditions are the conditions at the upper and lower surfaces of the glacier. At the upper surface the following are usually specified: the surface temperature T_H, and the magnitude and sign of the vertical velocity of ice motion through the interface $\mathbf{V}_H$ (the rate of accumulation of precipitation or the ablation rate). Here and in the following, a subscript H will refer to conditions at the upper surface of the glacier, and a subscript 0 will refer to the lower surface, in view of the fact that the coordinate origin will as a rule be on this surface.

If H is the thickness of the glacier, and u_H is the velocity of horizontal motion at the surface, then the boundary condition at the upper surface will be:

$$y = H; \; T = T_H; \quad u = u_H; \quad v = v_H. \tag{3.49}$$

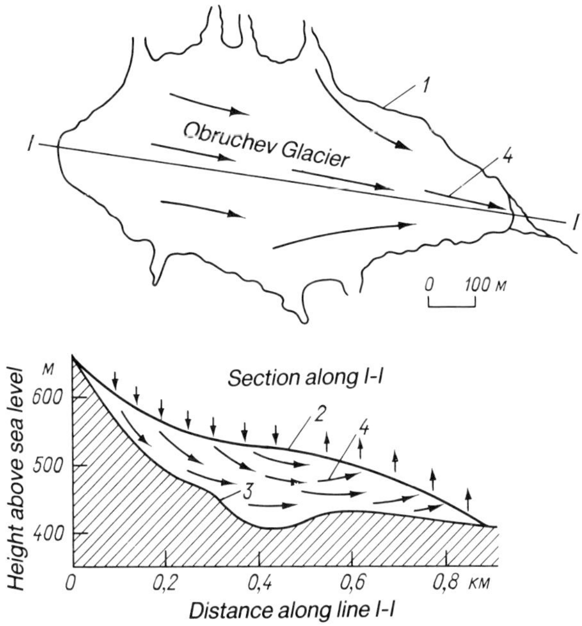

Fig. 3.8. One of the smallest mountain glaciers: Obruchev Glacier in the Polar Urals.

Notation the same as in Figure 3.6.

With regard to the lower surface, that is, the bed surface of an inland glacier or the boundary between an ice shelf and the sea, the following should be noted. In both cases either freezing or melting can occur at this surface, so that the velocity of vertical ice motion at this surface v_0 is in general unknown and must be found. In hydrodynamics the problem of the velocity along the interface usually does not come up, and this velocity is assumed to be zero. This condition, known as the adhesion hypothesis, is not valid for glaciers. Observations show that in many cases there is considerable sliding of the ice over the bed of an inland glacier ($v_0 \neq 0$). And, of course, this velocity will be equal to the velocity of the undisturbed ice at the lower surface of an ice shelf, where there is no resistance to the motion of the ice at the ice-seawater boundary.

With regard to conditions at the lower surface, note that for inland glaciers the temperature at this surface is usually unknown, and must be determined, on the basis of the equation describing the heat balance at the ice-bedrock interface. This will be considered in detail below, and suffice it to say here that a very simple equation for the heat balance at the glacier bed can be obtained by considering Fig. 3.9. This figure shows that at the lower surface of an inland glacier there will be a geothermal heat influx q_g and also a heat influx q_{fr} due to

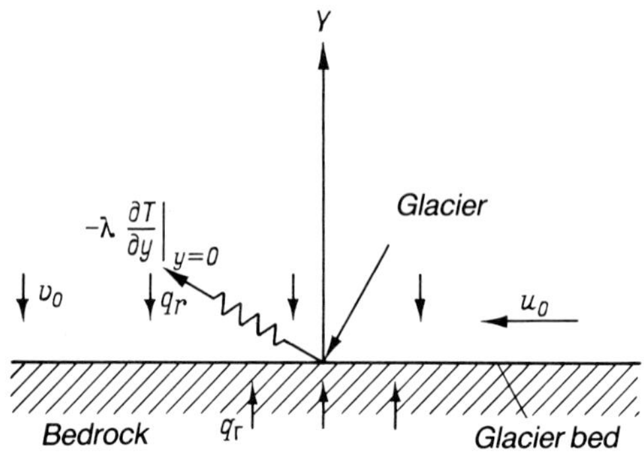

Fig. 3.9. Heat balance at lower surface of inland glacier, bedded on water-impermeable rock.

the friction between the glacier and the bed, if sliding occurs. In addition, some heat may be gained or lost by this surface because of the latent heat q_r associated with the freezing or melting at the glacier bed. The heat balance is rounded out in the steady state by the component of the heat removal through the ice to the upper regions of the glacier via thermal conduction ($-\lambda\ dT/dy\big|_{y=0}$):

$$\lambda \frac{\partial T}{\partial y}\Big|_{y=0} + q_g + q_{fr} + q_r = 0. \tag{3.50}$$

This condition can serve as the boundary condition at the lower surface, which would then have the form

$$y = 0;\ \lambda \frac{dT}{dy}\Big|_{y=0} + q_g + q_{fr} + q_r = 0;\ u = u_0;\ v = v_0. \tag{3.51}$$

However, when analyzing the conditions of formation of the temperature field in a glacier, we find it more convenient to assume that the temperature at the glacier bed is also specified, so that, knowing the distribution, and thus the dimensionless gradient, of the temperature at the bed, the true bed temperature can be found from the heat-balance equation for the ice-bed interface in the form (3.50). In this case the boundary condition at the bed becomes

$$y = 0;\quad T = 0;\quad u = u_0;\quad v = v_0. \tag{3.52}$$

3.5. Simplifying the Equations. Dimensionless Numbers.

Thus the temperature field and the motion of a glacier are defined by the system of equations of continuity (3.1), motion (3.41), and energy (3.47), with boundary conditions (3.49) and (3.51), and also the initial (time) conditions. Clearly, such a system of equations cannot be solved in its complete form. Thus it has to be simplified, by making a comparative assessment of the roles of the various members of the system and then dropping members which are of minor importance [76]. Let us make such an assessment. To do this, it will be advisable to change from the dimensional form of the equations to a dimensionless form. If this is done, then the same equation can be applied to an entire class of glaciers, the class being characterized by some given value of the dimensionless coefficients in front of the dimensionless derivatives of the equations.

In order to reduce the terms of an equation to dimensionless form, we must choose appropriate reduction scales. These may be quantities which are constant and typical of a glacier, so selected that the order of all the dimensionless quantities is equal to unity. For instance, if the length of the glacier varies from 0 to L along axis X, then we take L to be the characteristic size for reduction of x to dimensionless form. If the thickness varies from 0 to H along axis Y, then the characteristic size for y will be H, etc. If we limit ourselves to two dimensions (along axes X and Y), and if we assume the ice to be incompressible, then we obtain the following dimensionless coordinates:

$$\bar{x} = \frac{x}{L}; \quad \bar{y} = \frac{y}{H},$$

where L is the characteristic longitudinal size (length of glacier), and H is the characteristic thickness of the glacier (for ice caps the process often has to be considered using cylindrical coordinates).

The dimensionless velocities will be, correspondingly,

$$\bar{v} = \frac{v}{v_H}; \quad \bar{u} = \frac{u}{u_H},$$

where v_H is the characteristic longitudinal velocity of horizontal motion of the glacier, and u_H is the rate of accumulation of precipitation or ablation at the glacier surface.

Now let us find an expression for the dimensionless temperature. It should be so selected that its value at the lower boundary ($\bar{y} = 0$) is zero, even if its actual value there (T_0) is not zero. At the upper boundary the temperature $\bar{T}_H$ should be equal to unity for any actual value T_H of it. In order to satisfy these conditions, we represent the dimensionless temperature as:

$$\bar{T} = (T \quad T_0)/(T_H - T_0).$$

As our new time, let us take the quantity at/H^2, where a is the thermal diffusivity, equal to $a = \lambda/(c\rho)$. The units of a are length squared over time; for ice, for instance, $a \approx 34\ m^2/yr$. Therefore, the complex quantity at/H^2 is dimensionless and can be called the dimensionless time. It will be seen below how convenient it is to write the time in this way.

Now our dimensional coordinates can be related to their dimensional equivalents as follows:

$$x = L\bar{x}; \quad y = H\bar{y}; \quad v = v_H\bar{v}; \quad u = u_H\bar{u};$$

$$T = (T_H - T_0)\,\bar{T} - T_0; \quad t = \frac{H^2}{a}\,\bar{t}.$$

The dimensional and dimensionless derivatives of the energy equation then have the form:

$$\frac{\partial T}{\partial t} = \frac{\partial[(T_H - T_0)\bar{T} + T_0]}{\partial\left(\frac{H^2}{a}\,\bar{t}\right)} = \frac{(T_H - T_0)a}{H^2}\,\frac{\partial\bar{T}}{\partial\bar{t}};$$

$$\frac{\partial T}{\partial x} = \frac{\partial[(T_H - T_0)\bar{T} + T_0]}{\partial(L\bar{x})} = \frac{(T_H - T_0)}{L}\,\frac{\partial\bar{T}}{\partial\bar{x}};$$

$$\frac{\partial T}{\partial y} = \frac{(T_H - T_0)}{H}\,\frac{\partial\bar{T}}{\partial\bar{y}}; \quad \frac{\partial^2 T}{\partial x^2} = \frac{\partial^2[(T_H - T_0)T + T_0]}{\partial(L\bar{x})^2} =$$

$$= \frac{(T_H - T_0)}{L^2}\,\frac{\partial^2\bar{T}}{\partial\bar{x}^2}; \quad \frac{\partial^2 T}{\partial y^2} = \frac{(T_H - T_0)}{H^2}\,\frac{\partial^2\bar{T}}{\partial\bar{y}^2}.$$

By substituting these quantities into energy equaiton (3.47), we get

$$\frac{\rho c(T_H - T_0)a}{H^2}\,\frac{\partial\bar{T}}{\partial\bar{t}} + \frac{\rho c u_H(T_H - T_0)}{L}\,\bar{u}\,\frac{\partial\bar{T}}{\partial\bar{x}} +$$

$$+ \frac{\rho c v_H(T_H - T_0)}{H}\,\bar{v}\,\frac{\partial\bar{T}}{\partial\bar{y}} = \frac{\lambda(T_H - T_0)}{L^2}\,\frac{\partial^2\bar{T}}{\partial\bar{x}^2} +$$

$$+ \frac{\lambda(T_H - T_0)}{H^2}\,\frac{\partial^2\bar{T}}{\partial\bar{y}^2} + N\Phi'.$$

The dimensionless derivatives of all the terms in this equation are of the same order, the order or unity. This is because the corresponding values of the scales were appropriately chosen. Next we multiply all the terms of the equation by the reciprocal of the coefficient of the second term on the right-hand side, $H^2/[\lambda\,(T_H - T_0)]$. Now the dimensionless coefficient in front of each derivative denotes the relative value of the corresponding term of the equation, with respect to the second term on the right-hand side, the value of which we have taken to be unity. Then, after some simple transformations, we obtain

$$\frac{\partial\bar{T}}{\partial\bar{t}} + \frac{v_H H}{a}\frac{H}{L}\frac{u_H}{v_H}\bar{u}\frac{\partial\bar{T}}{\partial\bar{x}} + \frac{v_H H}{a}\bar{v}\frac{\partial\bar{T}}{\partial\bar{y}} = \frac{H^2}{L^2}\frac{\partial^2\bar{T}}{\partial\bar{x}^2} +$$

$$+ \frac{\partial^2\bar{T}}{\partial\bar{y}^2} + N\frac{H^2\Phi'}{\lambda(T_H - T_0)} \,. \qquad (3.53)$$

Apropos of this equation, note the following things:

(a) all the coefficients in each term of the equation are dimensionless;

(b) the quantity $v_H H/a$ is the Péclet number, a dimensionless criterion for the thermal similarity of processes which is familiar from heat physics;

(c) the dissipation function with its coefficient can be called the dimensionless dissipation function. Let us denote it by the symbol Φ':

$$N\frac{H^2}{\lambda(T_H - T_0)}\Phi' = \bar{\Phi}'.$$

(d) now it becomes clear why the dimensionless time was specified in the form of the complex at/H^2. Such a choice ensured that the coefficient of the time derivative in the formula was equal to unity, which simplified the equation considerably. In view of the foregoing remarks, we can now rewrite Equation (3.53) as

$$\frac{\partial\bar{T}}{\partial\bar{t}} + Pe\left(\frac{H}{L}\frac{u_H}{v_H}\right)\bar{u}\frac{d\bar{T}}{d\bar{x}} + Pe\bar{v}\frac{d\bar{T}}{d\bar{y}} = \frac{H^2}{L^2}\frac{\partial^2\bar{T}}{\partial\bar{x}^2} + \frac{\partial^2\bar{T}}{\partial\bar{y}^2} + \bar{\Phi}'.$$

The continuity equation for two-dimensional motion has the form $\partial u/\partial x + \partial v/\partial y = 0$, or in dimensionless form

$$\frac{u_H}{L}\frac{\partial\bar{u}}{\partial\bar{x}} + \frac{v_H}{H}\frac{\partial\bar{v}}{\partial\bar{y}} = 0.$$

However, since both terms of the continuity equation are of the same order of magnitude, it follows that $u_H/L \sim v_H/H$ or $u_H/v_H \sim L/H$. This means that the coefficient in parentheses in the second term on the left-hand side of the equation we are considering is of the order of unity.

Previously we gave some typical values for the thicknesses and lengths of various ice sheets, ice shelves, and mountain glaciers. The ratio H/L is usually of the order of 0.05 or less. Accordingly, the first term on the right-hand side of Equation (3.53), which gives the conductive heat transfer in the horizontal direction, turns out to be negligible and can be dropped. Thus the heat transfer in a glacier can be represented by the following dimensionless equation:

$$\frac{\partial \bar{T}}{\partial \bar{t}} + \mathrm{Pe}\,\frac{H}{L}\frac{u_H}{v_H}\bar{u}\,\frac{\partial \bar{T}}{\partial \bar{x}} + \mathrm{Pe}\bar{v}\,\frac{\partial \bar{T}}{\partial \bar{y}} = \frac{\partial^2 \bar{T}}{\partial \bar{y}^2} + \bar{\Phi}'. \tag{3.54}$$

Depending on the further simplifications made, we distinguish between classes of glacial formations.

A quantitative criterion of a 'glacier'. Previously we defined a glacier as a formation of natural ice in which substantial movement of the ice takes place. This definition is somewhat imprecise, however, since it is not clear when, at what velocity, the motion begins to affect the processes in a glacier. On the basis of equation (3.54), we can state a new definition: a glacier is a formation of natural ice for which the Péclet number is of the order of unity.

If for a glacial formation Pe << 1, then the processes of motion have practically no effect on the temperature field, so that the heat transfer in this ice body is described by the expression

$$\frac{\partial \bar{T}}{\partial \bar{t}} = \frac{\partial^2 \bar{T}}{\partial \bar{y}^2}. \tag{3.55}$$

This is the usual heat-conduction equation for a stationary medium, which is not a glacier.

From the definition of the Péclet number, taking its minimum value, when the term containing it still has some significance in the heat-conduction equation ($\mathrm{Pe} \sim 0.2$), we now obtain an expression defining a glacier:

$$v_H \geqq \frac{0.2a}{H}$$

This relation has been plotted in Figure 3.10, as the dividing line between a glacier and a formation which is not a glacier.

Quasi-steady-state glacier. Even though the first term in equation (3.54), representing the unsteady state, is of the same order as the terms describing the vertical heat transfer, still the effect of the unsteady state in a glacier can often be neglected. This is possible, for instance, when the characteristic time of the phenomenon responsible for the unsteadiness is very short, so that a temporary perturbation practically does not penetrate the glacier. If t_{char} is the characteristic time of the phenomenon, then in our dimensionless coordinates the **characteristic time** also has to be represented in dimensionless form:

$$\bar{t}_{char} = \frac{a t_{char}}{H^2} = \mathrm{Fo},$$

where Fo is a dimensionless time, known in heat physics as the Fourier number.

Thus it is clear that the very same dimensional time is significant for thin glaciers and negligible for thick glaciers. For instance, a perturbation time of 10 years for a glacier 30 m thick corresponds to a dimensionless time of about 0.35, that is, a time commensurable with unity and thus significant. At the same time, in its dimensionless form

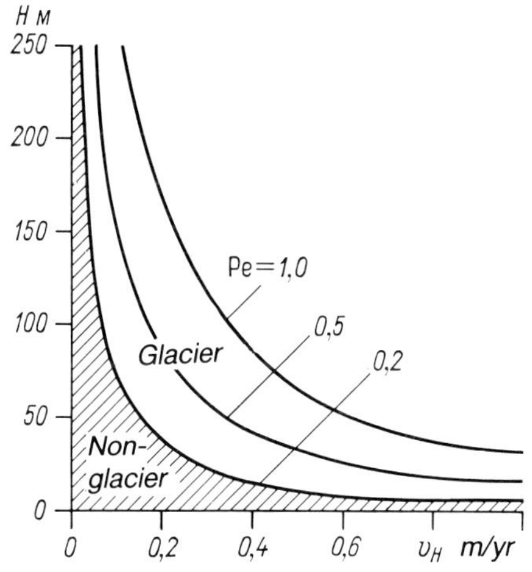

Fig. 3.10. A quantitative criterion for distinguishing a glacier from a nonglacier.

$\bar{t}_{char}$ for a glacier 300 m thick will be 3.5×10^{-3}, and for a glacier 1000 m thick $\bar{t}_{char} \approx 3.5 \times 10^{-5}$. For these two glaciers the dimensionless characteristic time is already so small that the thin layer of the glacier to which the unsteady component applies can be disregarded. Such glaciers will be called thick glaciers. For example, the annual temperature fluctuations (temperature amplitude of about 30 °C) for a glacier 12-15 m thick are only about 0.1 °C, so that in the following their effect will be ignored. Consequently, a layer 15 m thick can be excluded, and for a glacier more than 175 m thick, which can already be considered 'thick', this layer is less than 10% of the whole and thus is negligible for many calculations.

The temperature field in such a quasi-steady glacier is described by the equation

$$\mathrm{Pe}\,\frac{H}{L}\,\frac{u_H}{v_H}\,\frac{\partial \bar{T}}{\partial \bar{x}} + \mathrm{Pe}\,\frac{\partial \bar{T}}{\partial \bar{y}} = \frac{\partial^2 \bar{T}}{\partial \bar{y}^2} + \bar{\Phi}'. \qquad (3.56)$$

Glacier with block motion. The heat-transfer equation can be simplified considerably if we assume 'block motion' of the glacier ice.* In glaciology such motion usually refers to motion of a glacier for which its horizontal velocity is constant over the entire thickness. In this case it is advisable to use a moving coordinate system, with the origin located at the glacier bed and moving along it at the velocity of the 'block motion' of the glacier over the bed. The

* Also referred to as 'Blockschollen movement' or 'block flow' (translator).

horizontal velocity of the ice $u = u_H$ = const will then be zero over the entire thickness of the glacier, so that Equation (3.54) can be simplified to

$$\frac{\partial \bar{T}}{\partial \bar{t}} + Pe\bar{v}\frac{\partial \bar{T}}{\partial \bar{y}} = \frac{\partial^2 \bar{T}}{\partial \bar{y}^2} + \bar{\Phi}'. \tag{3.57}$$

Typical examples of such glaciers are all ice shelves.

It should be noted in particular that the unsteady term in (3.57) remains significant for 'thick' glaciers as well, for which the unsteadiness would not exist in a reference system at rest.

The unsteadiness appearing in our moving reference system is due to the transition of an observer moving with the system from one set of conditions (steady per se) to another set (also steady per se). In this sense the unsteadiness appearing in Equation (3.57) is none other than the substantial derivative with respect to time, a quantity widely used in hydrodynamics. In our case this derivative is

$$\frac{DT}{\partial t} = \frac{\partial T}{\partial t} + u\frac{\partial T}{\partial x}. \tag{3.58}$$

Consequently, since the local component ($\partial T/\partial t$) of the substantial derivative is zero, Equation (3.57) can be written as

$$\frac{H^2}{at_{char}}\bar{u}\frac{\partial \bar{T}}{\partial \bar{x}} + Pe\bar{v}\frac{\partial \bar{T}}{\partial \bar{y}} = \frac{\partial^2 \bar{T}}{\partial \bar{y}^2} + \bar{\Phi}'. \tag{3.59}$$

In spite of the absence of tangential shearing stresses in Equations (3.57) and (3.59), deformations of the ice particles in the glacier interior still exist.

Now let us go on to analyze the continuity equation for the block motion of a glacier: $\partial u/\partial x = -\partial v/\partial y$. Block motion implies constancy of the velocity of horizontal motion u over the glacier thickness at all times. This means that, if at some distance y' from the bed the velocity of this motion u varies along axis X by an amount $\partial u/\partial x$, then the variation in u above and below y' must be the same. If this were not the case, then the horizontal velocity would not remain constant over the glacier thickness, and the motion would no longer be block motion. Consequently, the mathematical condition for block motion is that the quantity $\partial u/\partial x$ remain the same (idem) along axis Y:

$$\frac{du}{dx} = \text{idem} \neq 0 \text{ along OY}.$$

Accordingly, from the continuity equation it follows that $\partial v/\partial y$ also remains the same along axis Y. Thus a necessary corollary of block motion of the glacier ice is

$$\frac{dv}{dy} = \text{const} = c_1, \tag{3.60}$$

from which

$$v = c_1 y + c_2. \tag{3.61}$$

From this it can be concluded that, in all glaciers characterized by block horizontal motion, the rate of vertical motion due to sinking of the ice during its longitudinal spreading or compression varies over the glacier thickness in a strictly linear manner.

'Flat block glacier'. In many cases of considerable practical and theoretical importance the variation in the parameters, for a coordinate system moving with the ice flow at a velocity u, is so gradual that parameter 1/Fo becomes very small. In this case the first term in Equation (3.57) or (3.59) can be neglected. This is permissible if velocity u is very low and if the thickness, mass-transfer rate, and surface temperature remain constant along axis X over a large distance, that is, if the glacier is 'flat'. As an example of this, we cite the conditions obtaining in the central flat region of an ice cap, and sometimes an ice shelf. The heat-transfer equation in this case has the form

$$\mathrm{Pe}\bar{v}\,\frac{\partial \bar{T}}{\partial \bar{y}} = \frac{\partial^2 \bar{T}}{\partial \bar{y}^2} + \bar{\Phi}'. \tag{3.62}$$

For the central region of an ice cap, where the velocities and their changes are very small, the quantity $\bar{\Phi}'$ can be neglected, so that the equation describing a flat block glacier simplifies to

$$\mathrm{Pe}\bar{v}\,\frac{\partial \bar{T}}{\partial \bar{y}} = \frac{\partial^2 \bar{T}}{\partial \bar{y}^2}\,. \tag{3.63}$$

An analysis of this very simple equation for the heat transfer in a glacier constitutes an important means of ascertaining general regularities, and such an analysis will be carried out below.

Significance of Péclet number. If we multiply the numerator and denominator of the expression for the Péclet number by the temperature differential across the glacier $(T_H - T_0)$, and if we expand the meaning of the thermal diffusivity a, then the definition of the Péclet number becomes

$$\mathrm{Pe} = \frac{v_H H}{a} = \frac{v_H c\rho (T_H - T_0)}{\frac{\lambda}{H}(T_H - T_0)}\,.$$

Here the numerator represents the heat flux transferred through the glacier by the vertical motion, and the denominator represents the heat flux which would be transferred during the same time by thermal conduction alone.

Now let us return to the very simple Equation (3.63) for the heat transfer in a flat block glacier, and let us also examine the meaning of the Péclet number. After some simple transformations Equation (3.63) becomes

$$c\rho v_H \bar{v} \frac{\partial \bar{T}}{\partial \bar{y}} = \lambda \frac{(T_H - T_0)}{H} \frac{d^2\bar{T}}{d\bar{y}^2} . \tag{3.64}$$

A comparison of the last two expressions indicates that the Péclet number is the ratio of the convective heat transfer because of vertical sinking of the ice to the transfer by thermal conduction. An evaluation of the significance of these quantities indicates three possible thermal regimes of a glacier [27]:

(1) $Pe \ll 1$. In this case the quantity representing the heat transfer due to sinking is negligible in comparison with the conductive transfer. The temperature field in such a glacier is described approximately by the equation $\lambda d^2T/dy^2 = 0$, from which it follows that $dT/dy = \text{const}$, that is, the temperature gradient in the glacier is constant over the thickness, while the temperature field is linear;

(2) $Pe \gg 1$. Heat transfer is mainly caused by the vertical transfer of mass in the glacier. The temperature throughout such a glacier is close to the temperature near one of the glacier surfaces, except for a comparatively thin region close to this glacier surface, where thermal conduction has an effect;

(3) $Pe \sim 1$. The heat transfer in this case is due equally to thermal conduction and to vertical transfer of the mass of the glacier. Thus both sides of the differential equation must be taken into account, since they are of the same order of magnitude.

Calculations carried out by the author in [27] demonstrated that both terms of Equation (3.64) have to be taken into account for $0.5 < Pe < 100$. If $Pe < 0.5$, then the deviation of the temperature field from linear over the whole glacier interior does not exceed 10% of the quantity $T_H - T_0$. For $Pe > 100$ the temperature throughout the glacier will, with an accuracy of up to 5%, be equal to the temperature near the upper surface of the glacier, except for a bottom layer with a thickness equal to less than 0.2 of the thickness of the glacier. Only in this layer will thermal conduction have an effect.

To conclude this section, let us just mention the surprising fact that most of the Earth's glaciers, from the great ice caps of Antarctica to the smallest mountain glaciers, have Péclet numbers for vertical heat transfer between 1 and 10. This means that in studies of heat-transfer processes in glaciers the thermal conduction and the heat transfer associated with the ice motion both always have to be considered.

Chapter 4

THE TEMPERATURE FIELD IN A FLAT TWO-DIMENSIONAL GLACIER WITH BLOCK SLIDING

4.1. A General Expression for the Temperature Field. Kinematics of a Glacier

In the preceding chapter it was shown that the simplest equation is one describing the heat transfer in a flat two-dimensional glacier which moves in such a way that the velocity of horizontal sliding is constant over the entire thickness of the glacier. Such a glacier will then have a constant thickness, temperature, and rate of accumulation or diminution of matter. Under conditions where the ice thickness is constant in time, while the coordinate origin is fixed to one of the glacier surfaces (to the bed in our case), the rates of accumulation or diminution of ice at the surfaces will be equal to the velocities of vertical ice motion through these surfaces.

The heat transfer in such a glacier is described by Equation (3.63), which in dimensional form in the coordinate system adopted by us becomes

$$\frac{d^2T}{dy^2} - \frac{v}{a}\frac{dT}{dy} = 0 \tag{4.1}$$

with the following boundary conditions at the upper and lower surfaces:

$$y = 0, \quad T = T_0;$$

$$y = H, \quad T = T_H. \tag{4.2}$$

For the further consideration of this problem we will follow [29].

Let us note again that T_H is usually unknown. The temperature at the lower surface will be specified below on the basis of an additional consideration.

Now, integrating (4.1) from 0 to y, we obtain

$$\frac{dT}{dy} = \frac{dT}{dy}\bigg|_{y=0} \exp\left(\int_0^y \frac{v}{a}\,dy\right). \tag{4.3}$$

Integration of (4.3) from 0 to H then gives an expression for the temperature gradient at the lower surface of the glacier:

$$\frac{dT}{dy}\bigg|_{y=0} = \frac{T_H - T_0}{\int_0^H \exp\left(\int_0^y \frac{v}{a}\,dy\right) dy}. \tag{4.4}$$

Substitution of (4.4) into (4.3) gives us the temperature gradient at any point in the glacier:

$$\frac{dT}{dy} = \frac{(T_H - T_0)\exp\left(\int_0^y \frac{v}{a}\,dy\right)}{\int_0^H \exp\left(\int_0^y \frac{v}{a}\,dy\right) dy}. \tag{4.5}$$

Finally, by integrating (4.5) from 0 to y, we obtain an equation uniquely determining the temperature field in a glacier:

$$\frac{T - T_0}{T_H - T_0} = \bar{T} = \frac{\int_0^y \exp\left(\int_0^y \frac{v}{a}\,dy\right) dy}{\int_0^H \exp\left(\int_0^y \frac{v}{a}\,dy\right) dy}. \tag{4.6}$$

The integrals in Equations (4.4), (4.5), and (4.6) can be expressed in the form of elementary or tabulated functions only for certain types of distribution of the vertical ice velocity over the glacier thickness. In connection with this, in [29] the possible types of velocity distributions were considered.

In the preceding chapter it was shown that velocity v in our glacier varies linearly over the glacier thickness according to law (3.61). Let us assume that the velocity of vertical ice motion is v_H at the upper surface and v_0 at the lower surface. Since axis Y is directed upward, we take this velocity to be positive when the motion is upward and negative when it is downward. Thus the boundary conditions for the velocities are:

$$y = 0, \quad v = v_0;$$

$$y = H, \quad v = v_H.$$

Coefficients c_1 and c_2 of solution (3.61) will then be

$$c_2 = v_0; \quad c_1 = \frac{v_H - v_0}{H}.$$

We next substitute the values of c_1 and c_2 into (3.61) and designate as $\bar{v}_0 = v_0/v_H$ the dimensionless velocity of vertical motion at the lower surface, to obtain

$$v = v_0 + \frac{v_H - v_0}{H}\,y = v_H\left(\bar{v}_0 + \frac{1 - \bar{v}_0}{H}\,y\right). \tag{4.7}$$

Figure 4.1, borrowed from [29], shows some possible shapes of the vertical-velocity distribution for the ice in a glacier. Distributions a, b, c, and d correspond to v < 0, when the heat transfer from the warmer bottom layers of the glacier to the colder surface is opposite to the downward moving masses of cold ice. Only a small portion of the heat leaving the bottom layers reaches the glacier surface, since almost all the heat goes to heat up the downward-moving ice. Thus for the cases being considered the temperature gradients in the upper ice layers, next to the surface, can be expected to decrease.

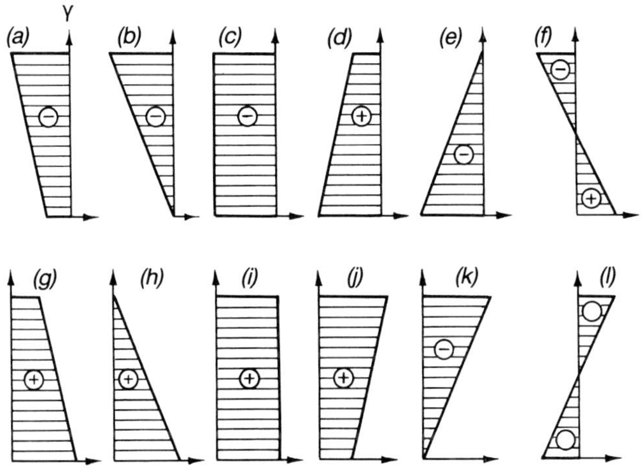

Fig. 4.1. Possible shapes of vertical-velocity distribution over glacier thickness under conditions of 'block motion' (after [29]).

(a-f) accumulation of precipitation at upper surface of glacier;
(g-l) ablation at this surface.

For the distribution shapes considered, except for case e, the velocity of vertical motion at both surfaces of the glacier is directed downward. Consequently, to ensure that the upper and lower surfaces of the glacier remain stationary, it must be assumed that an amount of substance arrives at the upper surface to compensate for the loss of substance due to the vertical velocity. For a flat glacier, in this case at the upper surface precipitation must accumulate at a rate equal to the vertical ice velocity at the surface. Similarly, at the lower surface, melting must occur at a rate equal to the vertical ice velocity there.

A completely different temperature distribution is to be expected when the ice motion is upward and the heat transfer from the warmer bottom regions of the glacier is due to both thermal conduction and ice motion (Figure 4.1 g-k). Here there is a cooling of the upward-moving ice particles and a release of the corresponding amount of heat; accordingly, the heat flux, and thus the temperature gradient, at the upper surface will exceed the flux and the gradient at the lower surface.

In cases g through k, at the upper surface of the glacier ablation will occur at a rate equal to the velocity of vertical ice motion at this surface; at the lower surface, on the other hand, freezing will occur at a rate equal to the velocity of motion of this surface.

Cases f and l in the figure correspond to conditions for which at both the upper and lower surfaces there is melting (case 1) or else a supply of material by accumulation of precipitation and freezing (case f).

For a more detailed analysis of the various cases shown in Figure 4.1, let us consider the nature of the variation in the horizontal velocities and also the shape of the streamlines corresponding to these cases in a flat block glacier.

Actually, it follows from the continuity equation that the rate of longitudinal expansion or contraction is

$$\frac{\partial u}{\partial x} = -\frac{\partial v}{\partial y} = -\frac{\partial}{\partial y}\left[v_H\left(\bar{v}_0 + \frac{1-\bar{v}_0}{H}y\right)\right] = -v_H\left(\frac{1-\bar{v}_0}{H}\right). \tag{4.8}$$

Then, paying attention to the signs, we find that cases a, b, f, g, and h in Figure 4.1 correspond to longitudinal expansion in the ice interior ($v_H - v_0 > 0$), while cases d, j, e, k, and l correspond to contraction ($v_H - v_0 < 0$). Cases c and i, for which $v_H = v_0$, are observed only in the absence of both spreading and contraction: $\partial u/\partial x = 0$.

Now let us consider the connection between the shape of the distribution in Figure 4.1 and the conditions of motion in the glacier, while noting especially the nature of the horizontal component of this motion.

By integrating (4.8) over x from some arbitrary zero, where the horizontal ice velocity is known and equal to $u_{x=0}$, to a point a distance L away from this zero, we get the following expression for the velocity at point L:

$$u_{x=L} = u_{x=0} - \int_0^L \frac{\partial v}{\partial y}\,dx = u_{x=0} - v_H\frac{1-\bar{v}_0}{H}L =$$

$$= u_{x=0} - \frac{v_H - v_0}{H}L. \tag{4.9}$$

Consequently, in a flat two-dimensional glacier with block motion, for which the vertical ice velocities do not vary along the length of the glacier, the horizontal velocity varies linearly along the direction of ice motion. In all cases where longitudinal expansion is observed, the velocity of horizontal motion increases:

$$-\frac{v_H - v_0}{H} > 0.$$

In the case of longitudinal compression, on the other hand, $-(v_H - v_0)/H < 0$, and the velocity decreases for downward motion along the flow.

When longitudinal expansion and contraction are both absent, the horizontal ice velocity remains constant over the whole glacier.

The nature of the ice flow is revealed in greater detail by the shape of the streamlines. It is a property of a streamline that each particle on it at a given time has a velocity directed along the tangent to the streamline. Thus the equation of the streamline is

$$\frac{dy}{v} = \frac{dx}{u} .$$

Substituting into this equation the velocities of vertical and horizontal motion from (4.7) and (4.9), we then obtain the following differential equation for the streamlines:

$$\frac{dy}{v_0 - \frac{v_H - v_0}{H} y} = \frac{dx}{u_{x=0} - \frac{v_H - v_0}{H} x} . \tag{4.10}$$

Consequently, the lines of flow in a flat two-dimensional glacier are described by the family of curves:

$$y = \frac{c_3}{y_{x=0} - \frac{v_H - v_0}{H} x} - \frac{v_0 H}{v_H - v_0} , \tag{4.11}$$

differing from one another only in the value of parameter c_3. Parameter c_3, the constant of integration, has units of the volume transport of ice in a glacier of unit width per unit time.

Thus the lines of ice flow in a flat two-dimensional glacier with block motion are hyperbolas, the shape of which depends considerably on the distribution of the vertical ice velocity over the glacier thickness. The simplest streamlines correspond to a vertical ice velocity which is constant over the thickness (no longitudinal expansion and no longitudinal contraction: $v_0 = 1$, Figure 4.1 c, g). The streamlines are then a family of straight lines, inclined to the glacier surfaces at an angle $\pm v_H/u_{x=0}$.

4.2. Accurate Solutions for the Temperature Field

4.2.1. Temperature Fields for Cooling Due to Ice Motion

In order to obtain working equations for the temperature field in a glacier, it is convenient to distinguish two regimes, depending on the direction of the vertical ice motion in the interior. If the vertical motion is downward, then the convective component of the transfer will function as a heat sink, distributed over the glacier thickness. The mean temperature of such a glacier will be lower than the mean temperature of the same ice mass in the absence of motion.

In order to arrive at specific expressions describing this process, let us consider the vertical-velocity distribution shown in Figure 4.1 a.

In this case precipitation accumulates at the upper surface of the glacier. The rate of vertical motion at this surface is negative and equal to the rate of accumulation. The velocity v_0 of ice motion at the lower surface is also negative, indicating melting at the bottom of the glacier. If $|v|$ is the absolute value (modulus) of the vertical ice velocity, then the distribution of vertical ice velocities in the given case can be written as

$$v = -|v_0| + \frac{|v_H| - |v_0|}{H} y = -|v_H|\left(\bar{v}_0 + \frac{1-\bar{v}_0}{H} y\right). \tag{4.12}$$

Substitution of (4.12) into (4.6) gives

$$\bar{T} = \frac{\int_0^y \exp\left[-\frac{|v_H|}{a}\left(\bar{v}_0 y + \frac{1-\bar{v}_0}{2H} y^2\right)\right] dy}{\int_0^H \exp\left[-\frac{|v_H|}{a}\left(v_0 y + \frac{1-\bar{v}_0}{2H} y^2\right)\right] dy}. \tag{4.13}$$

Let us express the numerator and denominator of (4.13) as error integrals. To do this, we write

$$f = \frac{1}{\sqrt{2}} \sqrt{\frac{|v_H|H}{a}}\left(\sqrt{1-\bar{v}_o}\,\frac{y}{H} + \frac{\bar{v}_0}{\sqrt{1-\bar{v}_0}}\right),$$

so that

$$\frac{|v_H|}{a}\left(\bar{v}_0 y + \frac{1-\bar{v}_0}{2H} y^2\right) = f^2 - \frac{1}{2}\,\frac{|v_H|H}{a}\,\frac{\bar{v}_0^2}{1-\bar{v}_0^2}.$$

Next in expression (4.13) we change to a new variable of integration:

$$\bar{T} = \frac{\int_\alpha^{\beta(y)} \exp(-f^2)\, df}{\int_\alpha^{\beta(H)} \exp(-f^2)\, df}. \tag{4.14}$$

Here α and β are new integration limits, corresponding to the new variable, which take the place of the old limits 0 and y or H, corresponding to the old variable of integration.

Passing to the new limits of integration, we note that the error function (erf) is defined by the expression

$$\mathrm{erf}(f) = \frac{2}{\sqrt{\pi}} \int_0^f \exp(-y^2)\, dy. \tag{4.15}$$

In addition, let us recall that y/H is the dimensionless coordinate $\bar{y}$, while the dimensionless complex $v_H H/a$ is none other than the Péclet number Pe. Taking this into account, as well as Equation (4.14), we obtain the following expression for the temperature field in a glacier:

$$\bar{T} = \frac{\operatorname{erf}\left[\sqrt{|Pe|}\left(\sqrt{1 - \bar{v}_0}\,\bar{y} + \frac{\bar{v}_0}{\sqrt{1 - \bar{v}_0}}\right)\right] + \operatorname{erf}\left(\sqrt{|Pe|}\frac{\bar{v}_0}{\sqrt{1 - \bar{v}_0}}\right)}{\operatorname{erf}\left[\sqrt{|Pe|}\left(\sqrt{1 - \bar{v}_0} + \frac{\bar{v}_0}{\sqrt{1 + \bar{v}_0}}\right)\right] - \operatorname{erf}\left(\sqrt{|Pe|}\,\frac{\bar{v}_0}{\sqrt{1 - \bar{v}_0}}\right)}. \quad (4.16)$$

If the rate of bottom melting is zero ($\bar{v}_0 = 0$, Figure 4.1 b), then Equation (4.16) simplifies to

$$\bar{T} = \frac{\operatorname{erf}(\sqrt{|Pe|}\,\bar{y})}{\operatorname{erf}(\sqrt{|Pe|})}; \quad \bar{v}_0 = 0. \quad (4.17)$$

This particular case was considered in detail by Robin [115].

If the velocity of vertical ice motion is constant over the entire glacier thickness, then the temperature will be

$$\bar{T} = \frac{\exp(-|Pe|\bar{y}) - 1}{\exp(-|Pe|) - 1}; \quad \bar{v}_0 = 1. \quad (4.18)$$

This case was studied by Wexler, as applied to ice shelves, and it is well-known in the theory of heat conduction [74].

In the absence of vertical ice motion ($v_H = v_0 = 0$) the temperature field in a glacier is linear:

$$\bar{T} = \bar{y}; \quad |Pe| = 0; \quad \bar{v}_0 = 0.$$

The straight lines in Figures 4.2 and 4.3 show the dimensionless temperature as a function of distance in the absence of ice motion. Below the lines are the curves for $\bar{T}$ as a function of $\bar{y}$ for two different values of the Péclet number. The curves were plotted for various values of the dimensionless ice velocity at the lower surface $\bar{v}_0$. The curves for $\bar{v}_0 = 0$ were constructed using formula (4.17), those for $v_0 = 1$ using formula (4.18), and those for $1 > \bar{v}_0 > 0$ using formula (4.16). These figures show that the values of Pe and $\bar{v}_0$ influence considerably the temperature distribution over the glacier interior.

If at the upper surface ice accumulates at a rate v_H, while at the bottom it melts at a rate v_0, but with $\bar{v}_0 > 1$ (melting rate higher than accumulation rate), then the process corresponds to conditions of longitudinal contraction of the ice ($\partial v/\partial y > 0$).

After some transformations which are essentially the same as those presented above and described in [29], we obtain the following cal-

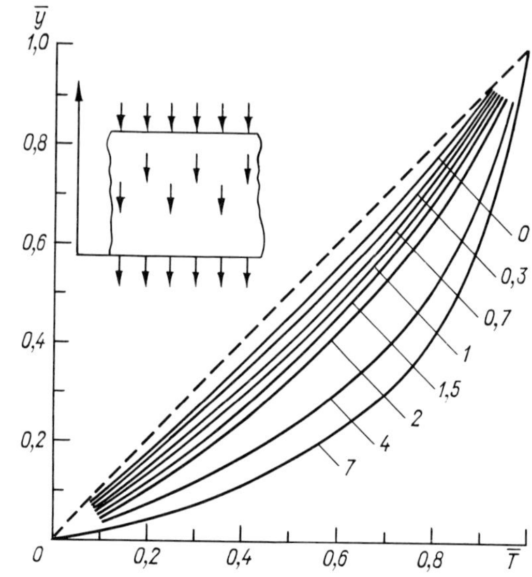

Fig. 4.2. Dimensionless temperature distribution in glacier for cooling due to ice motion. Péclet number Pe = 1. Numbers by curves correspond to values of dimensionless rate of melting at glacier bottom. Dashed line corresponds to absence of motion in glacier (Pe = 0).

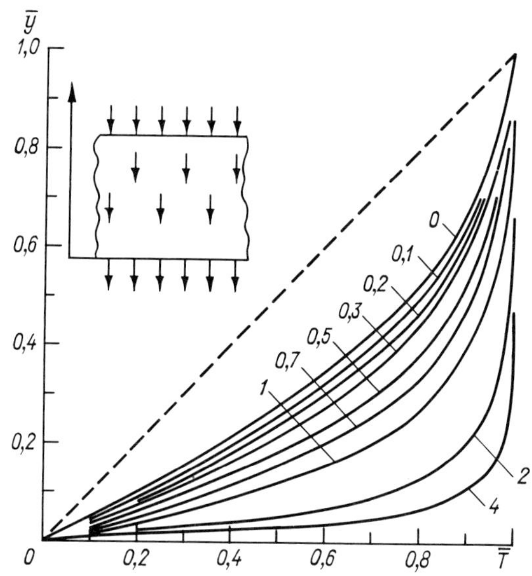

Fig. 4.3. The same as in Figure 4.2, except that here Pe = 6.

culated temperature field for this case:

$$\bar{T} = -\frac{\int_0^y \exp\left[\frac{|v_H|}{a}\left(\frac{\bar{v}_0 - 1}{2H} y^2 - \bar{v}_0 y\right)\right] dy}{\int_0^H \exp\left[\frac{|v_H|}{a}\left(\frac{\bar{v}_0 - 1}{2H} y^2 - \bar{v}_0 y\right)\right] dy}, \tag{4.19}$$

for which

$$\bar{T} = \frac{I\left[\sqrt{\frac{|Pe|}{2}}\left(\sqrt{\bar{v}_0 - 1}\,\bar{y} - \frac{\bar{v}_0}{\sqrt{\bar{v}_0 - 1}}\right)\right] + I\left(\sqrt{\frac{|Pe|}{2}}\frac{\bar{v}_0}{\sqrt{\bar{v}_0 - 1}}\right)}{I\left[\sqrt{\frac{|Pe|}{2}}\left(\sqrt{\bar{v}_0 - 1}\,y - \frac{\bar{v}_0}{\sqrt{\bar{v}_0 - 1}}\right)\right] + I\left(\sqrt{\frac{|Pe|}{2}}\frac{\bar{v}_0}{\sqrt{\bar{v}_0 - 1}}\right)}, \tag{4.20}$$

where $I(f) = \int_0^f \exp y^2\, dy$ is a tabulated function [81].

Figures 4.2 and 4.3 show $\bar{T}$ as a function of $\bar{y}$ for $n > 1$, which for the cases being considered corresponds to longitudinal contraction of the ice. As shown by these graphs, if longitudinal contraction of the glacier ice takes place, at a constant intensity, then even at very low Péclet numbers the temperature field for $\bar{v}_0 \gg 1$ may differ appreciably from linear. In all cases where the motion of the glacier causes ice to move downward, the glacier will be colder in its main part than it would be for a linear temperature distribution in the interior.

To conclude this subsection, we note that in some cases it is convenient to define Pe in terms of the vertical velocity at the lower surface. In this case Pe is designated as

$$Pe_0 = \frac{v_0 H}{a}.$$

An example of this case is shown in Figure 4.1e. The temperature field is then

$$\bar{T} = \frac{I\left[\sqrt{\frac{|Pe|}{2}}(\bar{y} - 1)\right] + I\left(\sqrt{\frac{|Pe|}{2}}\right)}{I\left(\sqrt{\frac{|Pe|}{2}}\right)}. \tag{4.21}$$

4.2.2. Temperature Fields for Warming Due to Ice Motion

Let us assume that melting takes place at the upper surface of a glacier, while freezing occurs at the lower surface. In this case ice particles move upward, carrying heat from the warmer lower ice layers to the upper cooler layers, as if aiding the thermal conduction. This effect is analogous to that which would be produced if internal heat sources were distributed over the glacier thickness; it is as if the glacier were heated by the ice motion.

Let us consider this effect quantitatively, on the basis of the results obtained in [29]. We begin with the conditions under which a longitudinal outflow of the glacier ice occurs (Figure 4.1g, h). In this case $n > 1$ and the distribution of vertical velocities has the form

$$v = |v_H|\left(\frac{\bar{v}_0 - 1}{H}\, y - \bar{v}_0\right). \tag{4.22}$$

As previously, we substitute (4.22) into (4.6) and carry out transformations of the type of (4.14) and (4.15), to obtain an expression for the temperature field corresponding to the distribution shown in Fig. 4.1g:

$$\bar{T} = \frac{\operatorname{erf}\left[\sqrt{\frac{|\mathrm{Pe}|}{2}}\left(\sqrt{\bar{v}_0 - 1}\,\bar{y} - \frac{\bar{v}_0}{\sqrt{\bar{v}_0 - 1}}\right)\right] + \operatorname{erf}\left(\sqrt{\frac{|\mathrm{Pe}|}{2}}\,\frac{\bar{v}_0}{\sqrt{\bar{v}_0 - 1}}\right)}{\operatorname{erf}\left[\sqrt{\frac{|\mathrm{Pe}|}{2}}\left(\sqrt{\bar{v}_0 - 1} - \frac{\bar{v}_0}{\sqrt{\bar{v}_0 - 1}}\right)\right] + \operatorname{erf}\left(\sqrt{\frac{|\mathrm{Pe}|}{2}}\,\frac{\bar{v}_0}{\sqrt{\bar{v}_0 - 1}}\right)} \tag{4.23}$$

If the rate of vertical ice motion is constant over the glacier thickness (rate of melting at upper surface exactly equal to rate of freezing at lower surface, Figure 4.1i), then the temperature field will be

$$\bar{T} = \frac{\exp(|\mathrm{Pe}|\bar{y}) - 1}{\exp(|\mathrm{Pe}|) - 1}\,. \tag{4.24}$$

The temperature field of a glacier in which the vertical velocity at the upper surface is zero, but with freezing at the lower surface ($v_H H/a = 0$, $\mathrm{Pe}_0 \neq 0$, Figure 4.1h), is described by the expression

$$\bar{T} = \frac{\operatorname{erf}\left[\sqrt{\frac{|\mathrm{Pe}_0|}{2}}\,(\bar{y} - 1)\right] + \operatorname{erf}\left(\sqrt{\frac{|\mathrm{Pe}_0|}{2}}\right)}{\operatorname{erf}\left(\sqrt{\frac{|\mathrm{Pe}_0|}{2}}\right)}\,. \tag{4.25}$$

Under conditions of longitudinal contraction of the glacier ice, with upward motion of the ice, the rate of ablation at the upper surface is greater than the rate of freezing at the lower surface (Figure 4.1j). Then the distribution of vertical velocities becomes

$$v = |v_H|\left(\bar{v}_0 + \frac{1 - \bar{v}_0}{H}\, y\right). \tag{4.26}$$

After carrying out similar calculations [29] for this case as well, we get

$$\bar{T} = - \frac{I\left[\sqrt{\frac{|Pe|}{2}}\left(\sqrt{1-\bar{v}_0}\,\bar{y} + \frac{\bar{v}_0}{\sqrt{1-\bar{v}_0}}\right)\right] - I\left(\sqrt{\frac{|Pe|}{2}}\frac{\bar{v}_0}{\sqrt{1-\bar{v}_0}}\right)}{I\left[\sqrt{\frac{|Pe|}{2}}\left(\sqrt{1-\bar{v}_0} + \frac{\bar{v}_0}{\sqrt{1-\bar{v}_0}}\right)\right] - I\left(\sqrt{\frac{|Pe|}{2}}\frac{\bar{v}_0}{\sqrt{1-\bar{v}_0}}\right)} \tag{4.27}$$

If the rate of freezing at the lower surface is zero ($\bar{v}_0 = 0$, Figure 4.1k), then expression (4.27) simplifies to

$$\bar{T} = \frac{I\left(\sqrt{\frac{|Pe|}{2}}\right)}{I\left(\sqrt{\frac{Pe}{2}}\right)}. \tag{4.28}$$

In Figures 4.4 and 4.5 the dashed lines correspond to the case in which the glacier ice does not move. The curves above the lines show the dependence of the dimensionless temperature $\bar{T}$ on the dimensionless $\bar{y}$, for values of the Péclet number typical of real glaciers.

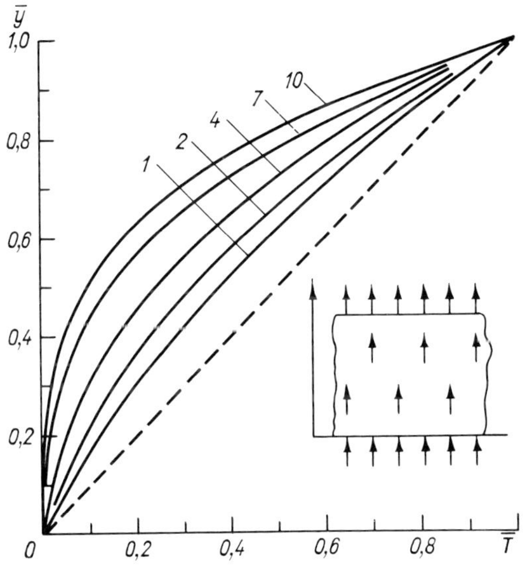

Fig. 4.4. Dimensionless temperature distribution in glacier, for warming due to ice motion. Pe = 1. Numbers by curves correspond to values of dimensionless freezing rate at lower surface. Dashed line corresponds to absence of ice motion in glacier (Pe = 0).

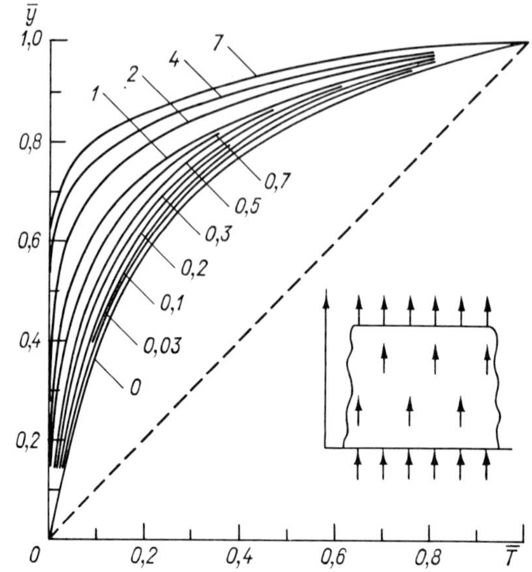

Fig. 4.5. Temperature distribution similar to that in Figure 4.4, except that here Pe = 6.

4.2.3. S-Shaped Temperature Fields

A glacier could possibly exist in which the directions of vertical ice motion are different in its upper and lower parts. For instance, consider an ice shelf with an accumulation of precipitation at its upper surface and melting at its lower surface. This case corresponds to distribution h in Figure 4.1. The upper part of the glacier is apparently being cooled off (concave curve of temperature distribution), while the lower part is being warmed up (curve of temperature distribution should be convex). In [29], where we were the first to point out the possibility of this type of temperature field, set up just by the ice motion in the glacier, such a distribution was called an S-shaped distribution.

A glacier having the vertical-velocity distribution shown in Figure 4.1f can be divided into two zones: an upper zone with a thickness $H/(1 + |\bar{v}_0|)$ in which the ice motion is downward, and a lower zone with a thickness $H|\bar{v}_0|/(1 + |\bar{v}_0|)$ in which the ice moves upward. The vertical-velocity distribution in the upper zone corresponds to cooling, and the distribution in the lower zone corresponds to warming. The point at which the vertical ice velocity is zero lies a distance $H|\bar{v}_0|/(1 + |\bar{v}_0|)$ from the lower surface of the glacier. If we call the temperature at this point T_h, then in the region from $\bar{y} = 0$ to $y = |\bar{v}_0|/(1 + |\bar{v}_0|)$ the temperature field in the glacier interior will be

$$\frac{T - T_0}{T_h - T^0} = \frac{\operatorname{erf}\left[\sqrt{\frac{|Pe|}{2}}\,\frac{\sqrt{|\bar{v}_0|}}{\sqrt{1 + |\bar{v}_0|}}\left(\frac{1 + |\bar{v}_0|}{|\bar{v}_0|}\,\bar{y} - 1\right)\right] + \operatorname{erf}\left(\sqrt{\frac{|Pe|}{2}}\,\frac{\sqrt{|\bar{v}_0|}}{\sqrt{1 + |\bar{v}_0|}}\right)}{\operatorname{erf}\left(\sqrt{\frac{|Pe|}{2}}\,\frac{\sqrt{|\bar{v}_0|}}{\sqrt{1 + |\bar{v}_0|}}\right)}, \tag{4.29}$$

and in the region from $\bar{y} = |\bar{v}_0|/(1 + |\bar{v}_0|)$ to $\bar{y} = 1$ it will be

$$\frac{T - T_0}{T_H - T_h} = \frac{\operatorname{erf}\left[\sqrt{\frac{|Pe|}{2}}\,\frac{1}{\sqrt{1 + |\bar{v}_0|}}\left(\bar{y} - \frac{|\bar{v}_0|}{\sqrt{1 + |\bar{v}_0|}}\right)\right]}{\operatorname{erf}\left(\sqrt{\frac{|Pe|}{2}}\,\frac{\sqrt{|\bar{v}_0|}}{\sqrt{1 + |\bar{v}_0|}}\right)}. \tag{4.30}$$

The value of T_h at the boundary between the upper and lower zones is found from the condition that there be a smooth transition of the temperature curve from zone to zone. The transition will be smooth if the temperature gradients corresponding to distributions (4.29) and (4.30) are equal at the point where the velocity of vertical ice motion is zero.

Next we differentiate (4.29) and (4.30), and after some transformations we obtain an expression giving T_h at the boundary between the upper and lower zones, that is, at the point where the vertical velocity is zero:

$$T_h = \frac{T_H \operatorname{erf}\left(\sqrt{\frac{|Pe|}{2}}\,\frac{\sqrt{|\bar{v}_0|}}{\sqrt{1 + |\bar{v}_0|}}\right) + T_0 \operatorname{erf}\left(\sqrt{\frac{|Pe|}{2}}\,\frac{1}{\sqrt{1 + |\bar{v}_0|}}\right)}{\operatorname{erf}\left(\sqrt{\frac{|Pe|}{2}}\,\frac{1}{\sqrt{1 + |\bar{v}_0|}}\right) + \operatorname{erf}\left(\sqrt{\frac{|Pe|}{2}}\,\frac{\sqrt{|\bar{v}_0|}}{\sqrt{1 + |\bar{v}_0|}}\right)}. \tag{4.31}$$

Expressions (4.29), (4.30), and (4.31) describe unambiguously the temperature field in such a glacier.

It is evident from Equation (4.31) that the temperature at the point where the vertical ice velocity is zero has to be considered a complex function of Pe and $\bar{v}_0$. It is interesting to note, however, that if $v_H = -v_0$ expression (4.31) no longer depends on Pe and simplifies to

$$T_h = \frac{T_H + T_0}{2} . \tag{4.32}$$

As an example, Figure 4.6 shows such a temperature field in the interior of a glacier. The curves were plotted for the case where the rates of accumulation of material at the upper and lower surfaces are identical ($v_H = -v_0$).

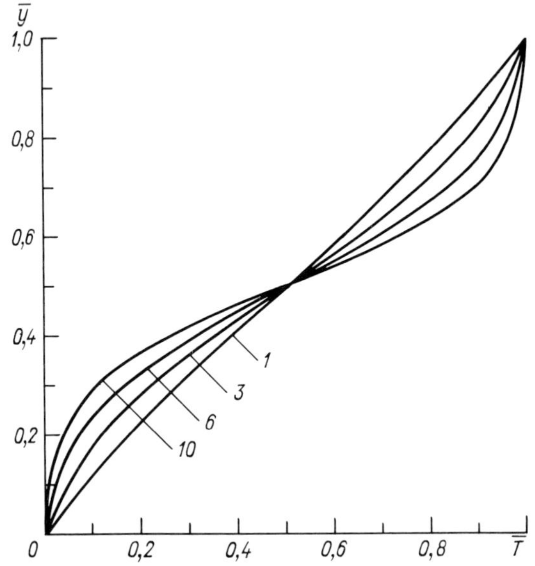

Fig. 4.6. Examples of S-shaped temperature field, where precipitation accumulates at the upper glacier surface and water freezes at the lower surface (after [29]). Rates of accumulation and freezing are equal. Numbers by curves give values of Péclet number.

Now let us consider a glacier in which melting takes place at its upper and lower surfaces. Such a glacier will be in a steady state if the longitudinal compression of the ice is such that an amount of mass is supplied to compensate for the mass lost through melting at the glacier surfaces. In this case the curve of the temperature field, which also has an inflection point, is the mirror image of an S-shaped curve. Let us find an analytical expression for the temperature field in this case, following [29]. As in the previous example, the interior of the glacier can be divided into two zones. In this case, in the region from $\bar{y} = 0$ to $\bar{y} = |\bar{v}_0|/(1 + |\bar{v}_0|)$, where the velocity of vertical ice motion is downward, the temperature field can be represented as

$$\frac{T - T_0}{T_h - T_0} = \frac{I\left[\sqrt{\frac{|Pe|}{2}}\,\frac{\sqrt{|\bar{v}_0|}}{\sqrt{1 + |\bar{v}_0|}}\left(\frac{1 + |\bar{v}_0|}{|\bar{v}_0|}\,\bar{y} + 1\right)\right] + I\left(\sqrt{\frac{|Pe|}{2}}\,\frac{\sqrt{|\bar{v}_0|}}{\sqrt{1 + |\bar{v}_0|}}\right)}{I\left(\sqrt{\frac{|Pe|}{2}}\,\frac{\sqrt{|\bar{v}_0|}}{\sqrt{1 + |\bar{v}_0|}}\right)}. \tag{4.33}$$

In the region from $\bar{y} = |\bar{v}_0|/(1 + |\bar{v}_0|$ to $\bar{y} = 1$, on the other hand, the temperature field is

$$\frac{T - T_h}{T_H - T_h} = \frac{I\left[\sqrt{\frac{|Pe|}{2}}\,\frac{\sqrt{|\bar{v}_0|}}{\sqrt{1 + |\bar{v}_0|}}\left(\frac{1 + |\bar{v}_0|}{|\bar{v}_0|}\,\bar{y} + 1\right)\right]}{I\left(\sqrt{\frac{|Pe|}{2}}\,\frac{1}{\sqrt{|\bar{v}_0| + 1}}\right)}. \tag{4.34}$$

As in the previous example, T_h at the boundary between the upper and lower zones is found from the conditions of smooth transition of the temperature curve from zone to zone. After some transformations analogous to those made above, we obtain

$$T_h = \frac{T_H I\left(\sqrt{\frac{|Pe|}{2}}\,\frac{\sqrt{|\bar{v}_0|}}{\sqrt{1 + |\bar{v}_0|}}\right) + T_0 I\left(\sqrt{\frac{|Pe|}{2}}\right)}{I\left(\sqrt{\frac{|Pe|}{2}}\,\frac{\sqrt{\bar{v}_0}}{\sqrt{1 + |\bar{v}_0|}}\right) + I\left(\sqrt{\frac{|Pe|}{2}}\right)}. \tag{4.35}$$

Expressions (4.31), (4.34), and (4.35) provide an unambiguous representation of the temperature field in a glacier in which ablation is observed at the upper and lower surfaces.

Just as in the previous case, for $v_H = - v_0$, when the rate of ablation at the upper surface equals the rate of bottom melting, the temperature at the point where the vertical velocity is zero does not depend on Pe and is given by the simple Equation (4.32).

4.3. Mean Temperature of a Glacier

In many cases the mean temperature of the glacier ice has to be known as well as the temperature distribution in the glacier. In the case being considered by us, where the ice has a constant density and specific heat over the entire glacier thickness, the mean temperature is

$$T_{av} = \frac{1}{H}\int_0^H T\,dy. \tag{4.36}$$

In the absence of glacier motion, when the temperature field of the glacier is linear, the mean temperature is defined as the halfsum of the temperatures at the upper and lower surfaces:

$$T_{av} = \frac{1}{H}\int_0^H \left(T_0 + \frac{T_H - T_0}{H}\,y\right) dy = \frac{T_H - T_0}{2}. \tag{4.37}$$

Thus the dimensionless mean temperature $\bar{T}_{av}$ in such a glacier will be

$$\bar{T}_{av} = \frac{T_{av} - T_0}{T_H - T_0} = \frac{1}{2}. \tag{4.38}$$

The dimensionless mean temperature of the ice in a glacier in which there is vertical ice motion differs from ½ by an amount K_T. Let us determine K_T, which we will call the coefficient of deformation of the temperature field due to ice motion. First we write

$$T_{av} = \frac{T_H - T_0}{2}K_T + T_0 = \frac{T_H - T_0}{2}\left(K_T + \frac{2T_0}{T_H - T_0}\right). \tag{4.39}$$

Then, in view of (4.38) and (4.39), we obtain the following expression for K_T:

$$K_T = \frac{2}{T_H - T_0}\,\frac{1}{H}\int_0^H T\,dy - \frac{2T_0}{T_H - T_0}, \tag{4.40}$$

or, taking (4.6) into account,

$$K_T = \frac{2}{H}\int_0^H \frac{\int_0^y \exp\left(\int_0^y \frac{v}{a}\,dy\right) dy}{\int_0^H \exp\left(\int_0^y \frac{v}{a}\,dy\right) dy}. \tag{4.41}$$

In the absence of vertical ice motion, when the temperature field is linear, $K_T = 1$, and when there is vertical motion $K_T \neq 1$.

Inspection of the curves in Figures 4.2 and 4.6 indicates that K_T should be greater than unity for downward motion of the ice in the glacier. Then the mean temperature of the ice will be lower than in a hypothetical glacier without vertical ice motion. Finally, K_T should be less than unity if the ice motion in the glacier is directed upward.

The integrals in Equation (4.41) can be represented in terms of elementary or tabulated functions for each of the types of vertical-velocity distributions depicted in Figure 4.1. Once again, we distinguish between the following two regimes: cooling of the glacier due to downward ice motion and warming due to upward ice motion.

4.3.1. Cooling of a Glacier Due to Ice Motion ($K_T > 1$)

Let us consider the motion of a glacier for longitudinal outflow of ice, when precipitation accumulates at its upper surface and melting takes place at its lower surface (Figure 4.1a). The temperature field in the glacier is then described by Equation (4.16). By substituting (4.16) into (4.40) and denoting, as previously,

$$\sqrt{\frac{|v_H|H}{2a}}\left(\sqrt{1-\bar{v}_0\bar{y}}+\frac{\bar{v}_0}{\sqrt{1+\bar{v}_0}}\right)=f,$$

we obtain a new variable of integration f, with the corresponding new limits of integration. Then

$$K_T=\frac{2\displaystyle\int_{\sqrt{\frac{|Pe|}{2}}\left(\sqrt{1-\bar{v}_0}\right)}^{\sqrt{\frac{|Pe|}{2}}\left(\sqrt{1-\bar{v}_0}+\frac{\bar{v}_0}{\sqrt{1-\bar{v}}}\right)}\operatorname{erf}(f)\,df-2\operatorname{erf}\left(\sqrt{\frac{|Pe|}{2}}\frac{\bar{v}_0}{\sqrt{1-\bar{v}_0}}\right)}{\operatorname{erf}\left[\sqrt{\frac{|Pe|}{2}}\left(\sqrt{1-\bar{v}_0}+\frac{\bar{v}_0}{\sqrt{1-\bar{v}_0}}\right)\right]-\operatorname{erf}\left(\sqrt{\frac{|Pe|}{2}}\frac{\bar{v}_0}{\sqrt{1-\bar{v}_0}}\right)}. \quad (4.42)$$

Note that

$$\int_y^\infty[1-\operatorname{erf}(f)]\,df=\int_y^\infty\operatorname{erfc}(f)\,df=\operatorname{ierfc}(y),$$

where ierfc (y) is a tabulated function [81].

Expression (4.42) in this case can be written as a ratio of tabulated functions:

$$K_T = \frac{2\left\{\operatorname{ierfc}\left[\sqrt{\frac{|Pe|}{2}}\left(\sqrt{1-\bar{v}_0} + \frac{\bar{v}_0}{\sqrt{1-\bar{v}_0}}\right)\right] - \operatorname{ierfc}\left(\sqrt{\frac{|Pe|}{2}}\frac{\bar{v}_0}{\sqrt{1-\bar{v}_0}}\right) - \sqrt{\frac{|Pe|}{2}}\sqrt{1-\bar{v}_0} \times \operatorname{erf}\left(\sqrt{\frac{|Pe|}{2}}\frac{\bar{v}_0}{\sqrt{1-\bar{v}_0}}\right)\right\}}{\sqrt{\frac{|Pe|}{2}}\sqrt{1-\bar{v}_0}\left\{\operatorname{erf}\left[\sqrt{\frac{|Pe|}{2}}\left(\sqrt{1-\bar{v}_0} + \frac{\bar{v}_0}{\sqrt{1-\bar{v}_0}}\right)\right] - \operatorname{erf}\left(\sqrt{\frac{|Pe|}{2}}\frac{\bar{v}_0}{\sqrt{1-\bar{v}_0}}\right)\right\}}. \tag{4.43}$$

For $\bar{v}_0 = 0$, that is, in the absence of melting at the lower surface of the glacier (Figure 4.1b), expression (4.43) simplifies to

$$K_T = \frac{2}{\sqrt{\frac{|Pe|}{2}}} \frac{\operatorname{ierfc}\left(\sqrt{\frac{|Pe|}{2}}\right) - \operatorname{ierfc}(0)}{\operatorname{erf}\left(\sqrt{\frac{|Pe|}{2}}\right)}. \tag{4.44}$$

If the velocity of vertically downward ice motion is constant over the entire glacier interior (Figure 4.1c), then K_T can be determined by substituting (4.18) into (4.40). After some transformations similar to those carried out above, we get

$$K_T = \frac{-2}{|Pe|} \frac{\exp(-|Pe|) + |Pe|}{\exp(-|Pe|) - 1}. \tag{4.45}$$

The cooling of the glacier due to vertical ice motion from the surface to the bottom regions is especially marked under conditions of longitudinal contraction of the ice, when the vertical velocity in the glacier interior exceeds the vertical velocity at the surface. This case corresponds to the vertical-velocity distribution type shown in Figure 4.1d. The coefficient characterizing the deformation of the temperature field can then be found by substituting (4.20) into (4.40). After some transformations similar to those carried out above, we obtain

$$K_T = \frac{2\left\{iI\left[\sqrt{\frac{|Pe|}{2}}\left(\sqrt{\bar{v}_0 - 1} + \frac{\bar{v}_0}{\sqrt{\bar{v}_0 - 1}}\right)\right] - iI\left(\sqrt{\frac{|Pe|}{2}}\,\frac{\bar{v}_0}{\sqrt{\bar{v}_0 - 1}}\right) + \sqrt{\frac{|Pe|}{2}} \times \sqrt{\bar{v}_0 - 1}\; I\left(\sqrt{\frac{|Pe|}{2}}\,\frac{\bar{v}_0}{\sqrt{\bar{v}_0 - 1}}\right)\right\}}{\sqrt{\frac{|Pe|}{2}}\sqrt{\bar{v}_0 - 1}\left\{I\left[\sqrt{\frac{|Pe|}{2}}\left(\sqrt{\bar{v}_0 - 1} + \frac{\bar{v}_0}{\sqrt{\bar{v}_0 - 1}}\right)\right] + I\left(\sqrt{\frac{|Pe|}{2}}\,\frac{\bar{v}_0}{\sqrt{\bar{v}_0 - 1}}\right)\right\}} \tag{4.46}$$

where

$$iI(m) = \int_0^m I(\zeta)\, d\zeta = \int_0^m \int_0^\zeta \exp(r^2)\, dr\, d\zeta.$$

If the velocity at the upper boundary of the glacier is zero (see Figure 4.1e), then K_T is found by substituting (4.21) into (4.40). Then

$$K_T = \frac{2\left[iI(0) - iI(-1) + \sqrt{\frac{|Pe_0|}{2}}\; I\left(\sqrt{\frac{|Pe_0|}{2}}\right)\right]}{\sqrt{\frac{|Pe_0|}{2}}\; I\left(\sqrt{\frac{|Pe_0|}{2}}\right)}\,. \tag{4.47}$$

4.3.2. Warming Due to Ice Motion ($K_T < 1$)

The vertical-velocity distributions in the glacier for this case are shown in Figure 4.1 g-k. For longitudinal spreading of the ice (see Figure 4.1g), the deformation coefficient is found by substituting (4.23) into (4.40). Then, after some transformations similar to those made in the derivation of (4.43), we obtain

$$K_T = \frac{2}{\sqrt{\frac{|Pe|}{2}}\sqrt{\bar{v}_0 - 1}} \times$$

$$\frac{\begin{gathered}\mathrm{ierfc}\left[\sqrt{\frac{|Pe|}{2}}\left(\sqrt{\bar{v}_0 - 1} - \frac{\bar{v}_0}{\sqrt{\bar{v}_0 - 1}}\right)\right] - \\ -\mathrm{ierfc}\left(\sqrt{\frac{|Pe|}{2}}\frac{\bar{v}_0}{\sqrt{\bar{v}_0 - 1}}\right) - 2\sqrt{\frac{|Pe|}{2}} \times \\ \times \sqrt{\bar{v}_0 - 1}\,\mathrm{erf}\left(\sqrt{\frac{|Pe|}{2}}\frac{\bar{v}_0}{\sqrt{\bar{v}_0 - 1}}\right)\end{gathered}}{\begin{gathered}\mathrm{erf}\left[\left(\sqrt{\frac{|Pe|}{2}}\left(\sqrt{\bar{v}_0 - 1} - \frac{\bar{v}_0}{\sqrt{\bar{v}_0 - 1}}\right)\right] - \\ - \mathrm{erf}\left(\sqrt{\frac{|Pe|}{2}}\frac{\bar{v}_0}{\sqrt{\bar{v}_0 - 1}}\right)\end{gathered}} . \tag{4.48}$$

Now, if the vertical ice velocity is constant over the glacier thickness. that is, if $\bar{v}_0 = 1$ (see Figure 4.1i), then K_T is determined by substituting expression (4.24) into (4.40):

$$K_T = \frac{2}{|Pe|} - \frac{\exp(|Pe|) - |Pe|}{\exp(|Pe|) - 1} . \tag{4.49}$$

If the vertical velocity at the upper surface is zero (see Figure 4.1h), then K_T is found by substituting expression (4.25) into (4.40). After some transformations, this gives

$$K_T = \frac{2\left[\mathrm{ierfc}\,(0) - \mathrm{ierfc}\,(1) + \sqrt{\frac{|Pe|}{2}}\,\mathrm{erf}\left(\sqrt{\frac{|Pe_0|}{2}}\right)\right]}{\sqrt{\frac{|Pe|}{2}}\,\mathrm{erf}\left(\sqrt{\frac{|Pe_0|}{2}}\right)} . \tag{4.50}$$

For longitudinal contraction of the ice as it moves vertically upward (see Figure 4.1j), the deformation coefficient is found by substituting (4.26) into (4.40), from which

$$K_T = \frac{2}{\sqrt{\frac{|Pe|}{2}}\sqrt{1-\bar{v}_0}} \times$$

$$\times \frac{iI\left[\sqrt{\frac{|Pe|}{2}}\left(\sqrt{1-\bar{v}_0}+\frac{\bar{v}_0}{\sqrt{1-\bar{v}_0}}\right)\right] + I\left(\sqrt{\frac{|Pe|}{2}} \times \frac{\bar{v}_0}{\sqrt{1-\bar{v}_0}}\right) - \sqrt{\frac{|Pe|}{2}}\sqrt{1-\bar{v}_0}\, I\left(\sqrt{\frac{|Pe|}{2}}\frac{\bar{v}_0}{\sqrt{1-\bar{v}_0}}\right)}{I\left[\sqrt{\frac{|Pe|}{2}}\left(\sqrt{1-\bar{v}_0}+\frac{\bar{v}_0}{\sqrt{1-\bar{v}_0}}\right)\right] - I\left(\sqrt{\frac{|Pe|}{2}}\frac{\bar{v}_0}{\sqrt{1-\bar{v}_0}}\right)}. \qquad (4.51)$$

If the rate of freezing at the lower surface is zero (Figure 4.1k), then expression (4.51) simplifies (since $\bar{v}_0 = 0$) to

$$K_T = \frac{2iI\left(\sqrt{\frac{|Pe|}{2}}\right)}{\sqrt{\frac{|Pe|}{2}}\, I\left(\sqrt{\frac{|Pe|}{2}}\right)} . \qquad (4.52)$$

The determination of the mean ice temperature in a glacier having an S-shaped temperature field does not differ from the cases described.

4.4. Temperature Gradients in the Glacier Interior

The field of temperature gradients, which governs the heat flows in the interior of a glacier, is a very important characteristic of the thermal regime of the glacier. In the absence of vertical ice motion in the glacier, and once again assuming that the physical parameters of the ice are constant over the glacier interior, we can write the temperature gradient, which is constant over the glacier thickness, as

$$\frac{dT}{dy} = \frac{T_H - T_0}{H} . \qquad (4.53)$$

If vertical ice motion is present in the glacier, then this gradient will be different at different distances from the glacier surface, and it will not be the same as the gradient in stationary ice.

In the general formulation of the problem, the temperature gradient in such a moving glacier is given by Equation (4.5). In order to obtain generalized relationships, it is convenient to introduce the dimensionless temperature gradient in the glacier. Let us define the dimensionless temperature gradient in a moving glacier $d\bar{T}/d\bar{y}$ at a distance $\bar{y}$ from the surface as a ratio of the temperature gradient in this moving glacier at a distance y to the gradient in ice which is at rest. Now, in view of (4.53) and (4.5), we have

$$\frac{dT}{dy} = \frac{d\bar{T}}{d\bar{y}} \frac{T_H - T_0}{H}.$$

The dimensionless temperature gradient is defined as

$$\frac{d\bar{T}}{d\bar{y}} = \frac{H \exp\left(\int_0^y \frac{v}{a}\, dy \right)}{\int_0^H \left[\exp\left(\int_0^y \frac{v}{a}\, dy \right) \right] dy} . \tag{4.54}$$

Just as before, the integrals in the numerator and denominator of (4.54) can be expressed in terms of elementary functions only for certain distributions of the vertical ice velocity over the glacier thickness. For the further analysis of this, it will once again be convenient to distinguish between regimes corresponding to cooling and warming of the ice due to its vertical motion.

Motion-caused cooling. The dimensionless temperature gradient is easy to determine in this case, provided that the dimensionless temperature distribution in the ice interior is known, since

$$\frac{d\bar{T}}{d\bar{y}} = \frac{\frac{dT}{dy}}{\left.\frac{dT}{dy}\right|_{v=0}} . \tag{4.55}$$

Thus, the dimensionless temperature gradient in a glacier with the vertical-velocity distribution depicted in Figure 4.1a will be found by differentiating expression (4.16) with respect to $\bar{y}$:

$$\frac{d\bar{T}}{d\bar{y}} = \frac{2\left\{\sqrt{\frac{|Pe|}{2}}\, \sqrt{1-\bar{v}_0}\, \exp - \left[\sqrt{\frac{|Pe|}{2}}\left(\sqrt{1-\bar{v}_0}\,\bar{y} + \frac{\bar{v}_0}{\sqrt{1-\bar{v}_0}}\right)\right]^2\right\}}{\sqrt{\pi}\left\{\operatorname{erf}\left[\sqrt{\frac{|Pe|}{2}}\left(\sqrt{1-\bar{v}_0} + \frac{\bar{v}_0}{\sqrt{1-\bar{v}_0}}\right)\right] - \operatorname{erf}\left(\sqrt{\frac{|Pe|}{2}}\, \frac{\bar{v}_0}{\sqrt{1-\bar{v}_0}}\right)\right]} . \tag{4.56}$$

If the rate of bottom melting is zero ($\bar{v}_0 = 0$, Figure 4.1b), then the dimensionless gradient is simpler:

$$\frac{d\bar{T}}{d\bar{y}} = \sqrt{\frac{|Pe|}{2}}\, \frac{\exp - \left(\sqrt{\frac{|Pe|}{2}}\, \bar{y}\right)^2}{\operatorname{erf}\left(\sqrt{\frac{|Pe|}{2}}\right)} . \tag{4.57}$$

The temperature gradient in a glacier with a downward velocity of vertical ice motion which is constant over the glacier thickness ($\bar{v}_0 = 1$, Figure 4.1c) is found by differentiating equation (4.18):

$$\frac{d\bar{T}}{d\bar{y}} = |Pe| \frac{\exp(-|Pe|\bar{y})}{1 - \exp(|Pe|)} . \tag{4.58}$$

If the ice moves under conditions of longitudinal contraction, that is, if the rate of bottom melting is higher than the vertical ice velocity at the upper surface ($\bar{v}_0 > 1$, Figure 4.1d), then the temperature gradient is given by the expression

$$\frac{d\bar{T}}{d\bar{y}} = \frac{\sqrt{\frac{|Pe|}{2}}\sqrt{\bar{v}_0 - 1}\,\exp\left[\sqrt{\frac{|Pe|}{2}}\left(\sqrt{\bar{v}_0-1}\,\bar{y} + \frac{\bar{v}_0}{\sqrt{\bar{v}_0-1}}\right)\right]^2}{I\left[\sqrt{\frac{|Pe|}{2}}\left(\sqrt{\bar{v}_0-1} + \frac{\bar{v}_0}{\sqrt{\bar{v}_0-1}}\right)\right] + I\left(\sqrt{\frac{|Pe|}{2}}\,\frac{\bar{v}_0}{\sqrt{\bar{v}_0-1}}\right)} . \tag{4.59}$$

Motion-caused warming. As shown above, if the vertical ice velocity in a glacier is directed upward, corresponding, for instance, to melting at the upper surface and freezing at the lower surface, then the mean temperature in the glacier will be higher than the mean temperature in the absence of ice motion. Differentiating (4.23)-(4.26), we can easily obtain expressions for the temperature gradients under various conditions of ice motion, attended by warming of the ice.

Under conditions of longitudinal spreading of the ice, when the melting rate at the surface is lower than the freezing rate at the bed ($\bar{v}_0 > 1$, Figure 4.1g), the dimensionless gradients are

$$\frac{d\bar{T}}{d\bar{y}} = \frac{2\sqrt{\frac{|Pe|}{2}}\sqrt{\bar{v}_0 - 1}\,\exp - \left[\sqrt{\frac{|Pe|}{2}}\left(\sqrt{\bar{v}_0-1}\,\bar{y} - \frac{\bar{v}_0}{\sqrt{\bar{v}_0-1}}\right)\right]^2}{\sqrt{\pi}\left\{\operatorname{erf}\left[\sqrt{\frac{|Pe|}{2}}\left(\sqrt{\bar{v}_0-1} - \frac{\bar{v}_0}{\sqrt{\bar{v}_0-1}}\right)\right] + \operatorname{erf}\left(\sqrt{\frac{|Pe|}{2}}\,\frac{\bar{v}_0}{\sqrt{\bar{v}_0-1}}\right)\right\}} . \tag{4.60}$$

If the rate of vertical ice motion is constant over the entire glacier interior ($v_0 = 1$, Figure 4.1i), then

$$\frac{d\bar{T}}{d\bar{y}} = \frac{|Pe|\exp(|Pe|\bar{y})}{\exp(|Pe|) - 1} . \tag{4.61}$$

For longitudinal contraction of the glacier ice, when the ablation rate at the upper surface is higher than the freezing rate at the bottom $\bar{v}_0 < 1$, Figure 4.1j), the expression for the temperature gradient is

$$\frac{d\bar{T}}{d\bar{y}} = \frac{\sqrt{\frac{|\mathrm{Pe}|}{2}}\sqrt{\bar{v}_0 - 1}\,\exp\left[\sqrt{\frac{|\mathrm{Pe}|}{2}}\left(\sqrt{1-\bar{v}_0}\,\bar{y} + \frac{\bar{v}_0}{\sqrt{1-\bar{v}_0}}\right)\right]^2}{I\left[\sqrt{\frac{|\mathrm{Pe}|}{2}}\left(\sqrt{1-\bar{v}_0} + \frac{\bar{v}_0}{\sqrt{1-\bar{v}_0}}\right)\right] - I\left(\sqrt{\frac{|\mathrm{Pe}|}{2}}\frac{\bar{v}_0}{\sqrt{\bar{v}_0 - 1}}\right)}. \tag{4.62}$$

4.5. The Effect of the Dissipation Function

As the results in Chapter 3 imply, the heat transfer in a flat two-dimensional glacier cannot in most cases be determined using Equation (3.63) or its dimensional equivalent (4.1), since these equations ignore the release of heat accompanying the motion of the glacier ice; this heat release is determined by the dissipation function Φ'. Consequently, to study the heat transfer in a glacier, we would have to analyze the heat-transfer equation in the form (3.62), or else in the form of its dimensional equivalent:

$$\frac{\partial^2 T}{\partial y^2} - \frac{v}{a}\frac{\partial T}{\partial y} = \frac{\Phi'}{\lambda}. \tag{4.63}$$

The rest of this section will follow our discussion in [29].

General case. By integrating equation (4.63) from 0 to y, we obtain

$$\frac{dT}{dy} = \exp\int_0^y \frac{v}{a}\,dy\left\{\int_0^y\left[\frac{\Phi'}{\lambda}\exp\left(-\int_0^y \frac{v}{a}\,dy\right)\right]dy + \left.\frac{dT}{dy}\right|_{y=0}\right\}. \tag{4.64}$$

Expression (4.64) describes the distribution of temperature gradients over the glacier thickness for a specified gradient at its lower surface. It is obvious that in the absence of heat sinks and sources equation (4.64) simplifies to expression (4.3).

The temperature gradient at the lower surface of the glacier can be found by integrating (4.64) from 0 to H:

$$\left.\frac{dT}{dy}\right|_{y=0} = \frac{T_H - T_0 - \int_0^H\left[\exp\left(\int_0^y \frac{v}{a}\,dy\right)\int_0^y \frac{\Phi'}{\lambda}\exp - \left(\int_0^y \frac{v}{a}\,dy\right)dy\right]dy}{\int_0^H \exp\left(\int_0^y \frac{v}{a}\,dy\right)dy}. \tag{4.65}$$

Substitution of (4.65) into (4.64) then gives the following equation for the temperature gradients in the glacier interior, in the presence of heat sinks and sources:

$$\frac{dT}{dy} = \exp\left(\int_0^y \frac{v}{a}\,dy\right)\left\{\int_0^y \frac{\Phi'}{\lambda}\exp\left(-\int_0^y -\frac{v}{a}\,dy\right)dy + \frac{T_H - T_0}{\int_0^H \exp\left(\int_0^y \frac{v}{a}\,dy\right)dy} - \frac{\int_0^H \exp\left[\int_0^y \frac{v}{a}\,dy \int_0^y \frac{\phi'}{\lambda}\exp\left(-\int_0^y \frac{v}{a}\,dy\right)dy\right]dy}{\int_0^H \exp\left(\int_0^y \frac{v}{a}\,dy\right)dy}\right\}. \qquad (4.66)$$

In the absence of sinks and sources this equation simplifies to (4.5).

The temperature field in a glacier with heat sinks and sources, determined by integrating (4.66) from 0 to y, is then

$$\bar{T} = \frac{1}{\lambda(T_H - T_0)}\int_0^y \exp\left[\int_0^y \frac{v}{a}\,dy \int_0^y \Phi' \exp\left(-\int_0^y \frac{v}{a}\,dy\right)dy\right]dy -$$

$$- \frac{\int_0^y \exp\left(\int_0^y \frac{v}{a}\,dy\right)dy \int_0^H \exp\left[\int_0^y \frac{v}{a}\,dy \int_0^y \Phi' \exp\left(-\int_0^y \frac{v}{a}\,dy\right)dy\right]dy}{\lambda(T_H - T_0)\int_0^H \exp\left(\int_0^y \frac{v}{a}\,dy\right)dy} +$$

$$+ \frac{\int_0^y \exp\left(\int_0^y \frac{v}{a}\,dy\right)dy}{\int_0^H \exp\left(\int_0^y \frac{v}{a}\,dy\right)dy}. \qquad (4.67)$$

A comparison of Equations (4.3), (4.5), (4.6), and (4.65)-(4.67) indicates that the presence of heat sources and sinks in the glacier makes the working equations extremely complicated. The problem is simplified somewhat, however, if we consider some particular cases.

For example, the simplest case is the distribution of vertical ice velocities shown in Figure 4.1c (downward vertical ice velocity which equals zero at the lower surface of the glacier). Then the vertical velocity v will be related very simply to the modulus of this velocity at the upper surface: $v = v_H = \text{const}$. Substitution of this relation into (4.67) then gives

$$\bar{T} = \frac{\int_0^y \exp\left(\frac{|v_H| y^2}{2Ha}\right) dy}{\int_0^H \exp\left(\frac{|v_H| y^2}{2Ha}\right) dy} \times$$

$$\times \left\{ 1 - \frac{\int_0^H \exp\left[-\frac{|v_H| y^2}{2Ha} \int_0^y \Phi' \exp\left(\frac{|v_H| y^2}{2Ha}\, dy\right)\right] dy}{\lambda(T_H - T_0)} \right\} +$$

$$+ \frac{1}{\lambda(T_H - T_0)} \int_0^y \exp\left[-\frac{|v_H| y^2}{2Ha} \int_0^y \Phi' \exp\left(\frac{|v_H| y^2}{2Ha}\, dy\right)\right] dy. \quad (4.68)$$

<u>Uniform heat release over glacier thickness</u>. Let us now assume that the heat sources have a constant strength over the glacier thickness. Then, after some transformations similar to those made above, we get

$$\bar{T} = \frac{\mathrm{erf}\left(\sqrt{\frac{|Pe|}{2}}\,\bar{y}\right)}{\mathrm{erf}\left(\sqrt{\frac{|Pe|}{2}}\right)} + \bar{\Phi}' \frac{E^*\left(\sqrt{\frac{|Pe|}{2}}\,\bar{y}\right)}{E^*\left(\sqrt{\frac{|Pe|}{2}}\right)} -$$

$$- \frac{\mathrm{erf}\left(\sqrt{\frac{|Pe|}{2}}\,\bar{y}\right)}{\mathrm{erf}\left(\sqrt{\frac{|Pe|}{2}}\right)} \frac{2E^*\left(\sqrt{\frac{|Pe|}{2}}\right)}{|Pe|}, \quad (4.69)$$

where

$$E^*(y) = \int_0^y \exp\left[-\varphi^2 \int_0^\varphi \exp(\varphi^2)\, d\varphi\right] d\varphi; \quad \bar{\Phi}' = \frac{\Phi H^2}{\lambda(T_H - T_0)}$$

is the dimensionless strength of the internal heat sources or sinks. In the case of heat sources emitting during ice motion, this quantity is the dimensionless dissipation function.

It is interesting to note that this very same expression for the dimensionless dissipation function was already obtained by us in the

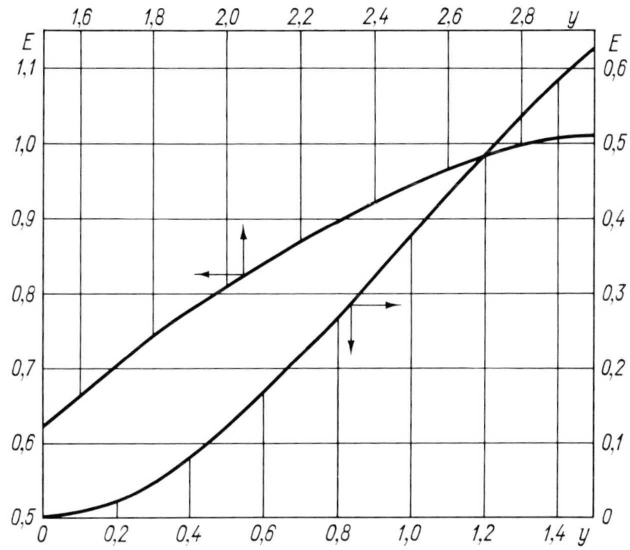

Fig. 4.7. Determining effect of heat sources distributed uniformly over glacier thickness. Source strengths E* in (4.69) for various values of independent variable.

preceding chapter, by a completely different means.

Note, too, that the dimensionless strength of the heat sources is the ratio of the amount of heat released over the entire glacier to the amount of heat passing through such a glacier, but without ice motion, just due to the difference between the temperatures at the upper and lower surfaces. This is obvious from the structure of the expression obtained for $\bar{\Phi}'$. The value of $E^*(y)$ have been tabulated by us and also plotted in Figure 4.7.

One of the two terms of expression (4.69) describes the temperature field in a glacier without heat sources or sinks (cf. Equation (4.17)), while the other term takes into account the effect of sources or sinks on the temperature field. The temperature difference caused by the presence of heat sources or sinks can be conveniently indicated by a dimensionless correction factor $\Delta\bar{T}_q$. If this correction is defined as $\Delta\bar{T}_q = (T - T_{q=0})/(T_H - T_0)$, where T is the temperature taking sources and sinks into account and $T_{q=0}$ is the temperature if sources and sinks are neglected, then for the case being considered the dimensionless correction for the effect of heat sources or sinks will be

$$\Delta\bar{T}_q = \bar{\Phi}' \frac{2E^*\left(\sqrt{\frac{|Pe|}{2}}\right)}{|Pe|} \left[\frac{E^*\left(\sqrt{\frac{|Pe|}{2}}\,\bar{y}\right)}{E^*\left(\sqrt{\frac{|Pe|}{2}}\right)} - \frac{\mathrm{erf}\left(\sqrt{\frac{|Pe|}{2}}\,\bar{y}\right)}{\mathrm{erf}\left(\sqrt{\frac{|Pe|}{2}}\right)}\right] . \quad (4.70)$$

As Equation (4.70) shows, this correction increases linearly with an increase in the dimensionless strength of the heat sources.

Chapter 5

APPROXIMATE METHODS FOR STUDYING THE TEMPERATURE FIELD IN A FLAT TWO-DIMENSIONAL GLACIER WITH BLOCK SLIDING

5.1. The Method of Polynomials

The complexity of the working formulas for the temperature field, even in the simplest case of a flat two-dimensional glacier with block motion, led us to examine the possibility of using approximate methods. One such method seeks, instead of an accurate solution of the equation, an approximate solution in the form of a linear combination of some system of linearly independent functions, satisfying the boundary conditions and certain other conditions. Below, the ideas of this method will be applied to analyze the temperature distribution in a glacier.

5.1.1. The Method of the 'Zeroth Moment'

As our linear combination of functions, it would be of interest to use a polynomial of low (for instance, fourth) degree [55]. Let us use this method to solve the previously considered problem of the temperature distribution in a glacier with accumulation at its upper surface. It will be assumed that neither melting nor freezing occurs at the lower surface. The heat-transfer equation will be used in its simplest form (3.64), but with the sign in front of the convective term changed. This is done so that, when analyzing a glacier with accumulation of precipitation at its surface (downward, and thus negative, vertical velocity), Pe will be positive (actually it is taken to be equal to its absolute value).

The boundary conditions are in this case:

$$\bar{y} = 0, \quad \bar{T} = 0; \quad \bar{y} = 1, \quad \bar{T} = 1. \tag{5.1}$$

Let us seek a solution of the equation in the form of a fourth-degree polynomial:

$$\bar{T} = A + B\bar{y} + C\bar{y}^2 + D\bar{y}^3 + E\bar{y}^4. \tag{5.2}$$

From the boundary conditions we get

$$A = 0; \quad B + C + D + E = 1. \tag{5.3}$$

In order to determine the coefficients of this polynomial, we must add three additional conditions to the above two boundary conditions. The first two of these are that polynomial (5.2) must satisfy Equation (3.64) at the upper and lower boundaries of the glacier. Thus we have

$$BPe + (6 + 3Pe)D + (12 + 4Pe)E = 0; \quad C = 0.$$

Loitsyanskii [55] has demonstrated that, in order to solve problems like these, we can use a whole system of additional integral conditions, obtained by multiplying both terms of the equation by successive degrees of the coordinates (0, 1, 2, ...) and then integrating over the entire thickness of the layer in which the process is being considered.

For the case at hand, we will use the integral condition for the zeroth degree of the coordinates, that is, the equation of the 'zeroth moment', to use Loitsyanskii's terminology:

$$\int_0^1 \left(\mathrm{Pe}\bar{y} \frac{d\bar{T}}{d\bar{y}} + \frac{d^2\bar{T}}{d\bar{y}^2} \right) d\bar{y} = 0. \tag{5.4}$$

Substituting (5.2) into (5.4), we now obtain

$$0{,}5\mathrm{Pe}B + (3 + 0.75\mathrm{Pe})D + (4 + 0.8\mathrm{Pe})E = 0.$$

Therefore, to determine the nonzero coefficients, we have a system of three equations:

$$\begin{aligned} &B + D + E = 1; \\ &B + D(6 + 3\mathrm{Pe}) + E(12 + 4\mathrm{Pe}) = 0; \\ &B + D(3 + 0.75\mathrm{Pe}) + E(4 + 0.8\mathrm{Pe}) = 0. \end{aligned} \tag{5.5}$$

Solution of these equations and substitution into expression (5.2) now yield the following equation for the temperature field in a glacier with the foregoing assumptions concerning the process:

$$\bar{T} = \frac{0.6\mathrm{Pe}^2 + 4.2\mathrm{Pe} + 12}{0.15\mathrm{Pe}^2 + 2.2\mathrm{Pe} + 12}\bar{y} - \frac{1.2\mathrm{Pe}^2 + 2\mathrm{Pe}}{0.15\mathrm{Pe}^2 + 2.2\mathrm{Pe} + 12}\bar{y}^3 + \\ + \frac{0.75\mathrm{Pe}^2}{0.15\mathrm{Pe}^2 + 2.2\mathrm{Pe} + 12}\bar{y}^4. \tag{5.6}$$

Figure 5.1 shows the temperature fields calculated using formula (5.1) for various values of Pe observed in real glaciers (Pe = 1, 3, 6, 10, 15). The abscissa is the dimensionless temperature $\bar{T}$, and the ordinate is the dimensionless distance $\bar{y}$ from the bottom of the glacier. The dashed curves in the figure show the temperature variations over the glacier thickness obtained on the basis of (4.17), corresponding to an accurate solution of Equation (3.64) with these same boundary conditions and Pe values.

The curves obtained using the approximate and precise methods are seen to be practically identical for Pe values from 1 to 6. It is precisely these values of Pe which are observed for the central regions of the Antarctic Ice Sheet, the large ice shelves, and many mountain glaciers. The discrepancy for high values of Pe, however, is quite considerable, as the figure demonstrates.

In view of the discrepancy at high Pe between the accurate solution and the solution obtained with the 'approximate zeroth-moment method', we decided to examine some other additional conditions which

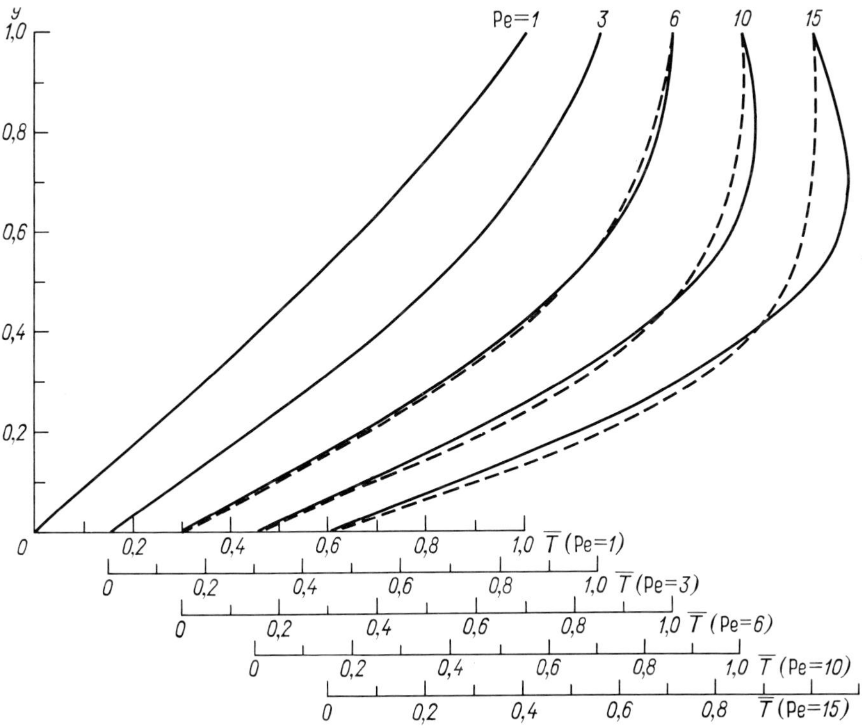

Fig. 5.1. Temperature distributions over glacier thickness, obtained using approximate method of 'zeroth moment' (solid curves) and from accurate solutions of equations (dashed curves).

could possibly be used to determine the coefficients in expression (5.3).

5.1.2. The Method of the 'Zeroth' and 'First' Moments

The above-mentioned three additional conditions required in order to find the coefficients in Equation (5.2) may differ from the conditions just used. For instance, instead of selecting the coefficients in (5.2) from the condition that Equation (3.64) be satisfied at the lower boundary of the glacier, as was done above, we can, in addition to the above equation of the 'zeroth moment' (5.4), construct and employ the equation of the 'first moment'. On the basis of the foregoing, we can write the equation of the 'first moment' as

$$\int_0^1 \left(\mathrm{Pe}\bar{y} \frac{d\bar{T}}{d\bar{y}} + \frac{d^2\bar{T}}{d\bar{y}^2} \right) \bar{y} \, d\bar{y} = 0. \tag{5.7}$$

Substitution of (5.2) into (5.7) then gives

$$0.3333\mathrm{Pe}B + D(2 + 0.6\mathrm{Pe}) + E(3 + 0.6667\mathrm{Pe}) = 0. \tag{5.8}$$

Consequently, we now have a new system of equations for determining the nonzero coefficients:

$$\begin{aligned} &B + D + E = 1; \\ &0.5PeB + D(0.3 + 0.75Pe) + E(1 - 0.8Pe) = 0; \qquad (5.9) \\ &0.3333PeB + D(2 + 0.6Pe) + E(3 + 0.6667Pe) = 0. \end{aligned}$$

The solution of this system and the substitution of the coefficients so obtained into expression (5.2) then yield the following equation for the temperature field in a glacier, with the previously made assumptions concerning the process:

$$\bar{T} = \frac{1 + 0.25Pe + 0.02Pe^2}{1 + 0.0833Pe + 0.0033Pe^2}\,\bar{y} - \frac{0.1667Pe + 0.0667Pe^2}{1 + 0.0833Pe + 0.0033Pe^2}\,\bar{y}^2 + \frac{0.05Pe^2}{1 + 0.0833Pe + 0.0033Pe^2}\,\bar{y}^4. \qquad (5.10)$$

Figure 5.2 shows curves for the variation of the dimensionless temperature $\bar{T}$ over the glacier thickness $\bar{y}$, constructed using the coefficients of (5.10) corresponding to the 'approximate method of the first moment'.

The dashed curves shown here were also plotted using the accurate solution. As before, the temperature distribution obtained using the 'approximate method of the first moment' shows good agreement with the accurate solution for Pe values from 1 to 6, but it diverges from the accurate solution for Pe = 10 to 15. It should be noted that for sufficiently high Pe (10-15) an 'inverse gradient' near the glacier surface is obtained, that is, the temperature in the upper ice layers is lower than the surface temperature. This is contrary to the physics of the process.

Consequently, both of these methods are recommended for use only at moderate (up to 6) values of Pe.

5.1.3. Improving the Method of Polynomials

It would be of interest to try to eliminate the 'inverse gradient' effect and to make the approximate method agree better with the accurate solution. One way of doing this is to increase the number of terms in polynomial (5.2). Let us seek an approximate description of the temperature field in the form of a fifth-degree polynomial with six terms:

$$\bar{T} = A + B\bar{y} + C\bar{y}^2 + D\bar{y}^3 + E\bar{y}^4 + F\bar{y}^5. \qquad (5.11)$$

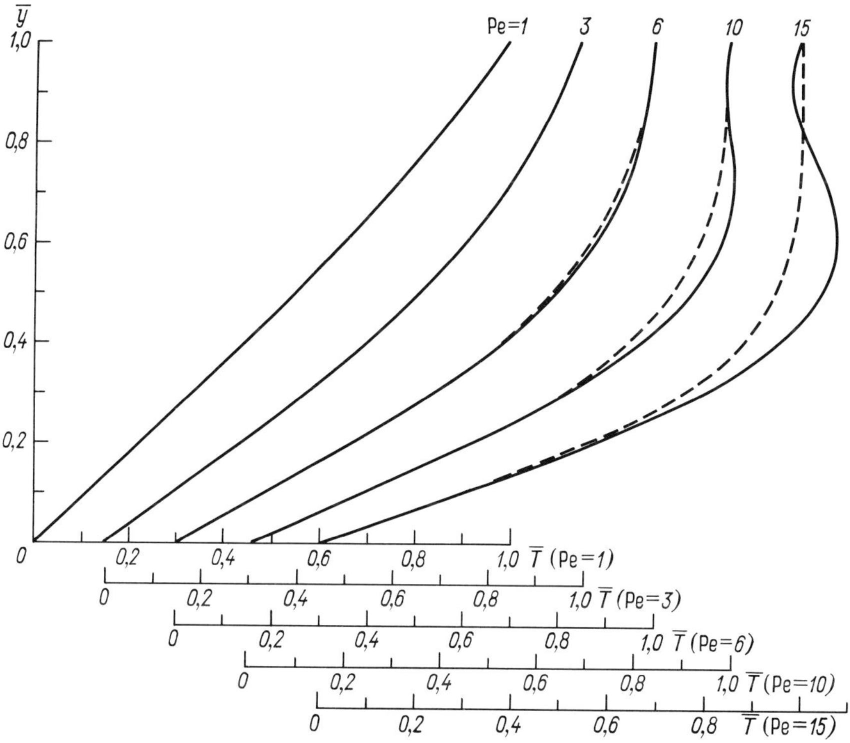

Fig. 5.2. Temperature distributions over glacier thickness, obtained using approximate method of 'zeroth and first moments' (solid curves) and from accurate solutions of equations (dashed curves).

We will take as the first five conditions to be satisfied by this polynomial the conditions which were used above to construct the 'approximate method of the zeroth moment'. As the sixth condition, let us add the equation of the 'first moment' (5.7).

Then, as previously, from the boundary conditions we have

$$A = 0; \; B + C + D + E + F = 1.$$

The conditions stating that (5.11) must satisfy Equation (3.64) at the boundaries then lead to

$$C = 0; \; B\mathrm{Pe} + D(6 + 3\mathrm{Pe}) + E(12 + 4\mathrm{Pe}) + F(20 + 5\mathrm{Pe}) = 0.$$

From the equations of the 'zeroth moment' and the 'first moment', respectively, we then get

$$\int_0^1 (6D\bar{y} + 12E\bar{y}^2 + 20F\bar{y}^3 + BPe\bar{y} + 3DPe\bar{y}^3 + $$
$$+ 4EPe\bar{y}^4 + 5Pe\bar{y}^5)\, d\bar{y} = 0;$$

$$\int_0^1 (6D\bar{y} + 12E\bar{y}^2 + 20F\bar{y}^3 + BPe\bar{y} + 3DPe\bar{y}^3 + $$
$$+ 4EPe\bar{y}^4 + 5Pe\bar{y}^5)\, \bar{y}\, d\bar{y} = 0.$$

After integration, the system of equations for the four nonzero coefficients of expression (5.11) can be written as

$$\begin{aligned} & B + D + E + F = 1; \\ & BPe + D(6 + 3Pe) + E(12 + 4Pe) + F(20 + Pe) = 0; \\ & B\frac{Pe}{2} + D(3 + \frac{3}{4}Pe) + E(4 + \frac{4}{5}Pe) + F(5 + \frac{5}{6}Pe) = 0; \\ & B\frac{Pe}{3} + D(2 + \frac{3}{5}Pe) + E(3 + \frac{2}{3}Pe) + F(4 + \frac{5}{7}Pe) = 0. \end{aligned} \tag{5.12}$$

Finally, after coefficients B, ..., F have been determined, the expression determining the temperature field and described by the fifth-degree polynomial has the form

$$\begin{aligned} \bar{T} = & \frac{2 + 0.62857Pe + 0.08333Pe^2 + 0.00476Pe^3}{2 + 0.29524Pe + 0.02857Pe^2 + 0.00063Pe^3}\bar{y} - \\ & - \frac{0.33333Pe + 0.10476Pe^2 + 0.03175Pe^3}{2 + 0.29524Pe + 0.02857Pe^2 + 0.00063Pe^3}\bar{y}^3 + \\ & + \frac{0.04762Pe^3}{2 + 0.29524Pe + 0.02857Pe^2 + 0.00063Pe^3}\bar{y}^4 - \\ & - \frac{0.02Pe^3 - 0.05Pe^2}{2 + 0.29524Pe + 0.02857Pe^2 + 0.00063Pe^3}\bar{y}^5. \end{aligned} \tag{5.13}$$

Although it turned out to be impossible to compare the curves based on Equation (5.13) with the accurate solution, in a diagram analogous to Figures 5.1 and 5.2, still the corresponding curves practically coincided for all Pe values typical of actual glaciers.

Consequently, it should be quite justified to describe approximately the temperature distribution in a flat steady-state glacier in terms of polynomials whose coefficients are specified by the boundary conditions and the equations of the moments.

The best description of the temperature distribution is given by a fifth-degree polynomial, whose coefficients are found from the boundary conditions, from the conditions of compliance with the heat-transfer equation, and from the equations of the zeroth and first moments.

We have calculated the values of the coefficients of polynomials (5.6) and (5.13) for Pe values from 0 to 15. These data, which can be used directly for computations, are given in Table 5.1.

Table 5.1. Values of coefficients of polynomials describing approximate temperature distribution in flat steady-state glacier in accumulation area.

Pe	Method of 'zeroth' moment			Method of 'zeroth' and 'first' moments			
	+B	-D	+E	+B	-D	+E	-F
0	1.0000	0.0	0.0	1.0	0.0	0.0	0.0
1	1.1707	0.2230	0.0523	1.1687	0.2021	0.0205	0.0129
2	1.3411	0.5176	0.1765	1.3390	0.4944	0.1406	0.0148
3	1.5037	0.8420	0.3383	1.5077	0.8860	0.4068	0.0285
4	1.6551	0.1724	0.5173	1.6724	1.3704	0.8284	0.1305
5	1.7943	1.4953	0.7010	1.8318	1.9330	1.3940	0.2928
6	1.9215	1.8039	0.8824	1.9850	2.5579	2.0833	0.5104
7	2.0374	2.0950	1.0576	2.1316	3.2291	2.8733	0.7758
8	2.1428	2.3673	1.2245	2.2716	3.9330	3.7420	1.0806
9	2.2389	2.6211	1.3822	2.4050	4.6580	4.6695	1.4165
10	2.3265	2.8571	1.5306	2.5321	5.3949	5.6392	1.7764
11	2.4067	3.0764	1.6697	2.6532	6.1361	6.6370	2.1541
12	2.4800	3.2800	1.8000	2.7685	6.8759	7.6517	2.5443
13	2.5474	3.4693	1.9219	2.8785	7.6099	8.6740	2.9426
14	2.6094	3.6454	2.0360	2.9834	8.3348	9.6970	3.3456
15	2.6667	3.8095	2.1428	3.0837	9.0481	10.7148	3.7504

The discrepancy between the results obtained using these appproximate methods and the results of the accurate solution for Pe values of 1, 3, 6, 10, 15 can be evaluated better from the data in Table 5.2, which gives the dimensionless temperature $\bar{T}$ as a function of the dimensionless distance $\bar{y}$ from the bottom of the glacier.

5.1.4. The Effect of Internal Heat Sources

Since the above-described approximate methods for calculating the temperature field in a flat steady-state glacier give a very good fit with the accurate solution, these methods can be recommended to estimate the effect of internal heat sources or sinks, and in particular to evaluate the dissipation function. The equation of heat transfer in a flat two-dimensional glacier in this case has the form (3.62). Once again, Pe will have positive values under conditions of accumulation at the glacier surface, corresponding to a downward motion of the ice.

Let us assume that the temperature field in this glacier is described by a fifth-degree polynomial (5.11). As previously, we stipulate that the coefficients of this polynomial satisfy the boundary

Table 5.2. Dimensionless temperature as function of dimensionless distance from glacier bottom. Comparison of two approximate descriptions of temperature field in flat steady-state glacier with temperature distribution based on accurate solution.

$\bar{y}$	Accurate solution	Method of zeroth and first moments (six terms)	Method of zeroth moment (five terms)
Pe = 1			
0.1	0.1166	0.1167	0.1169
0.2	0.2321	0.2321	0.2325
0.3	0.3453	0.3453	0.3457
0.4	0.4553	0.4552	0.4554
0.5	0.5608	0.5608	0.5608
0.6	0.6613	0.6612	0.6611
0.7	0.7569	0.7559	0.7557
0.8	0.8442	0.8441	0.8439
0.9	0.9255	0.9255	0.9254
Pe = 3			
0.1	0.1500	0.1499	0.1490
0.2	0.2956	0.2950	0.2946
0.3	0.4327	0.4316	0.4312
0.4	0.5581	0.5565	0.5563
0.5	0.6893	0.6675	0.6679
0.6	0.7650	0.7637	0.7643
0.7	0.8450	0.8444	0.8452
0.8	0.9099	0.9099	0.9105
0.9	0.9609	0.9611	0.9616
Pe = 6			
0.1	0.1963	0.1961	0.1912
0.2	0.3812	0.3796	0.3714
0.3	0.5453	0.5421	0.5350
0.4	0.6825	0.6784	0.6760
0.5	0.7906	0.7870	0.7906
0.6	0.8708	0.8688	0.8776
0.7	0.9268	0.9265	0.9384
0.8	0.9637	0.9645	0.9755
0.9	0.9866	0.9872	0.9928
Pe = 10			
0.1	0.2485	0.2484	0.2300
0.2	0.4736	0.4716	0.4448
0.3	0.6582	0.6553	0.6331
0.4	0.7953	0.7937	0.7868
0.5	0.8875	0.8885	0.9016
0.6	0.9436	0.9467	0.9769
0.7	0.9746	0.9775	1.0158
0.8	0.9901	0.9912	1.0251
0.9	0.9971	0.9970	1.0151
Pe = 15			
0.1	0.3015	0.3011	0.2630
0.2	0.5614	0.5602	0.5062
0.3	0.7548	0.7585	0.7143
0.4	0.8786	0.8993	0.8774
0.5	0.9472	0.9633	0.9908
0.6	0.9799	0.9928	1.0545
0.7	0.9933	0.9974	1.0704
0.8	0.9981	0.9943	1.0600
0.9	0.9995	0.9946	1.0280

conditions, Equation (3.62) at the boundaries, and also the equations of the zeroth and first moments. Then we obtain the following system of equations for the fourth unknown nonzero coefficients:

$$B_q + D_q + E_q + F_q = 1 - \frac{\bar{\Phi}'(0)}{2};$$

$$B_q Pe + D_q(6 + 3Pe) + E_q(12 + 4Pe) + F_q(20 + 5Pe) =$$

$$= \bar{\Phi}'(1) - \bar{\Phi}'(0) - Pe\bar{\Phi}'(0);$$

$$B_q \frac{Pe}{2} + D_q(3 + \frac{3}{4} Pe) + E_q(4 + \frac{4}{5} Pe) + F_q(5 + \frac{5}{6} Pe) =$$

$$= \int_0^1 \bar{\Phi}' \, d\bar{y} - \bar{\Phi}'(0) - \frac{Pe}{3} \bar{\Phi}'(0); \qquad (5.14)$$

$$B_q \frac{Pe}{3} + D_q(2 + \frac{3}{5} Pe) + E_q(3 + \frac{2}{3} Pe) + F_q(4 + \frac{5}{7} Pe) =$$

$$= \int_0^1 \bar{\Phi}' \, dy - \frac{\bar{\Phi}'(0)}{2} - \frac{Pe}{4} \bar{\Phi}'(0).$$

This system of algebraic equations uniquely determines the temperature field in a glacier with an arbitrary strength distribution of heat sources or sinks over the thickness. Earlier it was shown that the temperature at any point in the glacier, under the conditions specified by Equation (3.62), can be represented as the temperature in a glacier without internal sources $\bar{T}_{q=0}$ corrected by a factor $\Delta\bar{T}_q$ taking into account the effect of the sources:

$$\Delta\bar{T}_q = \frac{T - T_{q=0}}{T_H - T_0} .$$

For example, if heat sources are distributed uniformly over the glacier thickness, then correction $\Delta\bar{T}_q$ will have the form

$$\Delta\bar{T}_q = -\bar{\Phi}'\Bigg(\frac{1 + 0.147\ 62Pe + 0.0151Pe^2 + 0.000\ 04Pe^3}{2 + 0.295\ 24Pe + 0.028\ 57Pe^2 + 0.000\ 63Pe^3}\,\bar{y} -$$

$$- \frac{1}{2}\,\bar{y}^2 - \frac{0.166\ 67Pe + 0.280\ 95Pe^2 - 0.004\ 76Pe^3}{2 + 0.295\ 24Pe + 0.028\ 57Pe^2 + 0.000\ 63Pe^3}\,\bar{y}^3 +$$

$$+ \frac{0.166\ 67Pe + 0.547\ 62Pe^2 - 0.004\ 76Pe^3}{2 + 0.295\ 24Pe + 0.028\ 57Pe^2 + 0.000\ 63Pe^3}\,\bar{y}^4 -$$

$$- \frac{0.041\ 67Pe^2 - 0.000\ 28Pe^3}{2 + 0.295\ 24Pe + 0.028\ 57Pe^2 + 0.000\ 63Pe^3}\,\bar{y}^5$$

$$= \bar{\Phi}'(B_q\bar{y} + C_q\bar{y}^2 + D_q\bar{y}^3 + E_q\bar{y}^4 + F_q\bar{y}^5\Big). \qquad (5.15)$$

Table 5.3 gives corrections for the effect of heat sources, calculated for two values of Pe for $\bar{\Phi}' = 1$. For comparison, the table also gives the corrections obtained on the basis of accurate solution (4.70), for the very same conditions. The results yielded by the described approximate method show a good fit with the accurate solution.

Table 5.3. Corrections for effect of heat sources (sinks), calculated using approximate formula (5.15) and accurate solution (4.70).

	Pe = 1		Pe = 15	
$\bar{y}$	Approximate solution	Accurate solution	Approximate solution	Accurate solution
0.1	-0.0449	-0.0452	-0.0340	-0.0350
0.2	-0.0792	-0.0840	-0.0556	-0.0553
0.3	-0.1030	-0.1109	-0.0659	-0.0626
0.4	-0.1163	-0.1202	-0.0632	-0.0595
0.5	0.1193	-0.1112	-0.0539	-0.0504
0.6	-0.1124	-0.1104	-0.0405	-0.0392
0.7	-0.0963	-0.0872	-0.0266	-0.0275
0.8	-0.0716	-0.0780	-0.0147	-0.0172
0.9	-0.0392	-0.0348	-0.0062	-0.0080

Simpler expressions are obtained for the temperature field in a glacier, provided the internal sources are taken into account using a fourth-degree polynomial of form (5.2). In this case as well, the temperature inside the glacier can be represented in terms of the temperature in a glacier without internal sources corrected by a factor $\Delta\bar{T}_q$ for the source effect:

$$\Delta\bar{T}_q = B_q\bar{y} + C_q\bar{y}^2 + D_q\bar{y}^3 + E_q\bar{y}^4.$$

The values of $B_q, \ldots, E_q$ for a source distribution over the glacier thickness according to the $\bar{\Phi}'(\bar{y})$ law are given by the expressions:

$$B_q = \frac{-\bar{\Phi}'(0)\,\frac{B}{2} - \left[\int_0^1 \bar{\Phi}'(\bar{y})\,d\bar{y} - \bar{\Phi}'(0) - \frac{\Phi'(0)}{3}\,\mathrm{Pe}\right] \times (6 + \mathrm{Pe}) + [\bar{\Phi}'(1) - \bar{\Phi}'(0) - \mathrm{Pe}\bar{\Phi}'(0)]\left(3 + \frac{3}{4}\,\mathrm{Pe}\right)}{12 + 2.2\mathrm{Pe} - 0.15\mathrm{Pe}^2};$$

$$C_q = \frac{\bar{\Phi}'(0)}{2};$$

$$D_q = \bar{\Phi}'(0)\,\frac{D}{2} + \frac{\left[\int_0^1 \bar{\Phi}'(\bar{y})\,d\bar{y} - \bar{\Phi}'(0) - \frac{\bar{\Phi}'(0)}{3}\,Pe\right](12 + 3Pe) + [\bar{\Phi}'(1) - \bar{\Phi}'(0) - Pe\bar{\Phi}'(0)]\left(4 + \frac{3}{10}\,Pe\right)}{12 + 2.2Pe - 0.15Pe^2};$$

$$E_q = -\bar{\Phi}'(0)\frac{E}{2} - \frac{\left[\int_0^1 \bar{\Phi}'(\bar{y})\,d\bar{y} - \bar{\Phi}'(0) - \frac{\bar{\Phi}'(0)}{3}\,Pe\right](6 + 2Pe) - [\bar{\Phi}'(1) - \bar{\Phi}'(0) - Pe\bar{\Phi}'(0)]\left(3 + \frac{1}{4}\,Pe\right)}{12 + 2.2Pe - 0.15Pe^2}.$$

For $\bar{\Phi}'(\bar{y}) = \text{const} = \bar{\Phi}'$ these expressions simplify to:

$$B_q = \bar{\Phi}'\left(-\frac{B}{2} + \frac{5Pe + \frac{13}{12}\,Pe^2}{12 + 2.2Pe - 0.15Pe^2}\right);$$

$$C_q = \frac{\bar{\Phi}'}{2};$$

$$D_q = \bar{\Phi}'\left(\frac{D}{2} - \frac{Pe + \frac{3}{4}\,Pe^2}{12 + 2.2Pe - 0.15Pe^2}\right); \qquad (5.16)$$

$$E_q = \bar{\Phi}'\left(-\frac{E}{2} + \frac{\frac{5}{12}\,Pe^2 - Pe}{12 + 2.2Pe - 0.15Pe^2}\right).$$

5.1.5. A More General Statement of the Problem

In Chapter 4 we considered an accurate solution of the problem of the heat distribution in a flat steady-state glacier, in the general statement of this problem for ablation or accumulation at the surface and melting or freezing at the bottom, under conditions of longitudinal expansion or contraction of the ice. A total of 12 possible forms of the velocity distribution in a glacier were distinguished (Figure 4.1), and for these distributions working formulas were obtained giving the temperature distribution within the glacier.

Below we will consider an approximate method for calculating the temperature field in a glacier in this same general formulation.

Previously it was shown that the heat transfer in such a glacier under conditions of block [Blockschollen] movement is described by an equation which can be rewritten as

$$\frac{d^2\bar{T}}{d\bar{y}^2} + \bar{N}\bar{y}\,\frac{d\bar{T}}{d\bar{y}} + \bar{M}\,\frac{d\bar{T}}{d\bar{y}} = 0. \tag{5.17}$$

where $\bar{N} = Pe(1 - \bar{v}_o)$ and $\bar{M} = Pe\,\bar{v}_o$ for boundary conditions $\bar{y} = 0$, $\bar{T} = 0$, and $\bar{y} = 1$, $\bar{T} = 1$. Pe is taken to be positive for accumulation at the surface and negative for ablation.

For all the conditions considered, differing formally only in the values of Pe and $\bar{v}_0$, and their signs, let us apply the approximate method of describing the temperature field in terms of a polynomial of fifth degree (5.11). In addition to the boundary conditions, we will employ the same additional conditions as before, which our comparison of the approximate solution with the results of the accurate calculation showed to be so promising.

From the boundary conditions, we once again have

$$A = 0;$$

$$B + C + D + E + F = 1.$$

From the conditions that expression (5.11) must satisfy equation (5.17) at the boundaries, we get

$$C = B\,\frac{\bar{M}}{2};$$

$$B(\bar{N} - \bar{N}\bar{M} - \bar{M}^2) + D(6 + 3\bar{N} + 3\bar{M}) + E(12 + 4\bar{N} + 4\bar{M}) + \\ + F(20 + 5\bar{N} + 5\bar{M}) = 0.$$

Now, the substitution of expression (5.11) into the equations of the 'zeroth' and 'first' moments, obtained from Equation (5.17), gives

$$\begin{aligned} &B\left(\frac{\bar{N}}{2} - \frac{\bar{M}\bar{N}}{3} - \frac{\bar{M}^2}{2}\right) + D(3 + \frac{3}{4}\,\bar{N} + \bar{M}) + \\ &+ E(4 + \frac{4}{5}\,\bar{N} + \bar{M}) + F(5 + \frac{5}{6}\,\bar{N} + \bar{M}) = 0; \\ &B\left(\frac{\bar{N}}{2} - \frac{\bar{M}\bar{N}}{4} - \frac{\bar{M}^2}{2}\right) + D(2 + \frac{3}{5}\,\bar{N} + \frac{3}{4}\,\bar{M}) + \\ &+ E(3 + \frac{2}{3}\,\bar{N} + \bar{M}) + F(4 + \frac{5}{7}\,\bar{N} + \bar{M}). \end{aligned} \tag{5.18}$$

Consequently, after certain transformations we obtain the following system of equations for the coefficients:

$$B\left(1 - \frac{\bar{M}}{2}\right) + D + E + F = 0;$$

$$B(\bar{N} - \bar{M}\bar{N} - \bar{M}^2) + D(6 + 3\bar{N} + 3\bar{M}) + E(12 + 4\bar{N} + 4\bar{M}) +$$

$$+ F(20 + 5\bar{N} + 5\bar{M}) = 0;$$

$$B\left(\frac{\bar{N}}{2} - \frac{\bar{M}\bar{N}}{3} - \frac{\bar{M}^2}{2}\right) + D(3 + \frac{3}{4}\bar{N} + \bar{M}) +$$

$$+ E(4 + \frac{4}{5}\bar{N} + \bar{M}) + F(5 + \frac{5}{6}\bar{N} + \bar{M}) = 0; \qquad (5.19)$$

$$B\left(\frac{\bar{N}}{2} - \frac{\bar{M}\bar{N}}{4} - \frac{\bar{M}^2}{3}\right) + D(2 + \frac{3}{5}\bar{N} + \frac{3}{4}\bar{M}) +$$

$$E(3 + \frac{2}{3}\bar{N} + \frac{4}{5}\bar{M}) + F(4 + \frac{5}{7}\bar{N} + \frac{5}{6}\bar{M}) = 0.$$

Finally, by solving (5.18) and substituting the values obtained for the coefficients into (5.11), it is easy to arrive at a final expression for the temperature field in a glacier, for the conditions corresponding to the various profiles in Figure 4.1.

5.2. The Temperature Field in a Glacier with a Varying Surface Temperature

The problem of heat transfer in a glacier in which the temperature varies with time is one of the basic, and at the same time most difficult, problems of theoretical glaciology. As shown earlier, an accurate solution for the temperature distribution in a glacier with vertical ice motion and a temperature that varies with time can be obtained only for a semibounded body with a constant vertical velocity over its thickness and for conditions at the boundary which are very far from reality.

At the same time, a study of the heat transfer in a flat two-dimensional glacier with block [Blockschollen] movement and with a varying surface temperature is basic for the analysis of possible climatic changes on the basis of measurement data in deep boreholes in ice, as well as for the analysis of the reactions of a glacier to climatic change. In a number of cases such studies deal with the temperature field in a glacier moving from regions of cooler climate to regions of warmer climate. This is observed, for instance, in the peripheral regions of the Antarctic and Greenland Ice Sheets and other large ice sheets, and of island ice caps on long steep-dipping mountain glaciers.

Let us consider the heat transfer in a flat glacier within which the vertical ice velocity varies linearly over the glacier interior, while the temperature at one of the glacier surfaces (the upper surface, say) varies with time. We will confine ourselves to the accumulation region, on the assumption that the effect of bottom melting or freezing at the lower boundary of the glacier can be neglected. The effect of differences in the physical parameters over the glacier thickness will also be neglected here, as will the effect of internal heat sources and sinks.

The heat transfer in such a glacier is given by the following equation:

$$\frac{\partial \bar{T}}{\partial \bar{t}} - \mathrm{Pe}\bar{y}\,\frac{\partial \bar{T}}{\partial \bar{y}} - \frac{\partial^2 \bar{T}}{\partial \bar{y}^2} = 0. \tag{5.20}$$

Here, as in the previous section, Pe is positive for accumulation at the upper surface (downward ice motion in the glacier); $\bar{T} = (T - T_0)/(T_{H,t=0} - T_0)$ is the dimensionless temperature; $T_{H,t=0}$ is the initial temperature at the glacier surface; and T_0 is the temperature at the lower surface of the glacier, assumed to be invariable with time.

For simplicity, we assume that the dimensionless temperature at the lower glacier boundary is also constant and equal to zero ($\bar{y} = 0$, $\bar{T}_0 = 0$). The temperature at the upper surface, however, varies in time. Let us portray this temperature variation in terms of a polynomial of third degree:

$$\bar{y} = 1, \quad \bar{T} = \bar{T}_H(\bar{t}) = a_1 + b\bar{t} + c\bar{t}^2 + d\bar{t}^3. \tag{5.21}$$

The initial temperature distribution is assumed to be a function just of the physical parameters and the velocity of ice motion in the glacier:

$$\bar{t} = 0, \quad \bar{T}(\bar{y}, 0) = F(\mathrm{Pe}, \bar{y}), \tag{5.22}$$

where $\mathrm{Pe} = v_H H/a = \mathrm{const}$.

Since the fit is good between the accurate solution and the approximate descriptions of the temperature field in a glacier with the aid of a polynomial, let us use a polynomial to solve the problem at hand as well, taking its degree to be fourth. Thus we will seek a solution in the form (5.2), where A, B, C, D, E are functions of time.

Then, from the boundary conditions we get

$$A = 0,$$

$$a_1 + b\bar{t} + c\bar{t}^2 + d\bar{t}^3 = B + C + D + E. \tag{5.23}$$

Substitution of (5.22) and (5.20) gives

$$\dot{B}\bar{y} + \dot{C}\bar{y}^2 + \dot{D}\bar{y}^3 + \dot{E}\bar{y} - \mathrm{Pe}B\bar{y} - 2\mathrm{Pe}C\bar{y}^2 - 3\mathrm{Pe}D\bar{y}^3 -$$
$$- 4\mathrm{Pe}E\bar{y}^4 - 2C - 6D\bar{y} - 12E\bar{y}^2 = 0; \tag{5.24}$$

Here $\dot{B} = \partial B/\partial t$, $\dot{C} = \partial C/\partial t$, etc.

As the first of our additional conditions, we use the condition that the coefficient of expression (5.2) must satisfy equation (5.20) at the lower surface. Then, from (5.24) for $\bar{y} = 0$ we obtain

$$C = 0. \tag{5.25}$$

The equations of the 'zeroth' and 'first' moments from the heat-transfer equation are also taken as conditions for determining the coefficients. In this case these equations are

$$\begin{aligned} &\int_0^1 \left(\frac{\partial \bar{T}}{\partial \bar{t}} - \mathrm{Pe}\bar{y} \frac{\partial \bar{T}}{\partial \bar{y}} - \frac{\partial^2 \bar{T}}{\partial \bar{y}^2} \right) d\bar{y} = 0; \\ &\int_0^1 \left(\frac{\partial \bar{T}}{\partial \bar{t}} - \mathrm{Pe}\bar{y} \frac{\partial \bar{T}}{\partial \bar{y}} - \frac{\partial^2 \bar{T}}{\partial \bar{y}^2} \right) \bar{y}\, d\bar{y} = 0. \end{aligned} \tag{5.26}$$

Next, if we replace $\bar{T}$ in (5.26) by its approximate value (5.2), integrate, and substitute limits, we obtain

$$\begin{aligned} &\frac{1}{2}(\dot{B} - \mathrm{Pe}B - 6D) + \frac{1}{4}(\dot{D} - 3\mathrm{Pe}D) + \frac{1}{3}(\dot{E} - 4\mathrm{Pe}E) - 4E = 0; \\ &\frac{1}{3}(\dot{B} - \mathrm{Pe}B - 6D) + \frac{1}{5}(\dot{D} - 3\mathrm{Pe}D) + \frac{1}{6}(\dot{E} - 4\mathrm{Pe}E) - 3E = 0. \end{aligned} \tag{5.27}$$

Consequently, from (5.23), (5.25), and (5.27) after certain transformations we arrive at the following system of differential equations for the three nonzero coefficients B, D, E:

$$B + D + E = a_1 + b\bar{t} + c\bar{t}^2 + d\bar{t}^3;$$

$$\begin{aligned} &10\dot{B} - 10\mathrm{Pe}B - 60D - 80E + 5\dot{D} - 15\mathrm{Pe}D + 4\dot{E} - 16\mathrm{Pe}E = 0; \\ &10\dot{B} - 10\mathrm{Pe}B - 60D - 90E + 6\dot{D} - 18\mathrm{Pe}D + 5\dot{E} - 20\mathrm{Pe}E = 0. \end{aligned} \tag{5.28}$$

Subtracting the second Equation (5.28) from the third and denoting $a_1 + b\bar{t} + c\bar{t}^2 + d\bar{t}^3 = \bar{Q}$, we have

$$\begin{aligned} &-10E + \dot{D} - 3\mathrm{Pe}D + \dot{E} - 4\mathrm{Pe}E = 0; \\ &B = \bar{Q} - D - E. \end{aligned} \tag{5.29}$$

Next we substitute the quantity B from (5.29) into the second equation of expression (5.28), to obtain a system of two equations (the transformed second equation (5.28) and the first equation of system (5.29)):

$$\dot{D} + \dot{E} = 3\mathrm{Pe}D + (4\mathrm{Pe} + 10)E;$$

$$5\dot{D} + 6\dot{E} = -(60 + 5\mathrm{Pe})D - (80 + 6\mathrm{Pe})E + 10(\dot{\bar{Q}} - \mathrm{Pe}\bar{Q}). \tag{5.30}$$

Here $\dot{\bar{Q}} = \partial \bar{Q}/\partial \bar{t} = b + 2c\bar{t} + 3d\bar{t}^2$.

By eliminating $\dot{D}$ from the first Equation (5.30) and $\dot{E}$ from the second Equation (5.30), we obtain

$$\dot{E} = -(60 + 20Pe)D - (130 + 26Pe)E + 10(\dot{\bar{Q}} - Pe\bar{Q});$$

$$\dot{D} = (60 + 23Pe)D + (140 + 30Pe)E - 10(\dot{\bar{Q}} - Pe\bar{Q}).$$

To simplify the subsequent calculations, let us use the following notation:

$$\alpha_{11} = 60 + 23Pe; \quad \alpha_{21} = -(60 + 20Pe);$$

$$\alpha_{12} = 140 + 30Pe; \quad \alpha_{22} = -(130 + 26Pe); \tag{5.31}$$

$$\beta = 10(\dot{\bar{Q}} - Pe\bar{Q}) = 10(B + 2c\bar{t} + 3d\bar{t}^2 - Pea_1 -$$

$$- Peb\bar{t} - Pec\bar{t}^2 - Ped\bar{t}^3) = \beta_1 + \beta_2\bar{t} + \beta_3\bar{t}^2 + \beta_4\bar{t}^3$$

Therefore,

$$\beta_1 = 10(b - Pea_1); \quad \beta_2 = 10(2c - Peb);$$

$$\beta_3 = 10(3d - Pec); \quad \beta_4 = -10Ped.$$

The final form of the system of differential equations for coefficients D and E will then be

$$\frac{\partial D(\bar{t})}{\partial \bar{t}} = \alpha_{11}D(\bar{t}) + \alpha_{12}E(\bar{t}) - \beta(\bar{t});$$

$$\frac{\partial E(\bar{t})}{\partial (t)} = \alpha_{21}D(\bar{t}) + \alpha_{22}E(\bar{t}) + \beta(\bar{t}). \tag{5.32}$$

Coefficients α depend only on the physical properties of the ice, while coefficient β depend on both the nature of the temperature variation at the glacier surface and the physical properties of the ice.

5.2.1. A Particular Solution of the Nonhomogeneous System

Particular solutions for D and E can be described in terms of polynomials. Let us denote these coefficients by the symbols D_0 and E_0 and write them approximately as **third-degree polynomials**:

$$D_0 = \bar{m} + \bar{n}\bar{t} + \bar{q}\bar{t}^2 + \bar{r}\bar{t}^3;$$

$$E_0 = \bar{m}' + \bar{n}'\bar{t} + \bar{q}'\bar{t}^2 + \bar{r}'\bar{t}^3. \tag{5.33}$$

Then, substitution of (5.33) into (5.32) gives the following system of equations:

$$\begin{aligned}
&\bar{n} + 2\bar{q}\bar{t} + 3\bar{r}\bar{t}^2 = \alpha_{11}\bar{m} + \alpha_{11}\bar{n}\bar{t} + \alpha_{11}\bar{q}\bar{t}^2 + \alpha_{11}\bar{r}\bar{t}^3 + \\
&\quad + \alpha_{12}\bar{m}' + \alpha_{12}\bar{n}'\bar{t} + \alpha_{12}\bar{q}'\bar{t}^2 + \alpha_{12}\bar{r}'\bar{t}^3 - \beta_1 - \beta_2\bar{t} - \\
&\quad - \beta_3\bar{t}^2 - \beta_4\bar{t}^3; \\
&\bar{n}' + 2\bar{q}'\bar{t} + 3\bar{r}'\bar{t}^2 = \alpha_{21}\bar{m} + \alpha_{21}\bar{n}\bar{t} + \alpha_{21}\bar{q}\bar{t}^2 + \alpha_{21}\bar{r}\bar{t}^3 + \\
&\quad + \alpha_{22}\bar{m}' + \alpha_{22}\bar{n}'\bar{t} + \alpha_{22}\bar{q}'\bar{t}^2 + \alpha_{22}\bar{r}'\bar{t}^3 + \\
&\quad + \beta_1 + \beta_2\bar{t} + \beta_3\bar{t}^2 + \beta_4\bar{t}^3.
\end{aligned} \tag{5.34}$$

The right-hand and left-hand sides of each of these equations will be equal for any values of $\bar{t}$ provided that the sum of the coefficients for the same powers of $\bar{t}$ on the right-hand and left-hand sides are equal. By equating the sums of these coefficients, we arrive at the following eight equations for the quantities $\bar{n}, \ldots, \bar{r}; \bar{n}', \ldots, \bar{r}'$, for which polynomials (5.33) are particular solutions of system (5.32):

$$\begin{aligned}
&(1)\ \bar{n} = \alpha_{11}\bar{m} + \alpha_{12}\bar{m}' - \beta_1; \\
&(2)\ 2\bar{q} = \alpha_{11}\bar{n} + \alpha_{12}\bar{n}' - \beta_2; \\
&(3)\ 3\bar{r} = \alpha_{11}\bar{q} + \alpha_{12}\bar{q}' - \beta_3; \\
&(4)\ 0 = \alpha_{11}\bar{r} + \alpha_{12}\bar{r}' - \beta_4; \\
&(5)\ \bar{n}' = \alpha_{21}\bar{m} + \alpha_{22}\bar{m}' + \beta_1; \\
&(6)\ 2\bar{q}' = \alpha_{21}\bar{n} + \alpha_{22}\bar{n}' + \beta_2; \\
&(7)\ 3\bar{r}' = \alpha_{21}\bar{q} + \alpha_{22}\bar{q}' + \beta_3; \\
&(8)\ 0 = \alpha_{21}\bar{r} + \alpha_{22}\bar{r}' + \beta_4.
\end{aligned} \tag{5.35}$$

The first four equations of algebraic system (5.35) were derived from the first equation of system (5.34), and the last four were derived from the second equation of (5.34).

Since it is very difficult to solve a system of eight equations in eight unknowns, let us simplify things by reducing the system to fewer equations. Addition of the fourth and eighth equations of system (5.35) gives

$$(\alpha_{11} + \alpha_{21})\bar{r} + (\alpha_{12} + \alpha_{22})\bar{r}' = 0. \tag{5.36}$$

Equation (5.36) can be written more concisely as

$$i_0\bar{r} + i_0'\bar{r}' = 0, \tag{5.37}$$

where we have defined

$$\alpha_{11} + \alpha_{21} = i_0; \quad \alpha_{12} + \alpha_{22} = i_0'. \tag{5.38}$$

Next, using (5.36) to replace $\bar{r}'$ in the seventh equation of (5.35) by $\bar{r}$, we get

$$-3\bar{r} = \frac{\alpha_{12} - \alpha_{22}}{\alpha_{11} + \alpha_{21}} (\alpha_{21}\bar{q} + \alpha_{22}\bar{q}' + \beta_3).$$

If this equation is now added to the third equation of (5.35), then after some transformations we have

$$[(\alpha_{11} + \alpha_{21})\alpha_{11} + (\alpha_{12} + \alpha_{22})\alpha_{21}]\bar{q} + [(\alpha_{11} + \alpha_{21})\alpha_{12} + \\ + (\alpha_{12} + \alpha_{22})\alpha_{22}]\bar{q}' = -\beta_3(\alpha_{12} + \alpha_{22} + \alpha_{11} - \alpha_{21}). \tag{5.39}$$

For the sake of brevity, we designate the coefficients of $\bar{q}$, $\bar{q}'$, and β_3 as

$$[(\alpha_{11} + \alpha_{21})\alpha_{11} + (\alpha_{12} + \alpha_{22})\alpha_{21}] = i;$$

$$[(\alpha_{11} + \alpha_{21})\alpha_{12} + (\alpha_{12} + \alpha_{22})\alpha_{22}] = i'; \tag{5.40}$$

$$\beta_3(\alpha_{12} + \alpha_{22} - \alpha_{11} - \alpha_{21}) = j.$$

Then Equation (5.39) simplifies to

$$i\bar{q} + i\bar{q}' = j. \tag{5.41}$$

With the aid of (5.41), we can replace $\bar{q}'$ in the sixth equation of (5.35) by i, j, and $\bar{q}$. Then, the addition of this expression to the second equation of (5.35) gives, after some transformations,

$$-2J = (\alpha_{11}i + \alpha_{21}i')\bar{n} + (\alpha_{12}i + \alpha_{22}i')\bar{n}' - (i - i')\beta_2. \tag{5.42}$$

Using the notation:

$$(\alpha_{11}i + \alpha_{22}i') = i_1;$$

$$(\alpha_{12}i + \alpha_{22}i') = i_1'; \tag{5.43}$$

$$[2j = (i' - i)\beta_2] = j_1,$$

we can rewrite (5.42) as

$$i_1\bar{n} + i_1'\bar{n}' = -j_1. \tag{5.44}$$

Following the above procedure for the third time, that is, using (5.44) to replace $\bar{n}'$ by j_1, i_1' and $i, \bar{n}$ in the fifth equation of (5.35) and then adding this expression to the first equation of (5.35), after some transformations we arrive at the following equation:

$$-j_1 = (\alpha_{11} i_1 + \alpha_{21} i_1')\bar{m} + (\alpha_{12} i_1 + \alpha_{21} i_1')\bar{m} + \\ + (i_1' - i_1)\beta_1. \tag{5.45}$$

As previously, we define

$$(\alpha_{11} i_1 + \alpha_{21} i_1') = i_2;$$

$$(\alpha_{12} i_1 + \alpha_{21} i_1') = i_2'. \tag{5.46}$$

$$j_1 + (i_1' + i_1)\beta_1 = j_2.$$

Consequently, Equation (5.45) can be written as

$$i_2 \bar{m} + i_2' \bar{m}' = - j_2. \tag{5.47}$$

Now we have four equations (5.37), (5.41), (5.42), (5.43), with the aid of which it is easy to eliminate four unknowns ($\bar{m}'$, $\bar{n}'$, $\bar{q}'$, $\bar{r}'$) from the system of eight equations (5.35). This leaves us with a system of four equations for $\bar{m}$, $\bar{n}$, $\bar{q}$, $\bar{r}$:

$$\bar{n} = \alpha_{11}\bar{m} - \frac{\alpha_{12} j_2}{i_2'} - \alpha_{12}\frac{i_2}{i_2'}\bar{m} - \beta_1;$$

$$\bar{q} = \frac{\alpha_{11}\bar{n}}{2} - \frac{\alpha_{12} j_1}{2 i_1'} - \frac{\alpha_{12}}{2}\frac{i_1}{i_1'}\bar{n} - \frac{\beta_2}{2}; \tag{5.48}$$

$$\bar{r} = \frac{\alpha_{11}}{3}\bar{q} - \frac{\alpha_{12}}{3}\frac{j}{i'} - \frac{\alpha_{12}}{3}\frac{i}{i'}\bar{q} - \frac{\beta_3}{3};$$

$$0 = \alpha_{11}\bar{r} - \alpha_{12}\frac{i_0}{i_0'}\bar{r} - \beta_4.$$

Once again, for simplicity we use the notation:

$$\alpha_{11} - \alpha_{12}\frac{i_2}{i_2'} = \gamma; \quad \alpha_{12}\frac{j_2}{i_2'} + \beta_2 = \gamma';$$

$$\frac{\alpha_{11}}{2} - \frac{\alpha_{12}}{2}\frac{i_1}{i_1'} = \gamma_1; \quad \frac{\alpha_{12}}{2}\frac{j_1}{i_1'} + \frac{\beta_2}{2} = \gamma_1'; \tag{5.49}$$

$$\frac{\alpha_{11}}{3} - \frac{\alpha_{12}}{3}\frac{i}{i'} = \gamma_2; \qquad \frac{\alpha_{12}}{3}\frac{j}{i'} + \frac{\beta_3}{3} = \gamma_2';$$

$$\alpha_{11} - \alpha_{12}\frac{i_0}{i_0'} = \gamma_3. \tag{5.49}$$

The final system of equations is then:

$$\begin{aligned} \bar{n} &= \gamma\bar{m} - \gamma'; \\ \bar{q} &= \gamma_1\bar{n} - \gamma_1'; \\ \bar{r} &= \gamma_2\bar{q} - \gamma_2'; \\ 0 &= \gamma_3\bar{r} - \beta_4. \end{aligned} \tag{5.50}$$

The quantity $\bar{r}$ is found directly from the fourth equation of this system, after which it is easy to determine the remaining unknowns $\bar{q}$, $\bar{n}$, $\bar{m}$. The unknown coefficients $\bar{r}'$, $\bar{q}'$, $\bar{n}'$, $\bar{m}'$ can then be found directly from relations (5.37), (5.41), (5.44) and (5.47):

$$\bar{r} = \frac{\beta_4}{\gamma_3}; \quad \bar{q} = \frac{1}{\gamma_2}\left(\gamma_2' + \frac{\beta_4}{\gamma_3}\right); \quad \bar{n} = \frac{1}{\gamma_1\gamma_2}\left(\gamma_1'\gamma_2 + \gamma_2' + \frac{\beta_3}{\gamma_3}\right);$$

$$\bar{m} = \frac{1}{\gamma\gamma_1\gamma_2}\left(\gamma'\gamma_1\gamma_2 + \gamma_1'\gamma_2 + \gamma_2' + \frac{\beta_4}{\gamma_3}\right);$$

$$\bar{r}' = -\frac{i_0}{i_0'}\frac{\beta_4}{\gamma_3}; \quad \bar{q}' = -\frac{j}{i'} - \frac{i}{i'\gamma_2}\left(\gamma_2' + \frac{\beta_4}{\gamma_3}\right); \tag{5.51}$$

$$\bar{n}' = \frac{j_1}{i_1'} - \frac{i_1}{\gamma_1\gamma_2 i_1'}\left(\gamma_1'\gamma_2 + \gamma_2' + \frac{\beta_3}{\gamma_3}\right);$$

$$\bar{m}' = -\frac{j_2}{i_2'} - \frac{i_2}{i_2'\gamma\gamma_1\gamma_2}\left(\gamma'\gamma_1\gamma_2 + \gamma_1'\gamma_2 + \gamma_2' + \frac{\beta_4}{\gamma_3}\right).$$

Consequently, one of the particular solutions D_0 and E_0 of non-homogeneous equations (5.32) has been completely determined. In order to obtain the form of coefficients D and E corresponding to a complete solution of system (5.32) and defining the temperature field in a glacier, it is sufficient to add to the particular solution obtained the general solution of the corresponding homogeneous system of equations without the free terms.

5.2.2. The General Solution. Taking the Initial Condition into Account.

The general solution of the corresponding homogeneous system of equations (5.32) can be represented as

$$D = D_0 + \bar{\delta};$$

$$E = E_0 + \bar{\varepsilon},$$

where $\bar{\delta}$ and $\bar{\varepsilon}$ represent the general solution of the homogeneous system.

Consequently, the expressions for $\bar{\delta}$ and $\bar{\varepsilon}$ must satisfy the equations

$$\frac{d\bar{\delta}}{d\bar{t}} = \alpha_{11}\bar{\delta} + \alpha_{12}\bar{\varepsilon};$$
$$\frac{d\bar{\varepsilon}}{d\bar{t}} = \alpha_{21}\bar{\delta} + \alpha_{22}\bar{\varepsilon}. \tag{5.52}$$

This system is a normal system of the first order.

In order to find the general solution of this system, let us first construct its characteristic equation:

$$\begin{vmatrix} \alpha_{11} - \eta & \alpha_{12} \\ \alpha_{21} & \alpha_{22} - \eta \end{vmatrix} = 0. \tag{5.53}$$

Now, to each nonmultiple root η of the equation there will correspond a system of particular solutions:

$$\bar{\delta} = G_{1i} \exp(\eta_i\bar{t}); \quad \bar{\varepsilon} = G_{2i} \exp(\eta_i\bar{t}). \tag{5.54}$$

After evaluating the determinant, we can rewrite equation (5.53) as

$$\eta^2 - (\alpha_{12} + \alpha_{11})\eta + \alpha_{11}\alpha_{12} - \alpha_{21}\alpha_{12} = 0,$$

from which the value of η is specified by two expressions:

$$\eta_{1,2} = \frac{\alpha_{22} + \alpha_{11}}{2} \pm \sqrt{\left(\frac{\alpha_{11} + \alpha_{22}}{2}\right)^2 + \alpha_{21}\alpha_{12} - \alpha_{11}\alpha_{12}}. \tag{5.55}$$

Thus, the general solution of the homogeneous system has the form

$$\bar{\delta} = G_{11} \exp(\eta_1\bar{t}) + G_{12} \exp(\eta_2\bar{t});$$
$$\bar{\varepsilon} = G_{21} \exp(\eta_1\bar{t}) + G_{22} \exp(\eta_2\bar{t}). \tag{5.56}$$

It should be noted that for large $\bar{t}$ the importance of solution (5.56) in the general solution of the nonhomogeneous system (5.52) may well depend considerably on the value of Pe. For $\eta > 0$ the effect of terms $\bar{\delta}$ and $\bar{\varepsilon}$ in the general solution will be enhanced by an increase in the time $\bar{t}$ elapsing since the start of the process. This will not be the case only when η is negative.

By summing up the foregoing and substituting (5.56) and (5.33) into (5.52), we now obtain

$$D = \bar{m} + \bar{n}\bar{t} + \bar{q}\bar{t}^2 + \bar{r}\bar{t}^3 + G_{11}\exp(\eta_1\bar{t}) + G_{12}\exp(\eta_2\bar{t});$$
$$E = \bar{m}' + \bar{n}'\bar{t} + \bar{q}'\bar{t}^2 + \bar{r}'\bar{t}^3 + G_{21}\exp(\eta_1\bar{t}) + G_{22}\exp(\eta_2\bar{t}). \tag{5.57}$$

In the problem at hand, the temperature distribution can be represented, with the aid of (5.22) and (5.25), as a polynomial of three terms:

$$\bar{T}(\bar{t}, y) = B(t)\bar{y} + D(t)\bar{y}^3 + E(t)\bar{y}^4. \tag{5.58}$$

Previously (in Section 5.1) it was shown that the temperature field in a steady-state glacier can also be described approximately by a trinomial of the type

$$\bar{T} = B'\bar{y} + D'\bar{y}^3 + E'\bar{y}^4, \tag{5.59}$$

where B', D', E' are functions of Pe.

To be more explicit, let us assume that the initial condition can be represented in the form of (5.59). Then for the initial time $\bar{t} = 0$ distributions (5.58) and (5.59) can be equated to one another:

$$B(0)\bar{y} + D(0)\bar{y}^3 + E(0)\bar{y}^4 = B'\bar{y} - D'\bar{y}^3 + E'\bar{y}^4.$$

The coefficients of the powers of y can also be equated, respectively,

$$B(0) = B'; \quad D(0) = D' = \bar{m} + G_{11} + G_{12};$$
$$E(0) = E' = \bar{m}' + G_{21} + G_{22}. \tag{5.60}$$

The last two equations (5.60) can be used to determine coefficients G from the initial conditions. As our other two equations we can use Equation (5.53), taking (5.56) into account. Then for the initial time we have

$$(\alpha_{11} - \eta_1)G_{11} + (\alpha_{11} - \eta_2)G_{12} + \alpha_{12}G_{21} + \alpha_{12}G_{22} = 0;$$
$$(\alpha_{22} - \eta_1)G_{21} + (\alpha_{22} - \eta_2)G_{22} + \alpha_{21}G_{11} + \alpha_{21}G_{12} = 0. \tag{5.61}$$

Coefficients G_{11}, G_{12}, G_{21}, G_{22} are defined completely by the four equations (5.60) and (5.61) as

$$G_{11} = \frac{(\alpha_{11} - \eta_2)(D' - \bar{m}) + \alpha_{12}(E' - \bar{m}')}{\eta_1 + \eta_2};$$

$$G_{12} = \frac{(\alpha_{11} - \eta_1)(\bar{m} - D') + \alpha_{12}(\bar{m}' - E')}{\eta_1 - \eta_2};$$

$$G_{21} = \frac{(\alpha_{22} - \eta_2)(E' - \bar{m}') + \alpha_{21}(D' - \bar{m})}{\eta_1 - \eta_2}; \qquad (5.62)$$

$$G_{22} = \frac{(\alpha_{22} - \eta_1)(\bar{m}' - E') + \alpha_{21}(\bar{m} - D')}{\eta_1 - \eta_2}.$$

5.2.3. The Final Expression for the Temperature Field

As a result of the foregoing considerations, we have obtained all the parameters determing the values of coefficients A, C, D, E of expression (5.22). The unknown coefficient B can easily be found from the first equation of (5.28):

$$B = (a_1 - \bar{m} - \bar{m}') + (b - \bar{n} - \bar{n}')\bar{t} + (c - \bar{q} - \bar{q}')\bar{t}^2 + + (d - \bar{r} - \bar{r}')\bar{t}^3 - (G_{11} + G_{21})\exp(\eta_1\bar{t}) - - (G_{12} + G_{22})\exp(\eta_2\bar{t}).$$

To summarize the above study, it can be stated that the temperature field in a glacier whose surface temperature varies according to the law $\bar{t} = a_1 + b\bar{t} + c\bar{t}^2 + d\bar{t}^3$ is describable in terms of a polynomial of the form

$$\bar{T} = A + B\bar{y} + C\bar{y}^2 + D\bar{y}^3 + E\bar{y}^4,$$

in which the coefficients of the powers of $\bar{y}$ are time-dependent and defined as:

$$A = 0, \quad C = 0;$$

$$B = (a_1 - \bar{m} - \bar{m}') + (b - \bar{n} - \bar{n}')\bar{t} + (c - \bar{q} - \bar{q}')\bar{t}^2 + + (d - \bar{r} - \bar{r}')\bar{t}^3 - (G_{11} + G_{21})\exp(\eta_1\bar{t}) - - (G_{21} + G_{22})\exp(\eta_2\bar{t}); \qquad (5.63)$$

$$D = \bar{m} + \bar{n}\bar{t} + \bar{q}\bar{t}^2 + \bar{r}\bar{t}^3 + G_{11}\exp(\eta_1\bar{t}) + G_{12}\exp(\eta_2\bar{t});$$

$$E = \bar{m}' + \bar{n}'\bar{t} + \bar{q}'\bar{t}^2 + \bar{r}'\bar{t}^3 + G_{21}\exp(\eta_1\bar{t}) + G_{22}\exp(\eta_2\bar{t}).$$

The values of coefficients G in these formulas are found with the aid of Equations (5.62), and the values of η are given by formula (5.55). Coefficients $\bar{m}$, $\bar{n}$, $\bar{q}$, $\bar{r}$, $\bar{m}'$, $\bar{n}'$, $\bar{q}'$, $\bar{r}'$ are determined using expressions (5.51), and the quantities γ entering into equations (5.51) are in turn given by formulas (5.49). Parameters i, i', and j in (5.49) are determined using formulas (5.40), (5.43), and (5.46). The quantities α and β in these formulas are determined solely by the physical properties of the ice, the Péclet number, and the boundary conditions of formula (5.31).

Consequently, the calculation is begun by determining coefficients α and β, after which parameters i, i', and j are found and then γ is determined, and finally coefficients $\bar{m}$, $\bar{n}$, $\bar{q}$, $\bar{r}$, $\bar{m}'$, $\bar{n}'$, $\bar{q}'$, $\bar{r}'$, and G are found for the computations using formulas (5.63).

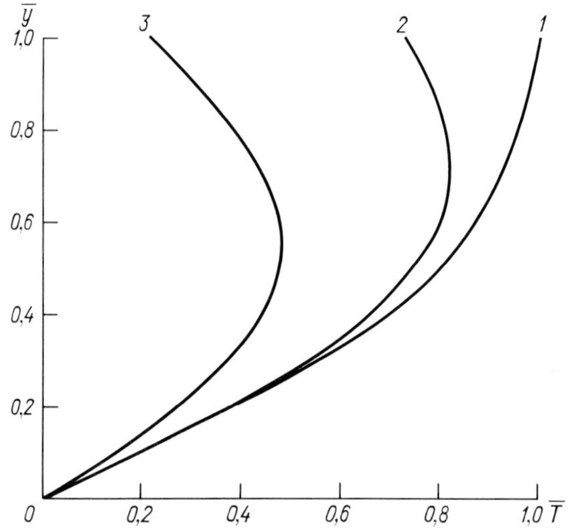

Fig. 5.3. Variation in time of temperature profile over glacier thickness, for vertical motion of glacier ice.
(1) initial distribution; (2, 3) typical temperature distribution when surface temperature varies according to law of third-degree polynomial with coefficients (5.64). Temperature fields correspond to glacier with Pe = 5, whose temperature distribution is described by a fifth-degree polynomial with coefficients (5.63).

As an example, we used the proposed method to study the temperature field in a glacier of constant thickness, the dimensionless surface temperature of which varies in time according to law (5.21). To be more definite, we assumed the following values for the coefficients in polynomial (5.21):

$$a_1 - 1; \quad b = 232 \cdot 10^{-3}; \quad c = 42; \quad d - 80. \tag{5.64}$$

The Péclet number is taken to be 5. The coefficient values in (5.64) correspond to the actual surface-temperature variations for the motion of an ice column from the central regions of Antarctica to the coast in the vicinity of the studies made in East Antarctica by Soviet expeditions. Figure 5.3 shows clearly the formation of the previously mentioned 'inverse gradient' caused by warming of the glacier surface.

Chapter 6

THE THERMAL REGIME OF GLACIERS

6.1. Thermophysical Studies of Accumulation-Ablation at the Upper and Lower Surfaces of Glaciers

Description of method. The calculation methods presented in Chapters 4 and 5 cannot be applied directly to study all glaciers. For instance, as our boundary condition at the lower surface of the glacier we assumed the temperature and the vertical velocity of the ice to be specified and the horizontal ice velocity to be constant over the glacier thickness. These conditions, however, appear to be valid only for floating ice shelves and icebergs, whose bottom temperatures can, as a first approximation, be assumed equal to the freezing temperature of seawater.

At the same time, for an ice shelf we do not usually know the magnitude (or even the direction) of the mass exchange at the lower surface (melting or freezing). Therefore, a calculation of the temperature field in the interior of an ice shelf and a comparison with the measured field still constitute about the only means of evaluating the intensity of the melting or freezing.

In an earlier work [32] we called the measurement of the temperature field in a glacier, and a comparison of this measured field with the observed field, the 'thermophysical method' of studying the melting or freezing at the glacier bottom or the ablation (accumulation) at the glacier top. Actually, as shown by Figures 4.2-4.6, the nature of the curve of the temperature field (convex, concave, S-shaped) already indicates the direction of the mass exchange going on at the upper and lower surfaces of the glacier. Inspection of these curves reveals that a variation of the Péclet number by unity, for the nominal Pe values of 3 to 6 typical of most present-day glaciers of the Earth, will appreciably alter the nature of the temperature field inside a glacier. Consequently, it is our opinion that the accuracy of the thermophysical method of determining the mean many-year values of the accumulation-freezing and ablation-melting can be equated to a difference of 1 between the values of the Péclet number. Taking into account the physical meaning of the Péclet number, we then obtain

$$\Delta v = \frac{a}{H}$$

or, if H is in meters,

$$\Delta v = 34 \frac{1}{H} \text{ m/yr}.$$

Accordingly, for a glacier about 300 m thick, which is the typical thickness of an ice shelf or iceberg, the accuracy in question will be close to 0.1 m/yr, or about 15% of its value for a glacier with Pe $\approx$ 6 and about 30% for a glacier with Pe $\approx$ 3. For inland glaciers with a thickness typical of Antarctica (3000 m), this accuracy will thus already be close to 0.01 m/yr, that is, considerably higher than that obtainable with direct measurements at the glacier surface [6]. With regard to the thermophysical method of studying mass-exchange processes, it should be noted that this method does not require repeated many-year observations of the accumulation-ablation, nor does it call for a meticulous manual processing of an ice core, so as to find the annual layers in the event of an accumulation of matter at the glacier surface. However, annual layers may not always be present; for instance, such layers may not form during the freezing of the bottom ice of a glacier. Another advantage of the thermophysical method is that, since the temperature field inside a glacier reflects factors acting for a long time, the mass-exchange characteristics obtained from a thermophysical study will represent mean values over many years, rather than instantaneous values, which may well be nonrepresentative.

This method is particularly important in cases where only single-shot remote observations are feasible, there being no possibility for observers to be present on the spot. This is why the thermophysical method was employed to investigate processes at the bottoms of ice shelves, which until recently have been inaccessible.

Use of the method to study glaciers on other planets. On the basis of the foregoing, it can be safely predicted that thermophysical studies of mass-exchange processes will be fundamental when investigating 'glaciers' of water, carbon dioxide, or ammonia on other planets. Chains of thermometers, connected to devices for the remote drilling of such 'ice' formations, by thermal drilling, for instance, can easily be lowered to the necessary depths without human aid, and then the data so obtained can be relayed back to Earth.

6.2. Boundary Conditions at the Lower Surface of an Inland Glacier

6.2.1. Heat Balance at the Lower Surface of a Glacier

An inland glacier moves over a surface either of bedrock, which is water-impermeable, or moraine or other sedimentary rock, which lets the water through. In the latter case the sedimentary rock bed may become saturated with meltwater, which may be completely or partly in a frozen state. Consequently, for generality, when considering the thermal regime of a sloping glacier, we have to take into account that it moves over a layer of partly ice-filled moraine or permafrost.

The temperature conditions at the boundary between the ice and the underlying frozen moraine will depend on the heat balance at this ice-moraine interface. Let us use the following notation: q_λ is the heat flux into the ice from the bottom of the glacier due to thermal conduction; q_{fr} is the heat flux arriving at the ice-moraine interface as a result of the thermal effects caused by the glacier motion; q_0 is the heat flux reaching the interface from the underlying moraine layers; and q_r is the heat removed per unit time to melt ice or released per unit time to freeze water at the ice-moraine interface (when such melting or freezing takes place). Then the heat balance at the boundary between the glacier and the underlying moraine can be represented as:

$$-\lambda \left.\frac{dT}{dy}\right|_{y=0} = q_\lambda = q_{fr} + q_0 - (q_r + q_w). \tag{6.1}$$

It is evident that expression (6.1) can be used directly as a boundary condition at the lower surface for the solution of Equation (4.1) [26, 28] or else to analyze more complicated conditions with the aid of numerical methods and a computer [5, 9, 86]. Two such cases are possible here:

(1) the temperature at the lower surface of the glacier is below the melting point (Figure 6.1 top). Then no melting-freezing processes will take place at the glacier bed, that is, the velocity of vertical ice motion at the bed is zero. Accordingly, term $(q_r + q_w)$ is also zero. The temperature at the lower surface of the glacier is in general unknown and must be determined independently:

(2) the temperature at the lower surface of the glacier is equal to the melting point of the ice (Figure 6.1 bottom). Only in this case are phase transitions at the lower surface possible. The rate of these transitions determines the intensity of the vertical ice motion. Thus, here $v|_{y=0} \neq 0$ and must be determined.

The heat flux $(q_r + q_w)$ will then be

$$q_r + q_w = \rho g[r + c(T_w - T_m)]v|_{y=0}. \tag{6.2}$$

Here T_w is the temperature to which the meltwater at the glacier bed is heated.

Since a glacier under steady-state conditions is being considered, it should be assumed that all the water formed by melting under the ice leaves the melting region. Water can flow off to the glacier periphery along its boundary with the bed or it can filter down into the subglacial moraine or sedimentary rock and then migrate sideward, to the periphery. Superheating of the water is possible if, before it leaves the melting region, it filters down deep into the sedimentary rock or moraine.

Such superheating can also occur if the water layer under the glacier is sufficiently large. For instance, suppose that subglacial lakes are present. In such a case, however, it is not likely that the superheating $T_w - T_m$ will be greater than 4 °C, since for higher temperature differentials the natural vertical convection of the lake water will start to have an effect, inhibiting the rise in T_w. The infiltration of water into sedimentary rock under a glacier can have a much greater effect. The temperature increase in the interior of a moraine is determined by the geothermic step, that is, it is close to 1 °C per 25 m of depth; therefore, if the water descends to a depth greater than 500 m before moving to the side of the melting region, then the value of q_w will become comparable with the latent heat of fusion (more than 10% of the latent heat) and must be taken into account.

Now, in view of Equation (6.1), we can represent the heat balance at the lower surface of the glacier as

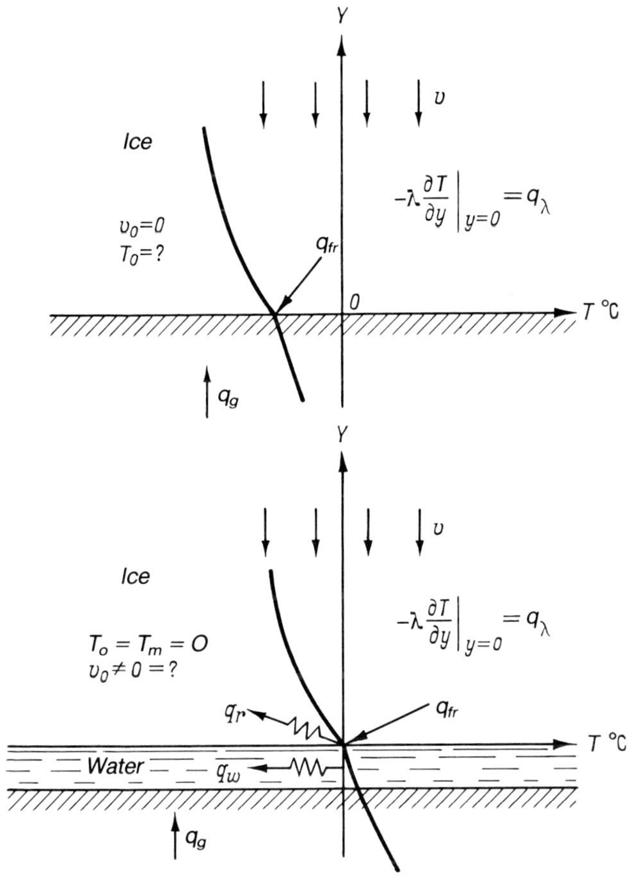

Fig. 6.1. Two types of boundary conditions at lower surface of inland glacier.

$$-\lambda \left.\frac{dT}{dy}\right|_{y=0} = q_{fr} + q_0 - \rho g[r + c(T_w - T_m)]v_0. \tag{6.3}$$

6.2.2. The Heat Flux Upward into the Glacier Due to Thermal Conduction

This flux will be uniquely determined if the temperature gradient at the glacier bed is known. The temperature gradient at the bed of a glacier with vertical ice flow is equal to the gradient in stationary ice corrected by a factor $\bar{\zeta}$, which depends on Pe and $\bar{v}_0$ and which takes into account the effect of the ice motion on the temperature field in the glacier:

$$\lambda \left.\frac{dT}{dy}\right|_{y=0} = \lambda \frac{T_H - T_0}{H} \bar{\zeta}.$$

Then the equation of heat balance (6.3) becomes

$$-\lambda \frac{T_H - T_0}{H} \bar{\zeta} = q_{fr} + q_H - \rho g[r + c(T_H - T_m)]v_0. \tag{6.4}$$

Note, however, that this correction factor is none other than the dimensionless temperature field at the lower surface of the glacier (see Section 4.3). Expressions for the dimensionless temperature gradient at the bed, for various cases of ice motion in a glacier, will be presented below. For instance, for the regime of ice cooling due to the downward ice flow when there is melting at the bed, we have

$$\bar{\zeta}_{\bar{v}_0 < 1,\downarrow} = \frac{2\left[\sqrt{\frac{|Pe|}{2}}\sqrt{1-\bar{v}_0}\, \exp - \left(\sqrt{\frac{|Pe|}{2}}\frac{\bar{v}_0}{\sqrt{1-\bar{v}_0}}\right)\right]^2}{\sqrt{\pi}\left\{\operatorname{erf}\left[\sqrt{\frac{|Pe|}{2}}\left(\sqrt{1-\bar{v}_0} + \frac{\bar{v}_0}{\sqrt{1-\bar{v}_0}}\right)\right] - \operatorname{erf}\left(\sqrt{\frac{|Pe|}{2}}\frac{\bar{v}_0}{\sqrt{1-\bar{v}_0}}\right)\right\}} \tag{6.5}$$

This regime corresponds to the accumulation region of the glacier. If the mass-exchange rate at the lower surface is zero, then

$$\bar{\zeta}_{\bar{v}_0 = 0} = \sqrt{\frac{|Pe|}{2}}\, \frac{1}{\operatorname{erf}\left(\sqrt{\frac{|Pe|}{2}}\right)} . \tag{6.6}$$

If the vertical velocity of ice flow is constant over the glacier thickness ($\bar{v}_0 = 1$), then we have

$$\bar{\zeta}_{\bar{v}_0 = 1,\downarrow} = \frac{|Pe|}{1 - \exp(|Pe|)} . \tag{6.7}$$

If the glacier movement corresponds to longitudinal compression ($\bar{v}_0 > 1$), then

$$\bar{\zeta}_{\bar{v}_0 > 1,\downarrow} = \frac{\sqrt{\frac{|Pe|}{2}}\sqrt{\bar{v}_0-1}\, \exp\left(\sqrt{\frac{|Pe|}{2}}\frac{\bar{v}_0}{\sqrt{\bar{v}_0-1}}\right)^2}{I\left[\sqrt{\frac{|Pe|}{2}}\left(\sqrt{\bar{v}_0-1} + \frac{\bar{v}_0}{\sqrt{\bar{v}_0-1}}\right)\right] + I\left(\sqrt{\frac{|Pe|}{2}}\frac{\bar{v}_0}{\sqrt{\bar{v}_0-1}}\right)} \tag{6.8}$$

If the vertical motion of the ice causes it to warm up, which occurs for ablation at the upper surface of the glacier, then for a surface melting rate which is less than the bottom freezing rate we get

$$\bar{\zeta}_{\bar{v}_0 > 1,\uparrow} = \frac{2\sqrt{\frac{|Pe|}{2}}\sqrt{\bar{v}_0-1}\,\exp - \left(\sqrt{\frac{|Pe|}{2}}\frac{\bar{v}_0}{\sqrt{\bar{v}_0-1}}\right)^2}{\pi\left\{\operatorname{erf}\left[\sqrt{\frac{|Pe|}{2}}\left(\sqrt{\bar{v}_0-1} - \frac{\bar{v}_0}{\sqrt{\bar{v}_0-1}}\right)\right] + \operatorname{erf}\left(\sqrt{\frac{|Pe|}{2}}\frac{\bar{v}_0}{\sqrt{\bar{v}_0-1}}\right)\right]}. \tag{6.9}$$

If $\bar{v}_0 = 1$, that is, if the surface melting rate equals the bottom freezing rate, then

$$\bar{\zeta}_{\bar{v}_0 = 1,\uparrow} = \frac{|Pe|}{\exp(|Pe|) - 1}. \tag{6.10}$$

If the rate of ablation at the upper surface is greater than the rate of bottom freezing, which implies longitudinal compression of the glacier ice, then we have

$$\bar{\zeta}_{0 < \bar{v}_0 < 1,\uparrow} = \frac{\sqrt{\frac{|Pe|}{2}}\sqrt{1-\bar{v}_0}\,\exp\left(\sqrt{\frac{|Pe|}{2}}\frac{\bar{v}_0}{\sqrt{1-\bar{v}_0}}\right)^2}{I\left[\sqrt{\frac{|Pe|}{2}}\sqrt{1-\bar{v}_0} + \frac{\bar{v}_0}{\sqrt{1-v_0}}\right)\Big] - I\left(\sqrt{\frac{|Pe|}{2}}\frac{\bar{v}_0}{\sqrt{1-\bar{v}_0}}\right)} \tag{6.11}$$

In the absence of bottom freezing, this expression simplifies to

$$\bar{\zeta}_{\bar{v}_0 = 0,\uparrow} = \frac{\sqrt{\frac{|Pe|}{2}}}{I\left(\sqrt{\frac{|Pe|}{2}}\right)}. \tag{6.12}$$

In the polynomial expressions for the temperature field which were derived in Chapter 5, factor $\bar{\zeta}$ is the coefficient of the first power of $\bar{y}$ in the polynomial. Consequently, using the approximate 'method of the zeroth moment' (Section 5.1), for cooling of the ice with zero mass exchange at the bed, we obtain

$$\bar{\zeta}_{\bar{v}_0 = 0,\downarrow} = B = \frac{0.6Pe^2 + 4.2Pe + 12}{0.15Pe^2 + 2.2Pe + 12}. \qquad (6.13)$$

Using the 'method of the zeroth and first moments', we have, correspondingly,

$$\bar{\zeta}_{\bar{v}_0 = 0,\downarrow} = B = \frac{1 + 0.25Pe + 0.02Pe^2}{1 + 0.0833Pe + 0.0033Pe^2} \qquad (6.14)$$

When the temperature field is approximated by a polynomial in five terms, we get

$$\bar{\zeta}_{\bar{v}_0 = 0,\downarrow} = \frac{2 + 0.62857Pe + 0.0833Pe^2 + 0.00476Pe^3}{2 + 0.29524Pe + 0.02857Pe^2 + 0.00063Pe^3} \qquad (6.15)$$

All the foregoing expressions will be valid if no heat sources are produced within the glacier by the ice motion. Such sources can be allowed for by introducing a correction factor $\Delta\bar{\zeta}_q$ to the factor $\bar{\zeta}$ obtained ignoring the effect of heat sources. The correction factor $\bar{\zeta}_q$ allowing for the effect of heat sources inside the glacier is then related as follows to correction factor $\bar{\zeta}$:

$$\bar{\zeta}_q = \bar{\zeta} + \Delta\bar{\zeta}_q \qquad (6.16)$$

Now the quantity $\Delta\bar{\zeta}_q$ for heat sources uniformly distributed over the glacier thickness can be evaluated by differentiating expression (4.70) or its approximation in the form of a fifth-degree polynomial (5.15). In the latter case $\overline{\Delta\zeta}_q$ is equal to coefficient B in the expression for the corresponding temperature correction or

$$\Delta\bar{\zeta}_q = -\bar{\Phi}' \frac{1 + 0.14762Pe + 0.0151Pe^2 + 0.00004Pe^3}{2 + 0.29524Pe + 0.02857Pe^2 + 0.00063Pe^3}. \qquad (6.17)$$

Here, as previously, $\bar{\Phi}'$ is the dimensionless intensity of the heat sources in the glacier.

6.2.3. Six Types of Thermal Regimes for a Glacier Bedded on a Water-Permeable Layer

Let us consider more carefully a glacier whose bed is a water-permeable layer, such as a layer of sedimentary rock or moraine of thickness H_M. Let T_M be the temperature at the lower surface of the moraine (at the boundary with the bedrock) and let λ_M be the thermal conductivity of the moraine. For simplicity, the melting temperature of the ice is taken to be T_m and assumed constant over the thickness of the glacier and the moraine.

Depending on a combination of various factors determining the thermal regime of this system, we can distinguish a number of typical cases, some of which can correspond to steady-state conditions, that is, they are invariable in time. Other cases may be transitional, nonsteady-state, but at the same time proceeding so slowly and gradually that they can be considered to be quasi-steady-state.

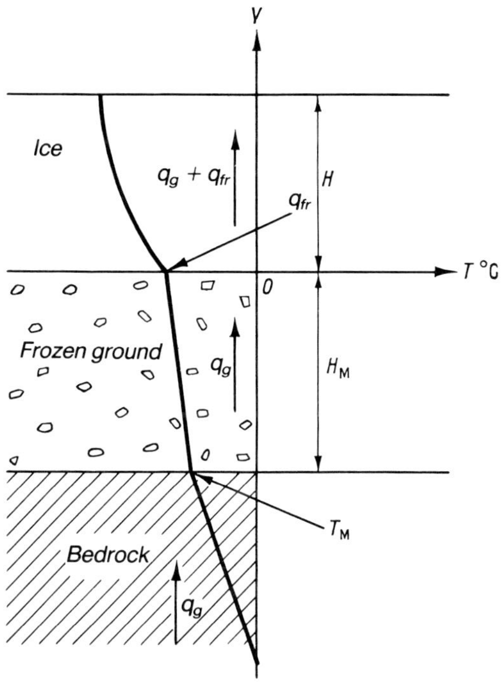

Fig. 6.2. First case of boundary conditions for glacier bedded on frozen moraine.

The first case is depicted schematically in Fig. 6.2. For it the temperature at the bottom of the moraine, at the boundary with bedrock, is lower than the melting temperature of the ice. In this case the upward heat flux through the moraine to the glacier-moraine interface can be written as

$$q_M = -\lambda_M \frac{T_0 - T_M}{H_M} .$$

By substituting this expression and (6.4) into (6.3) and taking into account that $v_0 = 0$, we arrive at the equation for the heat balance at the glacier-bed interface:

$$-\lambda \frac{T_H - T_0}{H} \bar{\zeta} = q_{fr} - \lambda_M \frac{T_0 - T_M}{H_M}. \tag{6.18}$$

If the temperature at the bottom of the moraine is lower than the melting point of ice, then the heat flux through the moraine equals the geothermal heat flux q_g:

$$-\lambda_M \frac{T_0 - T_M}{H_M} = q_g; \; T_M < T_m. \tag{6.19}$$

Now the substitution of (6.19) into (6.18) gives the following expression for the temperature at the glacier bed in this case:

$$T_0 = T_H - \frac{H(q_g + q_{fr})}{\lambda\bar{\zeta}}. \tag{6.20}$$

Finally, the temperature at the bottom of the moraine is found from (6.19) by a substitution of the type of (6.20):

$$T_M = T_H + q_g \frac{H_M}{\lambda_M} - \frac{H}{\lambda} \frac{q_g - q_{fr}}{\bar{\zeta}} = T_0 + q_g \frac{H_M}{\lambda_M}. \tag{6.21}$$

In the <u>second case</u> the temperature at the bottom of the moraine is equal to, or higher than, the melting temperature (Figure 6.3). Here it is always possible to set off an upper part of the moraine of thickness H_M^*, at the bottom of which the temperature is equal to the melting point of the ice. The heat flux through this upper part of thickness H_M^* can be lower than the geothermal flux by an amount q_{rM}, since melting is possible below this part of the moraine. It should be noted that the geothermal heat flux itself corresponds to the flow of heat given off when water freezes at a rate of only 0.5 cm/yr. Therefore, even melting at the moraine bottom at a rate only of the order of 0.1 cm/yr will already significantly affect the heat flux through the moraine, although it will just vary its thickness by 1 m in 1000 years, that is, it will hardly disturb the steady-state conditions at all. In this case the thermal regime of the upper part of the moraine is given by the equation

$$-\lambda_M \frac{T_0 - T_M^*}{H_M^*} = q_g - q_{rM}, \; T_M^* = T_m.$$

The temperature at the glacier bed, as in the previous case, is lower than the melting point of ice, and there will be no phase transitions at the bed ($v_0 = 0$).

In the <u>third case</u> the temperature at the glacier bed is equal to the melting point of the ice at the bed pressure. The following variations are possible:

(a) The entire moraine is ice-free, although it is saturated with water (Figure 6.4). Consequently, there are no phase transitions in the moraine. The heat balance in this case is given by Equation (6.3), in which the heat flux through the moraine is equal to the geothermal flux. Note that T_w will here depend on the moraine thickness and on the

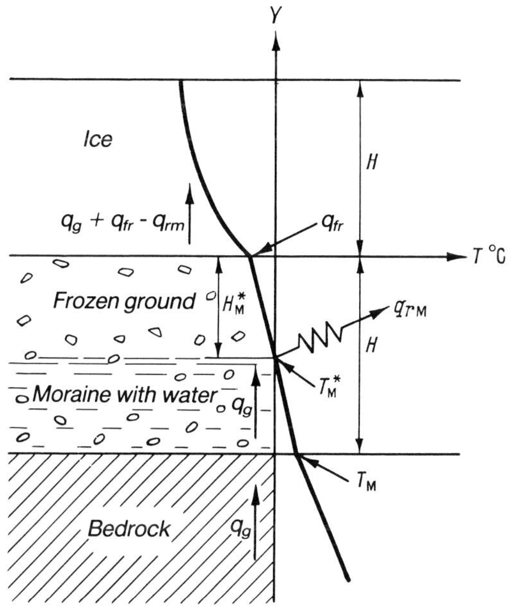

Fig. 6.3. Second case of boundary conditions for glacier on frozen moraine.

depth to which the water percolates into the moraine. In Equation (6.4) the expression for $\bar{\zeta}$ depends considerably on v_0, since the quantity $\bar{v}_0 = v_0/v_H$ enters into formulas (6.5), (6.7)-(6.11) determining $\bar{\zeta}$. However, since $\bar{v}_0$ is usually very low for inland glaciers (less than 0.1), the corresponding formulas (6.6) or their approximate equivalents (6.13), (6.14), or (6.15) can be used to find $\bar{\zeta}$. This was proposed in [30]; Figure 6.5 shows the coefficient M calculated in [30] as a function of the expected intensities of dimensionless melting at the glacier bed $\bar{v}_0$. M is defined as the ratio of $\bar{\zeta}$ in the presence of melting at the bed to $\bar{\zeta}$ for $\bar{v}_0 = 0$.

(b) If the temperature at the glacier bed is equal to the melting temperature of the ice, then the part of the moraine layer under the glacier, or even the entire moraine, may be filled with ice (Figure 6.6). The entire frozen layer will apparently have a temperature close to the melting point of ice, and thus it can be considered to be isothermal, that is, nonconducting. Hence, at the bottom of the moraine layer there must be continual melting, at a rate v_M defined as

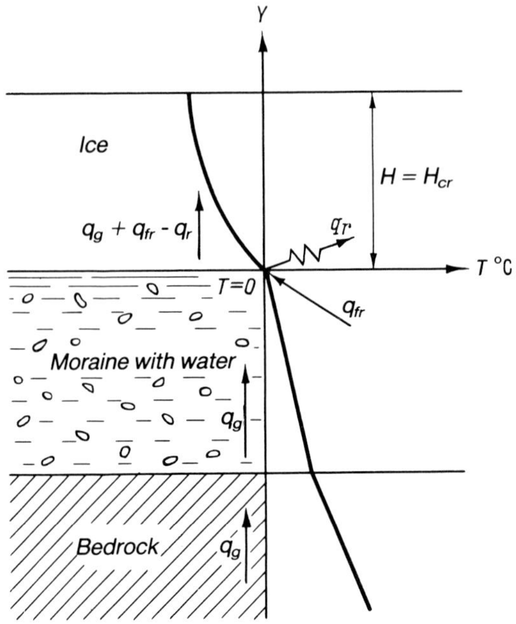

Fig. 6.4. Conditions at lower boundary of glacier bedded on water-saturated 'warm' moraine (third case of boundary conditions).

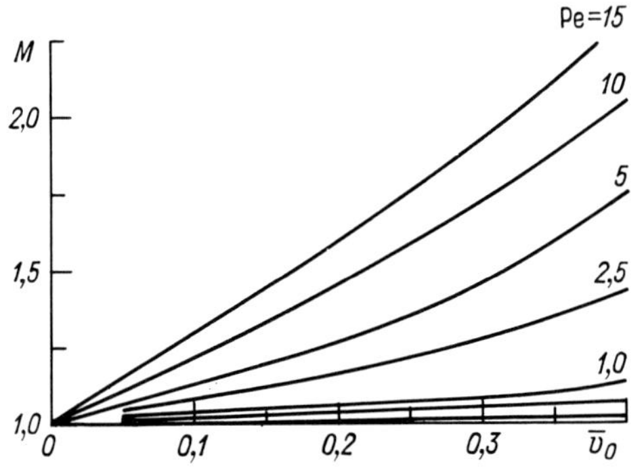

Fig. 6.5. Coefficient M showing effect of mass-exchange rate at glacier bottom on temperature gradient there (after [30]).

$$v_M = \frac{q_g}{\kappa_M \rho_M g(r + \Delta T_w c)} .$$

Here κ_M is the part by weight of the ice content in the moraine, and ΔT_w is the possible superheating of the water formed by the melting at the bottom of the frozen ground. This superheating has to be estimated

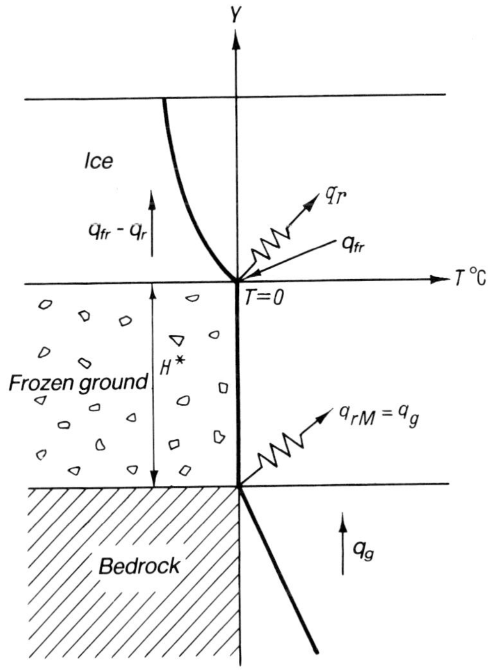

Fig. 6.6. Boundary condition at bottom of glacier bedded on ice-filled isothermal moraine (fourth case).

in the same way as the superheating of water for melting at the lower surface of a glacier bedded on a 'warm', that is, unfrozen, moraine.

For a geothermal heat flux close to the mean value over the Earth's surface and $\Delta T_w = 0$, the rate of melting at the bottom of a moraine containing 25% ice ($\kappa = 0.25$) will be close to 2 cm/yr. Thus in 1000 years of such melting the thickness of a frozen moraine layer will decrease by 20 m.

The equation for the heat balance at the bottom of such a glacier has the form

$$-\lambda \frac{T_H - T_m}{H} \bar{\zeta} = q_{fr} - \rho g[r + c(T_w - T_m)]v_0.$$

The rate of melting at the bed of the glacier is thus

$$v_0 = \left(\lambda \frac{T_H - T_m}{H} \bar{\zeta} + q_{fr}\right) \frac{1}{\rho g[r + c(T_w - T_m)]}. \tag{6.22}$$

Formula (6.22) shows that in some cases (when q_{fr} is low or when the temperature at the glacier surface is too low) rate v_0 can be 'negative'.

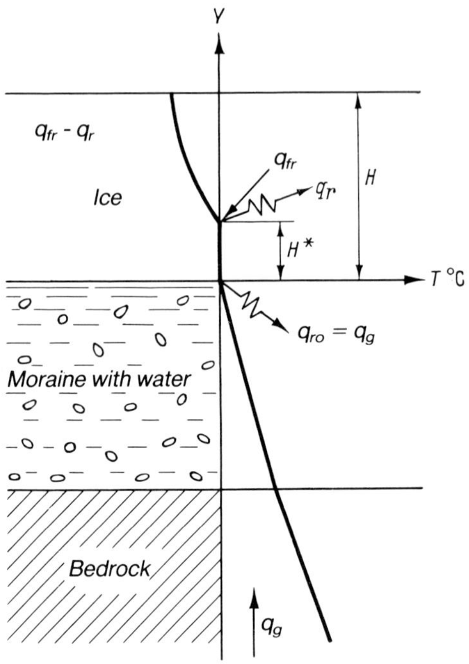

Fig. 6.7. Condition for existence of 'warm' isothermal layer in lower part of glacier (fifth case).

A negative rate corresponds to conditions whereby the regime can exist only for freezing at the lower surface of the glacier at a rate v_0. Such freezing is possible physically if, for instance, there is a vestigial subglacial lake under the glacier.

(c) The temperature in the glacier may be equal to the melting point of ice not only at the bed but also in part of the interior of the glacier (Figure 6.7). The glacier is then essentially divided into two zones, an upper zone where the temperature varies from T_H to the melting point (at the boundary with the lower zone), and a lower zone which is at the melting temperature throughout. A necessary condition for the stable existence of a lower isothermal zone is apparently the presence of a heat source at the boundary between the two zones, the source strength being high enough to compensate for the heat moving up through the zone interface into the cooler zone. Such a heat source may be, for instance, the friction created by the sliding of the ice of the upper zone over the stationary ice of the lower zone. If the friction heat developed at the interface between the zones is greater than the heat flux into the upper zone from below, then the excess friction heat will be expended to melt the ice at the interface. The temperature field in the upper zone, assumed to move relative to the lower stationary ice, is described by expressions determining the state shown in Figure 6.6, the thickness of the glacier being understood to be in this case the thickness of the upper zone. The rate of the possible melting at the zone boundary is given by expression (6.22), with the thickness of the upper zone ($H-H^*$) used as the glacier thickness.

(d) The limiting case of the thermal state of a glacier is the case in which the temperature throughout the glacier and the moraine is equal to the melting point of ice. Then, at the lower surface of the ice-filled part of the moraine, ice melts at the rate v_M. At the surface over which the ice slides, melting also takes place, at a rate

$$v_0 = \frac{q_{fr}}{\rho g[r + c(T_w - T_m)]}$$

Such glaciers, which are called 'temperate' by glaciologists, are often encountered in mountainous regions of the middle and lower latitudes. Their thermal regime is complicated considerably by heat exchange with the meltwater which permeates such glaciers. Some elements of this regime will be discussed below.

6.2.4. The Critical Thickness of a Glacier

In order to ascertain which case is observed in a specific glacier, let us analyze the conditions corresponding to a transition from one regime to another. For this it will be useful to define a critical glacier thickness H_{cr} at which the temperature in the glacier equals the melting point of ice only at the bed, but melting does not yet take place. If the glacier thickness is greater than critical, the third case is observed (melting at the bed or an isothermal warm zone in the lower part of the glacier). If, on the other hand, the glacier thickness is less than critical, the temperature at the bed will be below the melting point of ice and the thermal regime of the glacier will correspond to the first and second cases.

The critical thickness as an interface between regimes is found from Equation (6.4) with $v_0 = 0$ and with a temperature at the bed equal to the melting point:

$$-\lambda \frac{T_H - T_m}{H_{cr}} \bar{\zeta}_{\bar{v}_0 = 0} = q_{fr} + q_M, \tag{6.23}$$

which gives

$$H_{cr} = \frac{\lambda(T_m - T_H)}{q_{fr} + q_M} \bar{\zeta}_{\bar{v}_0 = 0}. \tag{6.24}$$

Here $\bar{\zeta}_{\bar{v}_0 = 0}$ is determined for the accumulation region by equation (6.6) or by its approximation in the form of polynomials (6.13), (6.14), (6.15). For the ablation region $\bar{\zeta}_{\bar{v}_0 = 0}$ is found using (6.12). By substituting these expressions into (6.24), we get the following cricital thicknesses for the accumulation region:

$$H_{cr\ ac} = \lambda \frac{T_m - T_H}{q_M + q_{fr}} \frac{2}{\sqrt{\pi}} \frac{\sqrt{\dfrac{v_H H_{cr}}{2a}}}{\mathrm{erf}\left(\sqrt{\dfrac{v_H H_{cr}}{2a}}\right)} \qquad (6.25)$$

and the ablation region:

$$H_{cr\ ab} = \lambda \frac{T_m - T_H}{q_M + q_{fr}} \frac{\sqrt{\dfrac{v_H H_{cr}}{2a}}}{I\left(\sqrt{\dfrac{v_H H_{cr}}{2a}}\right)} . \qquad (6.26)$$

If the ice thickness is greater than critical, continual melting takes place at the lower surface of the glacier.

It is clear from these formulas that the critical thickness is a function of the properties of the glacier ice, the surface temperature, the intensity of accumulation or ablation at the surface, and the heat flux arriving at the lower surface of the glacier.

The concept of the critical thickness of a glacier was introduced by N.N. Zubov [38], who considered only the simplest case of a glacier without sinking of the ice in its interior. Zubov's expression for H_{cr} is a particular case of (6.25) and (6.26), in the absence of ablation or accumulation at the surface. Actually, by evaluating the indeterminate form in formulas (6.25) and (6.26) for $v_H = 0$, we arrive at Zubov's expression:

$$H_{cr} = \frac{\lambda(T_m - T_H)}{q_M}. \qquad (6.27)$$

Figure 6.8 depicts the critical thickness of a glacier as a function of the surface temperature and the intensity of nourishment or ablation, according to formulas (6.25) and (6.26). The line for $v_H = 0$ was plotted using Zubov's formula (6.27). The curves above this line correspond to various accumulation intensities, and those below the line to ablation at the glacier surface. The calculations were carried out for ice with a thermal conductivity of 2.1 W/m °C, a heat capacity of 0.209 kJ N °C, and a specific weight of 8.98 kN/m^3. Flux q_M was taken to be the mean geothermal heat flux for the Earth's surface (the global mean q_g = 50 mW/m^2, and q_H values of 2 q_g and 3 q_g were used (see the various scales on the abscissa axis).

Inspection of these curves shows that even small amounts of accumulation or ablation for glaciers of great thickness will affect considerably the thermal regime of the glacier and its critical thickness, so that Equations (6.25) and (6.26) have to be used to find H_{cr}.

As was to be expected, under conditions of accumulation, when the subsidence of the glacier leads to its cooling, the critical thicknesses turn out to be greater than under ablation conditions, when the effect of the movement tends to heat the glacier.

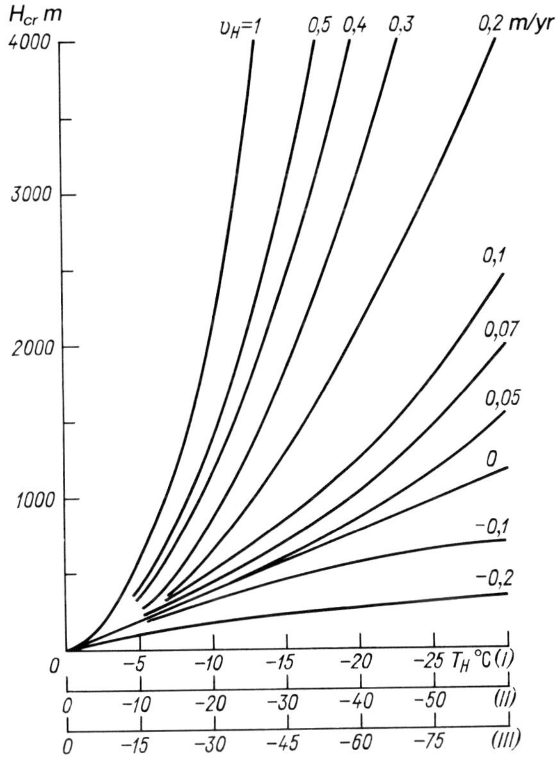

Fig. 6.8. Critical glacier thickness as function of temperature and vertical velocity of ice at upper surface (after [27]).

Note, by the way, that the same formulas can be obtained in the interpretation corresponding to expressions for $\bar{\zeta}_{\bar{v}_0} = 0$ in the form (6.13), (6.14), and (6.15). For example, using the more accurate approximation (6.15), we obtain

$$H_{cr\ ac} = \lambda \frac{(T_m - T_H)\left[2 + 0.62857\dfrac{|v_H|H_{cr}}{a} + 0.0833\left(\dfrac{|v_H|H_{cr}}{a}\right)^2 + 0.00476\left(\dfrac{|v_H|H_{cr}}{a}\right)^3\right]}{(q_{fr} + q_M)\left[2 + 0.29524\dfrac{v_H H_{cr}}{a} + 0.02857\left(\dfrac{v_H H_{cr}}{a}\right)^2 + 0.00063\left(\dfrac{v_H H_{cr}}{a}\right)^3\right]} . \tag{6.28}$$

If appreciable heat is emitted inside the glacier by internal sources, then we have

$$H_{cr} = \frac{\lambda(T_m - T_H)}{q_{fr} + q_H}(\bar{\zeta} + \Delta\bar{\zeta}_q).$$

Finally, by substituting quantities $\bar{\zeta}$ and $\Delta\bar{\zeta}_q$ from (6.15) and (6.17) into this expression, we obtain

$$H_{cr\ ac} = \lambda \frac{(T_m - T_H)\left[1 + 0.48095\frac{|v_H|H_{cr}}{a} + 0.0682\left(\frac{|v_H|H_{cr}}{a}\right)^2 + 0.00472\left(\frac{|v_H|H_{cr}}{a}\right)^3\right]}{(q_{fr} + q_H)\left[2 + 0.29524\frac{|v_H|H_{cr}}{a} + 0.02857\left(\frac{|v_H|H_{cr}}{a}\right)^2 + 0.00063\left(\frac{|v_H|H_{cr}}{a}\right)^3\right]} \tag{6.29}$$

Critical temperature differential in a glacier bedded on an ice-filled moraine. In order to decide whether melting takes place at the lower surface of a moraine filled with ice, or whether the temperature at this surface is below the melting point and there is no submorainal melting, let us introduce the concept of the critical temperature differential in a glacier bedded on a moraine. This critical temperature differential is defined as the difference between the surface temperature of the glacier and the melting temperature of the ice at the moraine-bedrock boundary, for which the temperature at this boundary is apparently the melting temperature but there is still no melting there. If the difference between the surface temperature of a real glacier and the melting temperature at the bottom of the moraine exceeds the critical differential, then the temperature at the moraine-bedrock boundary will be lower than the melting point of the ice. If the critical differential ΔT_{cr} is greater than the difference between the surface temperature of the glacier and the melting point of the ice at the bottom of the moraine, then melting will take place at the moraine-bedrock boundary. Substituting $T_M = T_m$ into (6.20), we obtain the following expression for the critical temperature differential:

$$\Delta T_{cr} = T_H - T_m = \frac{H}{\lambda}\frac{q_g + q_{fr}}{\bar{\zeta}_{\bar{v}_0} = 0} - \frac{H_M}{\lambda_M}q_g. \tag{6.30}$$

6.3. Heat Sources Participating in the Heat Balance at the Bottom of a Glacier

6.3.1. The Traditional Representation of the Motion-Caused Heat Flux Arriving at the Bed

The motion of ice in a glacier takes place under the influence of mass forces. Therefore, for a glacier in which inertial forces can be neglected, the amount of heat given off during the motion is determined by the work done by the ice particles as they move downward in the gravitational field, that is, by the work of subsidence. Roughly speaking, the heat produced in the glacier interior during the ice motion equals the heating effect of the work of subsidence performed by the ice in the gravitational field.

Let us calculate the heating effect of the subsidence work done in the gravitational field by an ice column of thickness H with a base area of unity, assuming a linear variation of the vertical ice velocity from a value equal to the accumulation rate (at the surface) to zero (at the bottom). For a glacier with a constant ice density over its thickness, this heating effect will be

$$q_Q = N\rho g \int_0^H v\, dy = \frac{\rho g H v_H}{2}, \tag{6.31}$$

where N is the thermal equivalent of work.

For a glacier about 3000 m thick and a mean vertical velocity of about 0.1 m/yr, this quantity equals the global mean value of the geothermal heat flux and thus must be taken into account when analyzing the heat-transfer processes in a glacier. At the same time, in the opinion of many [14, 27, 115], the heat given off during the ice movement may be delivered in toto to the ice-bed boundary, not being distributed over the whole mass of the glacier. This possibility arises because the near-bottom parts of the glacier experience a very great shear force, while the upper layers have only a very small shear acting on them. In this case the shear work done at the bed per unit time, expressed in thermal units, will also constitute a source of heat caused by the glacier movement (q_{fr}). If τ_b is the shear stress at the bed and u_0 is the velocity of horizontal motion, then

$$q_{fr} = N\tau_b u_0. \tag{6.32}$$

The ice velocity at the bed for the glacier model of constant thickness being considered here is found from the continuity equation.

Now let us assume that the thickness and accumulation of the glacier both vary as a function of the distance x from the ice divide, according to laws H(x) and $v_H(x)$. Then the velocity of horizontal motion is

$$u = \frac{1}{H}\int_0^H v_H\, dx. \tag{6.33}$$

Correspondingly, for axisymmetric ice outflow in a glacier whose thickness and accumulation rate vary with the distance R from the outflow center according to laws H(R) and v(R), the velocity of horizontal ice motion will be

$$u(R) = \frac{\int_0^R v(R)\, dR}{H(R)\, dR}\ . \qquad (6.34)$$

At one time Equation (6.31) was used to evaluate the heat flux arriving at the glacier bed [108], and Nye stated in [109] that the shear stress at the bottom of many glaciers, or at any rate of the large ice sheets, can be taken to be constant and equal to about 98 kN/m^2. This appeared to be an acceptable value, since it agreed with Glen's data, according to which at low shear stresses a slow ice flow develops and only beginning at some $\tau = \tau_0 = 98\ kN/m^2$ does the flow velocity rise sharply and a slight stress increment give rise to an appreciable increase in the deformation rate. It was natural to assume, on the basis of the theory of ideal plasticity used by Nye, that the shear stresses along the glacier bed are constant and equal to τ_0. It was assumed that the limiting resistance to shear τ_0 in this case can be likened to a regulator of the glacier motion [15]. As soon as the stress along the bed "exceeds τ_0, the velocity of ice motion increases sharply and prevents an excessive increase in the glacier thickness as well as a further rise in τ, which thus cannot exceed τ_0 by much. A state is therefore reached which is similar to the plastic state, and the glacier begins to slide over its bed as a solid block" [14].

Subsequent studies have not, however, verified the hypothesis of a constant tangential stress at the bed of a glacier or ice sheet. The quantity τ_b in (6.32) was found from the conditions for equilibrium of the ice column, using the approximate expression

$$\tau_b = \rho g H \operatorname{tg} \alpha_s,$$

where α_s is the inclination of the glacier surface to the horizontal. This quantity was found to be different for different parts of a glacier or ice sheet [4]. The problem of the true value of the stress and the distribution of heat sources at the bed and in the interior of a glacier still awaits solution. Let us now consider one possible, in our opinion quite promising, approach to this problem.

6.3.2. Determination of the Heat Developed at the Glacier Bed

As previously, let us consider the horizontal ice velocity at the surface u_H, the glacier thickness H, and the horizontal velocity u_y at a distance $\bar{y}$ from the glacier bed. Then we have

$$u = f[\bar{y}, H(x)].$$

We next put this relation into dimensionless form and assume similarity of the velocity fields of the glacier along the x axis. Apparently such an assumption is valid, at any rate for sufficiently large regions of the glacier. In this case the dimensionless velocity u/u_H will be a function only of the dimensionless glacier thickness $\bar{y}$:

$$\bar{u} = f(\bar{y}).$$

In the last chapter it was shown that a dependence of this sort can always be represented in terms of a polynomial of sufficiently high degree. Let us seek it in the form of such a polynomial, but let us limit ourselves to the first three terms:

$$\bar{u} = b_1 + b_2\bar{y} + b_3\bar{y}^2. \tag{6.35}$$

We note once again that the number of terms may be greater; only three terms are taken here simply because this is the number of terms corresponding to our present knowledge about the ice motion in a glacier.

The coefficients in polynomial (6.35) will be determined from the following conditions typical of any glacier:

(1) at the surface ($\bar{y} = 1$) the tangential stresses are zero, so that the velocity gradient over the glacier thickness will also be zero:

$$\bar{y} = 1; \quad \frac{d\bar{u}}{d\bar{y}} = b_1 + 2b_2 = 0; \quad b_2 = -\frac{b_1}{2}; \tag{6.36}$$

(2) at the glacier bed the tangential stress τ will be related to the ice deformations by an experimental rheological expression. If this relation corresponds to Glen's law [99], then we have

$$\bar{y} = 0; \quad \left.\frac{du}{dy}\right|_{y=0} = B\tau_b^n. \tag{6.37}$$

Here B and n are the experimental constants of Glen's law.

Now, by substituting (6.35) into (6.37), we get for $\bar{y} = 0$

$$b_1 = \frac{BH}{u_H}\tau_b^n; \tag{6.38}$$

(3) the velocity at the surface is in dimensionless form equal to unity, so that from (6.35) we have

$$\bar{y} = 1; \; b + b_1 + b_2 = 1. \tag{6.39}$$

Then coefficients b, b_1, and b_2 are given by system of Equations (6.36), (6.38), (6.39) and they are:

$$b = 1 - \frac{H}{2u_H} B\tau_b^n; \quad b_1 = \frac{H}{u_H} B\tau_b^n; \quad b_2 = \frac{1}{2}\frac{H}{u_H} B\tau_b^n. \tag{6.40}$$

Note in particular that the three above conditions were obtained without making any special assumptions concerning the nature of the processes in the glacier, and thus they are general conditions, valid for any glacier.

Consequently, if the true curve of the velocity distribution over the glacier thickness can be approximated by a second-degree polynomial, then this polynomial has been found by us and it will have the coefficients given in (6.40).

It should be repeated that the representation of the velocity field in terms of a second-degree polynomial corresponds to the level of our present knowledge, and that a more accurate analysis of the velocity field is always obtainable by increasing the number of terms in the polynomial. The tangential stress at the bed, which enters into expressions (6.40), is related to the thickness and surface inclination of the glacier by an integral condition.

For a glacier with a flat horizontal bed this condition, first derived by Nye, has the form

$$\frac{dH}{dx} = \frac{\tau_b}{\rho g H} = \mathrm{tg}\ \alpha_s. \tag{6.41}$$

The analogous relation for a glacier with a sloping bed is somewhat more complicated [14]. However, to simplify the calculations, the main purpose of which is to demonstrate the method of analysis, in the following we will assume that the glacier bed is horizontal.

By substituting condition (6.41) into (6.40), we obtain the final expression for b, b_1, b_2 characterizing the velocity distribution inside the glacier:

$$\begin{aligned} b &= 1 - (\rho g)^n B \frac{H^{n+1}}{2u_H} (\mathrm{tg}\ \alpha_s)^n; \\ b_1 &= (\rho g)^n B \frac{H^{n+1}}{u_H} (\mathrm{tg}\ \alpha_s)^n; \\ b_2 &= -(\rho g)^n B \frac{H^{n+1}}{2u_H} (\mathrm{tg}\ \alpha_s)^n. \end{aligned} \tag{6.42}$$

Accordingly, the variation in the ice velocity over the glacier thickness can be expressed as

$$\frac{u}{u_H} = 1 + \frac{BH^{n+1}}{u_H} (\rho g\ \mathrm{tg}\ \alpha_s)^n \left(-\frac{1}{2} + \bar{y} - \frac{1}{2}\bar{y}^2\right). \tag{6.43}$$

The quantity $B \frac{H^{n+1}}{u_H} (\rho g\ \mathrm{tg}\ \alpha_s)^n$ determining the nature of the velocity variation over the glacier thickness is dimensionless. Let us call this the characteristic of the glacier motion G*.

The quantity b characterizes the ratio of the sliding velocity at the bed to the velocity of ice motion at the surface. Coefficient b is also dimensionless and may with complete justification be called the dimensionless parameter of basal sliding of the glacier, or, for short, simply the sliding parameter. The velocity of sliding over the bed is given by the expression

$$u_{sl} = u_H b = u_H[1 - (\rho g)^n B \frac{H^{n+1}}{2u_H} (tg\ \alpha_s)^n]. \quad (6.44)$$

It should be noted that in some cases the velocity at the glacier surface is unknown. For example, this may happen when parts of large ice sheets quite far away from outcrops of the stationary bedrock are studied, direct measurements of the ice velocity being difficult in this case because of the absence of reference points (central regions of Antarctica). Another instance of this is the reconstruction of ancient ice sheets. However, in these cases it is usually possible to determine the mean velocity of ice motion over the glacier thickness from the continuity conditions (by calculational means). From the definition of the mean velocity, we have

$$u_{av}H = \int_0^H u\ dy.$$

By substituting (6.35) into this expression, we obtain the following relation between the velocity at the surface (to be substituted into our formulas) and the mean velocity:

$$u_H = \frac{u_{av}}{b + \frac{b_1}{2} + \frac{b_2}{3}}.$$

If the variation of the velocity over the glacier thickness is known, then we can determine the thermal effects of the motion. It is convenient to divide these effects into (1) the heat released at the boundary between the ice and the bed, and (2) the heat sources produced by internal friction, which are distributed over the glacier thickness.

The heat produced at the ice-bedrock boundary is equal to the work doen as the ice slides over the bed, expressed in thermal units:

$$A_{fr} = \tau_b u_{sl}; \quad q_{fr} = N\tau_b u_{sl}. \quad (6.45)$$

Here A_{fr} is the work done, and q_{fr} is the heat released, per unit area of the bed per unit time, as the glacier slides with a velocity u_{sl}.

Finally, by substituting into (6.45) the expressions for the sliding velocity (6.44) and the tangential stress (6.41), we obtain

$$q_{fr} = N\rho g H u_H b \text{ tg } \alpha_s = N\rho g H u_H \left[1 - B(\rho g)^n \frac{H^{n+1}}{2u_H} (\text{tg } \alpha_s)^n\right] \text{tg } \alpha_s. \tag{6.46}$$

6.3.3. Determination of the Heat of Internal Friction for a Glacier

This heat is produced by the work of the internal-friction forces. This work A_{in} is assumed to be the difference between the entire work expended on the motion of the glacier ice A_Σ and that part of the work needed to overcome the sliding friction at the bed A_{fr}. These quantities will be referred to unit time and to a column of glacier ice of thickness H with a base area of unity. Thus we have

$$A_{in} = A_\Sigma - A_{fr}. \tag{6.47}$$

The work expended on the ice motion in a column H of unit cross section is the work of the motion in the gravitational field of all the ice particles in this column:

$$A_\Sigma = \int_0^H \rho g v \, dy. \tag{6.48}$$

Let us assume the ice density to be constant over the glacier thickness. The velocity components v and u for such a two-dimensional glacier will be related as follows: $\partial u/\partial x = -\partial v/\partial y$, so that in view of (6.35) we have

$$dv = \frac{\partial u_H}{\partial x}\left(b + b_1 \frac{y}{H} + b_2 \frac{y^2}{H^2}\right) dy.$$

Integration of this expression over y then gives

$$v = H \frac{\partial b}{\partial x} \bar{y} + \frac{H}{2} \frac{\partial b_1}{\partial x} \bar{y}^2 + \frac{H}{3} \frac{\partial b_2}{\partial x} \bar{y}^3 + C,$$

where C is the constant of integration.

Substitution of expressions (6.42) into the latter then gives

$$v = H \frac{\partial}{\partial x}\left[(u_H \bar{y} + \rho g)^n BH^{n+1} (\text{tg } \alpha_s)^n \left(- \frac{1}{2} \bar{y} + \right.\right.$$

$$\left.\left. + \frac{\bar{y}^2}{2} - \frac{1}{6} \bar{y}^3\right)\right] + C,$$

or, after certain transformations,

$$v = H\left\{\frac{\partial u_H}{\partial x}\bar{y} + (\rho g)^n B\left[(\text{tg}\ \alpha_s)^n \frac{\partial H^{n+1}}{\partial x} + H^{n+1}\frac{\partial(\text{tg}\ \alpha_s)^n}{\partial x}\right] \times\right.$$

$$\left.\times\left(-\frac{1}{2}\bar{y} + \bar{y}^2 - \frac{1}{6}\bar{y}^3\right)\right] + C. \tag{6.49}$$

If we neglect bottom melting or freezing, then

$$\bar{y} = 0; \quad v_0 = 0; \quad C = 0$$

Now, by substituting the expression obtained for v into (6.48), we get, after integration and transformations, the following formula:

$$A_\Sigma = \rho g H^2\left\{\frac{\partial u_H}{2\partial x} - \frac{3}{24}(\rho g)^n\left[(\text{tg}\ \alpha_s)^n \frac{\partial H^{n+1}}{\partial x} + \right.\right.$$

$$\left.\left. + H^{n+1}\frac{\partial(\text{tg}\ \alpha_s)^n}{\partial x}\right]\right\} = \frac{\rho g H(v_H - u_H\ \text{tg}\ \alpha_s)}{2} - \tag{6.50}$$

$$- \frac{3}{24}\rho g H^2 (\rho g)^n\left[(\text{tg}\ \alpha_s)^n \frac{\partial H^{n+1}}{\partial x} + H^{n+1}\frac{\partial(\text{tg}\ \alpha_s)^n}{\partial x}\right].$$

For ice under conditions of block [Blockschollen] motion ($b = 1$, $b_1 = b_2 = 0$), this expression simplifies to the familiar formula:

$$A_\Sigma = \frac{1}{2}\rho g H(v_H - u_H\ \text{tg}\ \alpha_s). \tag{6.51}$$

Finally, if we substitute (6.50) into (6.47), then taking (6.45) into account we arrive at the following equation for the total heat produced by energy dissipation due to internal friction in an ice column of unit thickness and unit cross section:

$$A_{in} = \frac{\rho g H(v_H - u_H\ \text{tg}\ \alpha_s)}{2} - \frac{3}{24}(\rho g)^n\left[(\text{tg}\ \alpha_s)^n \frac{\partial H^{n+1}}{\partial x} + \right.$$

$$\left. + H^{n+1}\frac{\partial(\text{tg}\ \alpha_s)^n}{\partial x}\right] - \rho g H u_H(\text{tg}\ \alpha_s)\left[1 - B(\rho g)^n \frac{H^{n+1}}{2u_H}(\text{tg}\ \alpha_s)^n\right]. \tag{6.52}$$

6.3.4. Source-Strength Distribution over Glacier Thickness

The strength of the heat sources inside a glacier is an important characteristic. In order to ascertain the distribution of the source strength over the glacier thickness, let us consider the ice flow in the glacier, using the Navier-Stokes equations for an ideal liquid.

Previously it was shown that Glen's experimental law for the flow of ice as a non-Newtonian liquid can be put into the form of Newton's law of friction for an ideal liquid, provided that the quantity (3.25) is taken to be the apparent viscosity μ_a. Once this apparent viscosity is used in Glen's law, it can be written formally like Newton's law of friction:

$$\tau = \mu_a \frac{\partial u}{\partial y}.$$

We assume for simplicity that the ice flow in the glacier satisfies the Navier-Stokes equation with an apparent ice viscosity that is constant over the glacier thickness. Actually, this will be strictly valid only for temperate glaciers. In cold glaciers the apparent viscosity increases with the distance from the glacier bed, but apparently the mean viscosity over the thickness can be so selected that the error introduced by this simplification will be minimal.

In this case the release of internal-friction energy characterizing the volume source strength is given by dissipation function (3.44). For a flat glacier the dissipation function is

$$\Phi = 2\mu_a\left[\left(\frac{\partial v}{\partial y}\right)^2 + \left(\frac{\partial u}{\partial x}\right)^2\right] + \mu_a\left(\frac{\partial u}{\partial y} + \frac{\partial v}{\partial x}\right)^2. \tag{6.53}$$

We can simplify this expression by evaluating the relative order of each term for specific glacier conditions and by dropping terms of second order of smallness. To do this, we put the coordinates in the expression for the dissipation function into dimensionless form, as we did in Chapter 2: $x/L = \bar{x}$; $y/H = \bar{y}$; $v/v_H = \bar{v}$; $u/u_H = \bar{u}$.

For large glaciers it is usually true that

$$\frac{H}{L} \sim \delta \ll 1; \quad \frac{v_H}{u_H} \sim \delta \ll 1.$$

Next we place over each dimensionless complex preceding each derivative the order of the complex:

$$\Phi = 2\mu_a \overset{1}{\left(\frac{v_H}{H}\right)^2} \left\{\left[\overset{1}{\left(\frac{\partial \bar{v}}{\partial \bar{y}}\right)^2} + \overset{(\delta/\delta)^2}{\left(\frac{u_H}{L}\frac{H}{v_H}\right)^2} \left(\frac{\partial \bar{u}}{\partial \bar{x}}\right)^2\right] + \right.$$

$$+ \frac{1}{2}\left[\overset{1/\delta^2}{\left(\frac{u_H}{H}\frac{H}{v_H}\right)^2} \left(\frac{\partial \bar{u}}{\partial \bar{y}}\right)^2 + 2\overset{\delta/\delta}{\left(\frac{u_H}{L}\frac{H}{v_H}\right)} \left(\frac{\partial \bar{u}}{\partial \bar{y}}\frac{\partial \bar{v}}{\partial \bar{x}}\right) + \right.$$

$$\left.\left. + \overset{\delta^2}{\left(\frac{v_H}{L}\frac{H}{v_H}\right)^2} \left(\frac{\partial \bar{v}}{\partial \bar{x}}\right)^2\right]\right\}.$$

Then a comparison of the relative orders of the marked terms enables us to represent the dissipation function for the conditions obtaining in large glaciers as

$$\Phi = \mu_a \left(\frac{v_H}{H}\right)^2 \left(\frac{\partial \bar{u}}{\partial \bar{y}}\right)^2 = \mu_a \left(\frac{\partial u}{\partial y}\right)^2 . \tag{6.54}$$

Accordingly, the strength of the heat sources arising due to the dissipation of energy because of internal friction is

$$q_{in} = N\mu_a \left(\frac{\partial u}{\partial y}\right)^2 . \tag{6.55}$$

After substituting (6.35) into (6.55), we obtain after certain transformations the following equation for the distribution over the glacier thickness of the strength of the volume heat sources due to internal friction:

$$q_{in} = N\mu_a u_H^2 \frac{b_1^2}{H^2}(1 - 2\bar{y} + \bar{y}^2). \tag{6.56}$$

The apparent viscosity, averaged over the glacier thickness, can be selected correctly provided we recall that the total strength of the heat sources, distributed over the glacier thickness according to law (6.56), has to equal the amount of heat specified earlier as the heating effect of the work of glacier motion in the gravitational field, excluding the heating effects of the sliding of the glacier over the bed. Then,

$$\mu_a \frac{u_H^2 b_1^2}{H^2} \int_0^H \left(1 - 2\frac{y}{H} + \frac{y^2}{H^2}\right) dy = A_\Sigma - A_{fr},$$

where $A_\Sigma - A_{fr}$ is given by expression (6.52). The apparent viscosity is then

$$\mu_a = 3\,\frac{A_\Sigma - A_{fr}}{H}\,\frac{H^2}{u_H^2 b_1^2}. \tag{6.57}$$

Consequently, the strength of the heat sources of internal friction can be represented in a form not containing the apparent viscosity:

$$q_{in} = 3N\,\frac{A_\Sigma - A_{fr}}{H}\,(1 - 2\bar{y} + \bar{y}^2). \tag{6.58}$$

It should be noted that the quantity $(A_\Sigma - A_{fr})/H$ is the volume strength of the heat sources, assumed to be distributed uniformly over the glacier thickness. Let us call this quantity q_m^*. Then the distribution of the dimensionless source strength $\bar{q}_m = q_m/q_m^*$ will have a form which is the same for all the glaciers being considered:

$$\bar{q}_m = \frac{q_{in}}{q_m^*} = 3 - 6\bar{y} + 3\bar{y}^2. \tag{6.59}$$

Chapter 7

THE THERMAL REGIME OF INLAND ICE CAPS

Ice caps are appropriate objects of study using the methods developed, since their shapes and main features allow simpler generalizations, thus making it possible to use calculation schemes which are always simpler than nature. On the other hand, the Antarctic Ice Sheet alone, which is actually a system of ice caps, accounts for as much as 90% of the entire volume of glacier ice in the world [32]. This in itself is an indication of the enormous significance, for theory as well as practice, of the processes going on in ice caps.

As an example, let us consider some elements of the thermal regime of the inland Antarctic Ice Sheet.

7.1. The Thermal Regime of the Antarctic Ice Sheet

7.1.1. A Model of a Steady-State Flat Glacier with Block Sliding and Melting at the Bed (Central Antarctica)

Robin [115] greatly aided the development of ideas about the temperature conditions inside large ice caps when he proposed using a very simple model of a steady-state flat glacier with block sliding (Section 3.5) to analyze the processes in the central regions of such formations. With the aid of this model, and the analysis method described above, in Chapters 4 and 6, as applied to the central region of the Antarctic Ice Sheet, Robin was able to show that in the entire central region of an antarctic ice cap the temperature at the bottom of the glacier is equal to the melting point of ice and that the ice at the bed melts continually.

This conclusion was reached by determining the critical glacier thicknesses according to the method described in the previous chapter and then comparing these values with the actual ice thicknesses. The calculations were carried out for a glacier resting on bedrock ($q_M = q_g$), which apparently corresponds to the actual conditions beneath an antarctic glacier. The geothermal flux was taken to be close to 80 mW/m^2 data of [25, 105]).

As an example, Figure 7.1 shows the variations in the temperature and accumulation rate at the glacier surface, as well as the height of this surface and the location of the glacier bed, over the profile of the Mirny, Vostok, and Amundsen-Scott (South Geographic Pole) Stations. Inspection of the figure shows the critical thickness of the ice sheet in its entire central region to be less than the true thickness; thus, over the whole central part of the ice sheet melting occurs at the bed.

Close to the margin of the ice sheet there is a region where the actual thickness is reduced, to the same degree that the critical thickness exceeds the true thickness. This region, which apparently forms a

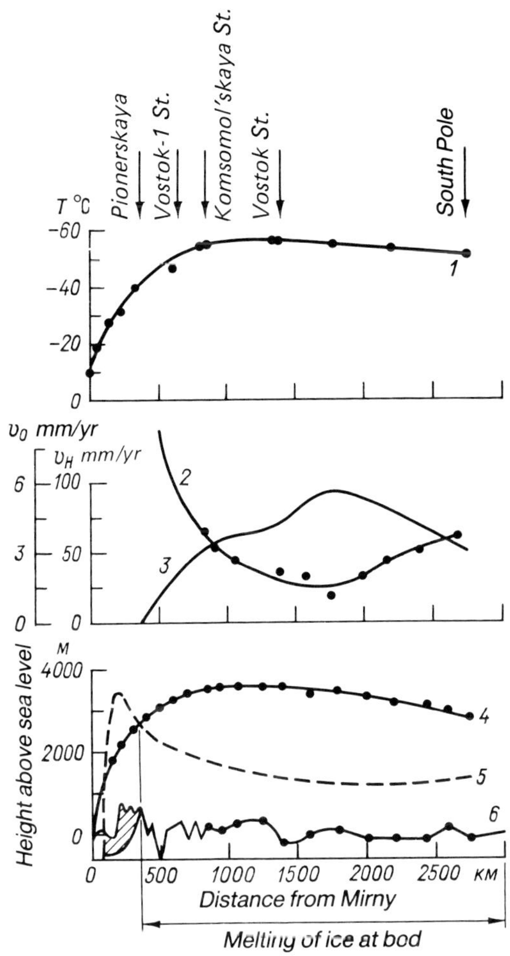

Fig. 7.1. Determination of critical thickness, region of basal melting, and rate of this melting in central region of ice sheet, according to profile for Mirny, Vostok, and South-Pole Stations (after [27]). Dots show observational data: (1) surface temperature of ice sheet T; (2) rate of nourishment v_H; (3) rate of melting at bed v_0; (4) surface of ice sheet; (5) surface of glacier of critical thickness; (6) subglacial bed.

ring all around Antarctica, is a region in which the temperature at the bed of the ice sheet is below the melting point. Estimates indicate that the bedrock is frozen through in this region down to a depth of as much as 500 m. Closer to the antarctic coast there is a band about 50 - 100 km wide in which melting at the glacier bottom is once again typical, but now mainly due to the heat released by the glacier movement. Figure 7.1 shows that the velocity of this motion, calculated from the continuity conditions, increases abruptly toward the edge of the ice sheet.

Figure 7.2 depicts the boundaries of the region of subglacial melting in the central part of the Antarctic Ice Sheet, obtained in [133] on the basis of a steady-state model of a flat glacier with block sliding [28]. Region 1 corresponds to a heat balance at the bed for which the heat flux arriving at the glacier bottom is a purely geothermal flux, assumed to be 50 mW/m^2. Region 2 corresponds to an increase in the area of bottom melting, because of the additional supply of heat to the glacier bottom by reason of the ice movement. This additional heat flux q_{fr} is assumed to be 50 mW/m^2 as well. If, as a first approximation, the tangential stress at the glacier bed is taken to be 98 kN/m^2 (see Chap. 3), then under conditions of block motion the assumed heat flux due to the motion will correspond to a velocity of glacier movement close to 15 m/yr.

The melting rate in the central regions, as shown by calculations based on the equation of heat balance in water (6.3) and on the expressions for the temperature field (4.16) and the temperature gradient at the bed (4.56), did not exceed 4 or 5 mm/yr. The total amount of water formed as a result of this melting is estimated to be 20 km^3/yr. This meltwater is apparently forced outward to the edges of the melting region, where it refreezes at the bottom of the glacier, subsequently moving with the latter as an ice layer of water origin. In some places there may also be a runoff of this meltwater without any refreezing in the seas around Antarctica, right up to the outer margin of the ice sheet.

The subglacial melting in central Antarctica revealed by thermophysical analysis has also been verified by observations.

In 1960 near Mirny Station the author succeeded in observing an event of the breaking away (calving) of the edge of the Antarctic Ice Sheet. The iceberg so formed turned upside down in the water, revealing its lower surface. The bulk of the iceberg was composed of ordinary glacier ice, containing air bubbles, but the bottom 5 m of it turned out to be completely transparent 'bottle' ice, separated from the main bulk of glacier ice by a definite boundary (interface). An analysis carried out by B.A. Savel'ev showed this lower layer to be ice formed by the freezing of fresh water, rather than by the metamorphism which created the upper part of the formation. Apparently, the lower ice layer was formed by the refreezing of water arriving from the central regions [27].

Estimates carried out in [27] indicate that the thickness of the refrozen ice layer corresponds exactly to the case where all the water formed by bottom melting in central Antarctica freezes into a uniform layer along the entire inner boundary of the central melting zone.

The existence of subglacial melting was proven conclusively by a flow of fresh water into a deep borehole near Byrd Station, after the subglacial bed had been reached at a depth of more than 2000 m from the surface [124].

Despite the slow rate of the subglacial melting, its importance is great. Because the rate of vertical ice flow is not zero at the glacier bed, it follows that some air, compressed to the pressure at the bed, makes its way to the bottom of the glacier from the surface.

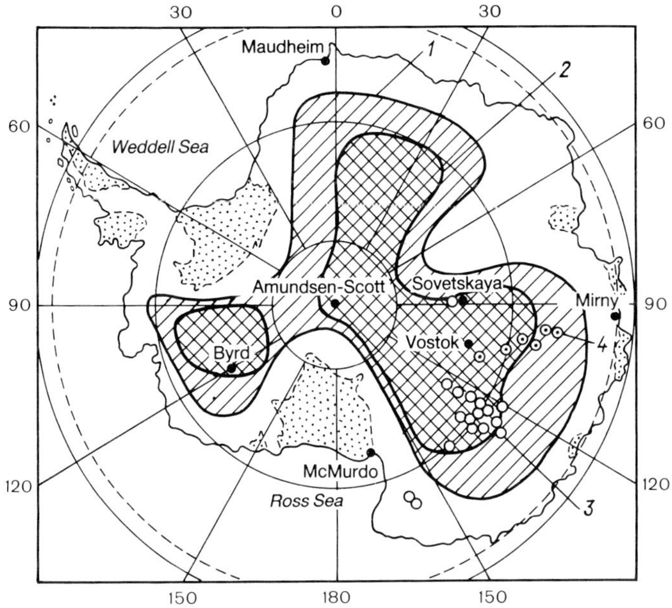

Fig. 7.2. Map of Antarctica showing regions of melting at glacier bed in central regions of ice sheet. Steady-state model.

(1) boundary of bottom melting region, assuming that throughout this region $q_0 = q_g = 52$ mW/m^2 = const; (2) the same, for $q_0 = q_g = 104$ mW/m^2 = const (after [44]); (3) sites of subglacial lakes (interpretation of radar data); (4) sites of 'lakes' at surface of ice sheet, discovered by low-flying aircraft.

The surface layer of an ice sheet is known to be very porous [1]. As particles of snow and firn sink downward, these pores are closed, and the air trapped in the pores subsequently moves downward along with the ice. The pressure in the air bubbles rises correspondingly. As a first approximation, this is equal to the hydrostatic pressure in the ice. For instance, at a depth of 4000 m below the surface the air pressure in the bubbles would be about 39 MPa. It is known that air bubbles are no longer connected with one another in the ice of central Antarctica at a depth of around 150 m, for an ice density at this depth of about 8 kN/m^3. Consequently, for a mean melting rate of about 3 mm/yr, about 20,000 m^3 of air will reach every square kilometer of the glacier bed. This means that the vertical motion of the ice from the glacier surface to the bed, with a subsequent melting of this ice, acts as a kind of compressor, raising the pressure of the trapped air to that of the glacier bed. According to the data of [95], all this air becomes completely dissolved in the meltwater. Each liter of water will thus have about 0.1 liter of air dissolved in it (at the normal pressure). It should be mentioned, too, that a certain amount of organic debris, pollen, and spores, borne into central Antarctica by the winds, also continually makes its way to the lower surface of the glacier [1].

On the basis of the foregoing, we put forth in [32] a hypothesis that some forms of life may exist in the subglacial meltwater of central Antarctica. These can be living organisms, carried down to the bed from the surface of the glacier, which managed to survive their extended stay in the cold ice as it moved down to the bed. These could also be organisms remaining alive at the bed from the time when the region was not covered with ice. In the former case the life forms would not have been in contact with the above-ground surroundings for hundreds of thousands of years, and in the latter case for several millions of years (the present age of the Antarctic Ice Sheet). In [32] we also discussed a mechanism for the possible separation of air during the refreezing of meltwater at the periphery of the region of subglacial melting. If there is indeed such a separation of air, then at some places along the glacier bed pockets of compressed air may be present at the boundary between the ice and the meltwater.

7.1.2. Subglacial Lakes

In [27] we considered the consequences of the conclusion, obtained by thermophysical analysis, that ice is continually melting under the ice cap of Antarctica in the central region of the cap. It appeared to us that at certain places, in depressions of the subglacial bed, the meltwater should be able to collect and form lakes.

The first confirmation of this was obtained only later, in 1972. Analyzing the results of radar soundings of the bed of the ice sheet in central Antarctica, Robin and his coworkers [111] noted that in some places the reflected signal revealed the presence of a smooth, perfectly horizontal lower boundary of the ice. In these places the intensity of the reflected signal was found to increase abruptly and considerably. Now, just such an increase in the intensity of the reflected signal would be observed for a transition from the ice-bedrock boundary to an ice-water boundary, beyond which the water layer is quite deep (more than a few meters deep, that is, greater than the wavelength of the radar signal). Thus, these investigators concluded that the observed reflection was from a subglacial lake. The white circles in Figure 7.2 indicate places for which the radar reflections were interpreted as reflections from subglacial lakes.

Later measurements indicated that these lakes may have diameters as great as tens of kilometers. Their depths are apparently considerably less than the depths of the closed depressions in the water-impermeable subglacial bedrock, which serve as the initial reservoirs for these lakes. Actually, during the lifetimes of these lakes, which appear to be about the same as the lifetime of the ice sheet itself (of the order of ten million years), over each of the lakes an ice column 10,000 m thick must have melted, assuming a melting rate at the lower surface of just 1 mm/yr.

Since the ice at the bed contains a certain amount of morainal and other foreign (vis-à-vis the ice) material, which will settle down to the lake bottom, it follows that a sedimentary layer must form there. In addition, some morainal material and other, for instance, meteoritic, material will continually find its way to the lake because of the horizontal movement of the ice and this will become distributed over the lake bottom. For example, if the concentration of foreign matter in the glacier is around 0.001, then the thickness of the sediment layer due to melting of the 'roof' of the subglacial lake will be of the order of 10 m. If the total thickness of the moraine layer transported by the glacier is about 1 cm, while the velocity of the ice movement is of the order of 10 m/yr, it follows that during the lifetime of a lake 10 km in diameter a moraine layer up to 100 m thick may be fed into the lake. The thickness of the sedimentary layer at the bottom of a subglacial lake may, by the way, give us valuable information about the past history of the glacier. Such information is gained by studying the material content and stratification of the sedimentary layer. This layer may have a number of sublayers, corresponding to different periods in the life of the ice cap. A schematic diagram of a subglacial lake of the described type is shown in Figure 7.3.

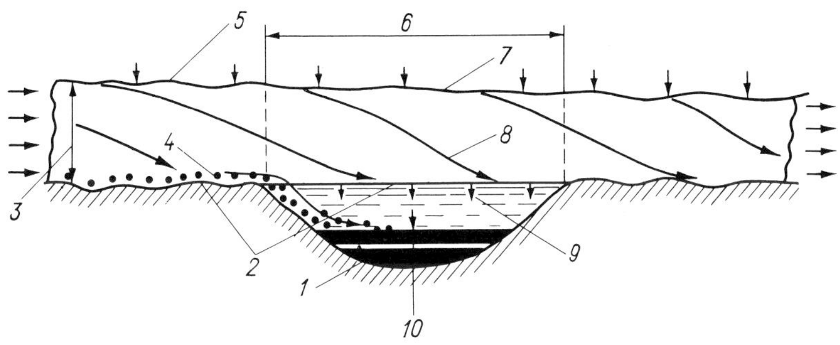

Fig. 7.3. Schematic representation of subglacial lake in central part of thick ice cap (type of Antarctica).

(1) depression in subglacial bedrock (lake reservoir); (2) lower surface of glacier; (3) glacier thickness; (4) particles carried along by glacier at its lower surface; (5) upper surface of glacier (off-lake part); (6) 'lake' boundary projected to glacier surface; (7) upper surface of glacier above lake; (8) lines of flow inside glacier; (9) (9) lake; (10) sedimentary layer at lake bottom.

Finally, with regard to subglacial lakes, note that for a glacier thickness of about 3 km and a lake diameter of several tens of kilometers it should be possible to distinguish between the upper surface of a glacier over a lake and the surface of a glacier resting on solid ground. Thus, certain stable, visually distinguishable (at low viewing angles) areas several kilometers in diameter which have been detected with low-flying aircract are of considerable interest, since these, in our opinion, are surface signs of subglacial lakes. In Figure 7.2 the sites of such lakes, detected during flights on the Mirny Station to Komsomol'skaya Station route, are denoted by circles with dots in them. However, when an aircraft flew right over a lake, and the glacier surface was viewed at an angle close to the vertical, it was impossible to distinguish visually between the surface over the lake and the surface over bedrock. No such visual differences in the glacier surface were detected by ground-based radar observations either.

7.1.3. The Temperature Distribution in an Ice Cap

A sizeable amount of data has been collected so far about the temperature state inside a glacier. In the Arctic data have been obtained on the temperature inside the ice caps of Franz Josef Land, Devon Island, and Vavilov Island, and also data on the temperature over the entire thickness of the Greenland Ice Sheet in the vicinity of Camp Century. The largest number of observations have been carried out in Antarctica, where deep boreholes were drilled near Mirny and Byrd Stations, and later near Vostok Station.

In the central region of the ice caps of Greenland and Antarctica the temperature increases as we descend from the upper surface of the glacier to the bed. This temperature rise is very small in the upper part of the glacier and it increases with the approach to the bed, as is to be expected under conditions of cooling by the vertical downward motion of the ice for accumulation at the glacier surface.

In the marginal zone of an ice cap the temperature drops with increasing distance from the upper surface, and only in the lower part of the glacier does it once again rise, as the bed is approached.

The first theoretical data on the temperature over the interior of an ice cap were obtained by Robin, on the basis of a steady-state model in which the heat transfer in the central regions of the glacier is determined by the heat conduction and the vertical ice movement, whereas in the marginal regions it is determined by the additional heat transfer in the horizontal direction, which creates an inverse gradient in the surface layers. This gradient is caused by the descent of ice from the colder upper parts of the ice cap to the lower warmer regions. Robin thereby obtained the first isotherm chart of the interior of the Greenland ice cap, demonstrating that this ice sheet is frozen through to the very bottom.

V.N. Bogoslovskii used a hydrointegrator for this same model to obtain the temperature distribution in the marginal zone of the antarctic ice cap in the vicinity of Mirny Station. His results confirmed that the horizontal heat transfer due to ice movement from cold regions to warmer regions plays the main role in setting up the inverse gradient revealed by observations. Zotikov employed the same model, but complicated somewhat by the effect of bottom melting, to arrive at the first theoretical temperature distributions for the central region of the Antarctic Ice Sheet (vicinity of Vostok and Komsomol'skaya Stations); he used for this the heat-transfer equations for a steady-state model of a flat glacier with block sliding. This turned out to be possible because of the (somewhat surprising) constancy of the Péclet number along the lines of outward spreading of the glacier, for the major portion of either an inland antarctic ice cap or an ice shelf [32, 76]. This constancy of Pe is explained by the fact that it is the product of two quantities: the annual accumulation of snow and the glacier thickness. At the center of the ice sheet, you see, the ice thickness is a maximum and the annual accumulation is a minimum. In the marginal part of the ice sheet, on the other hand, the accumulation is very great, while the ice thickness is small. The two quantities thus vary in such a way from the center of the glacier to its margin that their product remains almost constant.

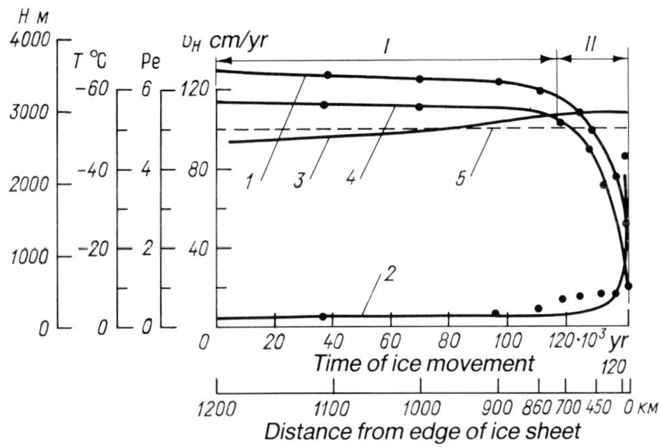

Fig. 7.4. Explanation of constancy of Pe along outflow line of East Antarctic Ice Sheet (after [32]).

(1) glacier thickness H; (2) annual accumulation v_H; (3) Péclet number ($Pe = v_H H/a$); (4) surface temperature; (5) mean value of Pe over entire profile. Dots indicate measured values.

Figure 7.4 shows the accumulation rate, the glacier thickness, and the surface temperature as functions of the time of ice movement from the vicinity of the ice divide to the margin of the East Antarctic Ice Sheet. It is seen that, from the central region of the ice sheet right out to places 400 to 500 km from the edge, not only the Péclet number but also the surface temperature of the glacier are constant in time. Consequently, for this zone the heat transfer in the glacier was described using a coordinate system whose origin moves horizontally with the ice along a line of outward flow. The analysis was based on the solution of the simplest equation describing the heat transfer in a flat steady-state glacier with block motion [27]. It is seen from Figure 7.4 that further toward the glacier margin the situation changes abruptly. Although the Péclet number once again remains almost invariable for ice movement along a flow line, the surface temperature of the glacier varies sharply with time. Thus, in order to analyze the temperature field in this zone, it is necessary to solve a heat-transfer equation complicated by a time-dependent term taking into account the unsteadiness connected with the variation in time of the surface temperature of a vertical ice column moving with the glacier from cooler regions to warmer regions.

Peripheral region II (Figure 7.4), in which this unsteadiness manifests itself, is called the marginal zone, while region I, the 'steady-state' region, is the central zone of the antarctic ice cap.

A study of more complicated models of the heat transfer in ice sheets, with boundary conditions closer to reality, called for the use of numerical modeling with the aid of high-speed computers. The first to do this were Jenssen and Radok [103, 104], who in 1963 considered the movement of a vertical column of ice along a flow line. They used a more complicated equation for the heat transfer in the vertical direction, allowing for conduction and convection, which included a term taking the unsteadiness into account.

Since the direct solution of Equation (3.57) is fraught with serious difficulties, even for $\bar{\phi}' = 0$, Jenssen and Radok resorted to numerical methods of solving the heat-transfer equation. This equation was put into the form

$$a \frac{\partial^2 T}{\partial y^2} + \left(2 \frac{\partial a}{\partial y} - v\right)\frac{\partial T}{\partial y} = \frac{\partial T}{\partial t}, \qquad (7.1)$$

which corresponds to Equation (3.57) for $\bar{\phi}' = 0$, but complicated by allowing for the variation of the thermal diffusivity over the glacier thickness. The rate of vertical ice movement over the glacier thickness was assumed to vary linearly.

The temperature specified for the glacier surface, which varied with time, was taken as the boundary conditions. The time variation of this temperature corresponds to the movement of a vertical ice column from the central part of the ice sheet to its warmer peripheral regions. At the lower boundary the condition stipulated a balance of the heat arriving at and leaving the ice-bed interface. The initial condition was taken to be a steady-state distribution of temperature at the center of the ice sheet.

Equation (7.1) was represented in terms of finite differences and solved with a computer for a number of conditions close to these obtaining in ice sheets. The calculation results verified that the inverse gradient in the marginal regions of an ice sheet thinning down toward its edges is indeed related to the ice movement from the central regions to the periphery.

It should be noted that, in all the problems associated with the movement of a vertical ice column, the thermal effects due to energy dissipation during viscoplastic motion were assumed to be concentrated near the glacier bed and they were calculated as a point heat source at the bed. This assumption follows directly from the physical model of the process. The assumption of a linear variation of the vertical ice velocity over the glacier thickness corresponds to a constant horizontal velocity over this thickness. Actually, from the continuity equation for a flat glacier we have $\partial u/\partial x = -\partial v/\partial y$ but $\partial v/\partial y = \text{const}$, so that $\partial u/\partial x = \text{const}$ over the entire glacier thickness. Thus, the originally vertical column retains its shape as it moves, and there is no shear and heat release in the interior (all the shear is concentrated just at the glacier bed).

The work of Budd, Jenssen, and Radok [86] contributed significantly to an understanding of the thermal regime of the Antarctic Ice Sheet. These investigators considered the transfer of heat in a coordinate system at rest. They rejected the assumption of a linear variation of the vertical velocity over the glacier thickness and assumed instead that

$$v = \frac{v_H}{H} y \frac{u}{u_H}, \qquad (7.2)$$

where u_H is the velocity of horizontal ice movement at the glacier surface. The equation describing the heat transfer in the interior, as given in [86], then has the form

$$a \frac{\partial^2 T}{\partial y^2} - \frac{v_H}{H} y \frac{u}{u_H} \frac{\partial T}{\partial y} + u \frac{\partial T}{\partial x} + \rho g \alpha (H-y) \frac{\partial T}{\partial y} = \frac{\partial T}{\partial t}. \tag{7.3}$$

The last term on the left-hand side of this equation takes into account the thermal effects of internal friction. This equation was solved simultaneously with the equation of motion, which gives the distribution of velocities v over the glacier thickness. Budd, Jenssen, and Radok then considered the temperature distribution using a coordinate system moving with the vertical ice column; assuming a linear distribution of the vertical velocity over the glacier thickness, from v_H at the surface to zero at the bed, they obtained the following heat-transfer equation:

$$a \frac{\partial^2 T}{\partial y^2} - \frac{v_H}{H} y \frac{\partial T}{\partial y} = \frac{\partial T}{\partial t} . \tag{7.4}$$

In order to simplify the solution of this equation, the authors of [86] made an important assumption, with regard to $\partial T/\partial t$, which for a steady-state glacier can be identified with the rate of heating of the ice column as it moves from the cooler central regions to the warmer peripheral regions. They took this temperature variation with time to be constant over the column thickness and equal to the rate S of the variation of the surface temperature as the ice column moves downstream along the ice flow. Consequently,

$$\frac{\partial T}{\partial t} = u_H \frac{\partial T_H}{\partial x} = u_H \frac{\partial T_H}{\partial H_T} \frac{\partial H_T}{\partial x} = u_H \alpha K_T = S = \text{const.}$$

Here H_T is the height of the glacier surface above sea level, and K_T is the rate of variation of the surface temperature with height.

Then the equation describing the heat transfer over the glacier thickness becomes

$$a \frac{d^2 T}{dy^2} - \frac{v_H}{H} y \frac{\partial T}{\partial y} = S. \tag{7.5}$$

Although Equations (7.5) and (7.4) are very different, still evaluations made in [86] indicate that Equation (7.5) correctly describes the main features of the temperature distribution obtained by solving Equation (7.4).

The heat-transfer model corresponding to Equation (7.3) is undoubtedly more accurate than the ice-column model (moving coordinate system) based on the assumption of block flow. However, the law of ice flow is such that even small errors in determining the stresses or temperatures lead to large errors in evaluating the rates of displacement, and thus in evaluating the velocities of the glacier ice as well.

Moreover, the flow law itself is by no means accurately known, especially for low temperatures, small stresses, and low rates of displacement, and for ice of oriented structure.

In [86] calculations of the temperature field, made with a computer and based on Equation (7.4), were compared with calculations for a moving vertical ice column, based on data corresponding to several profiles of the Antarctic Ice Sheet. Budd, Jenssen, and Radok felt that the discrepancy due to imprecise knowledge of the fundamental parameters (accumulation, ice thickness, ice velocity, conditions at bed, etc.) influences the calculation results much more than do any differences between the initial models described by Equations (7.3), (7.4), and (7.5). Accordingly, they used model (7.5) to calculate the temperature state of the Antarctic Ice Sheet.

More recent progress toward understanding the thermal regime of glaciers and ice sheets has involved gathering new temperature data and calculating the temperature state of the glacier interior under specific conditions.

Some totally new information was obtained by determining the ratio of the stable isotopes ^{16}O and ^{18}O over the glacier thickness. This ratio, other conditions being the same, is determined bv the temperature

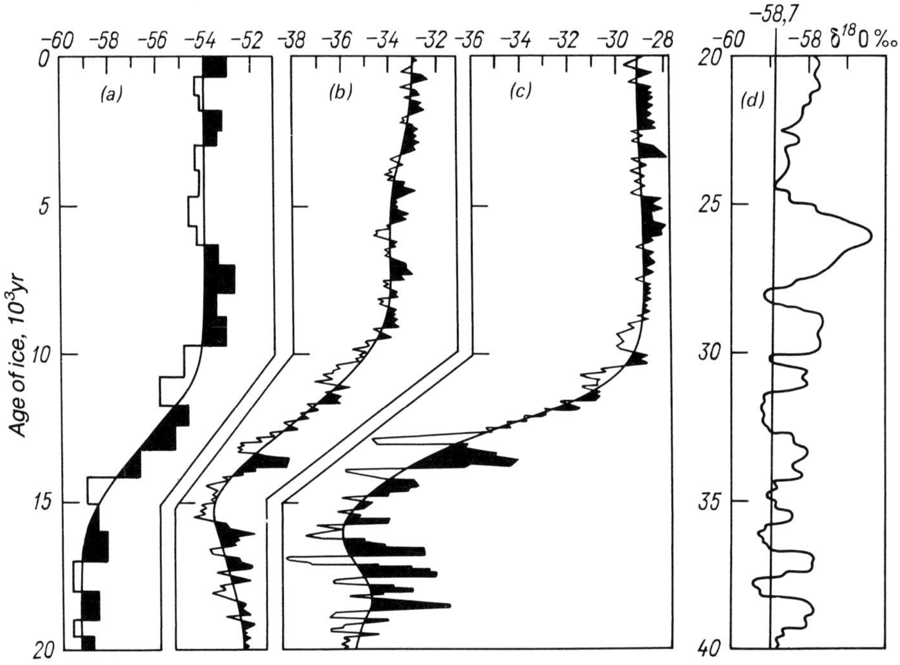

Fig. 7.5. Variation of ^{18}O isotope content with time, according to data of ice cores from deep boreholes.

(a) Vostok Station, 0 - 20,000 yr; (b) Byrd Station, 0 - 20,000 yr; (c) Camp Century, 0 - 20,000 yr; (d) Vostok Station, 20,000 - 40,000 yr [8, 43, 44].

of the snow or ice particles at the time of their fall onto the surface. Therefore, if we know the age of the ice and the ratio of the oxygen isotopes at each level, we can deduce the variation of the temperature of the glacier surface with time.

Figure 7.5 shows the variations of the ^{18}O isotope content in the ice sheets of Antarctica and Greenland during the last 20,000 years, according to ice-core data from deep boreholes. The graph clearly reveals an abrupt isotopic shift in ice layers 11,000 to 15,000 years old. This corresponds to the boundary between the Pleistocene and the Holocene, that is, to the lifetime of the last Quaternary inland ice sheet. As a first approximation, this shift is equivalent to a temperature change of 5 °C at Vostok Station, 8 °C at Bryd Station, and 12 °C at Camp Century, since a 1 ‰ variation of the ^{18}O content in the ice is, as a first approximation, equivalent to a temperature variation of 1 to 1.5 °C [44].

The data presented can be interpreted either as a variation of the surface temperature due to a change in the climate or else as a variation of the surface temperature associated with a change in the thickness of the glacier ice, and thus with a change in the elevation of the glacier surface above sea level. Studies of the variation in surface height based on measurements of the pressure in air bubbles trapped in the glacier ice [118] indicate that the elevation of the Greenland Ice Sheet at the time when the Valdai Ice Sheet covered Europe was 1300 m higher than at present. Correspondingly, the elevation of the West Antarctic Ice Sheet was 500-600 m higher than the present elevation. The height of the ice sheet in East Antarctica, in the vicinity of Vostok Station, was also different.

7.2. Numerical Modeling of the Temperature Field of an Ice Cap

7.2.1. A New Statement of the Problem

The apparent unsteadiness of the surface temperature of an ice sheet, revealed by measurements of the ^{18}O isotope content in cores from deep boreholes, calls for a closer study of the heat-transfer equation for ice sheets, taking into account the unsteady term. It is also important to get away from the assumption of heat propagation in a vertical ice column which moves from the central regions of an ice cap to its periphery, always with its side walls vertical and flat. Such an assumption is valid only for completely block flow, which is apparently almost nowhere observed in reality. It would also be desirable to dispense with the hypothesis of a linear variation of the vertical velocity with depth, not by introducing artificial laws of this variation but rather on the basis of an analysis of the true continuity equation, under conditions where the velocity of horizontal outward flow of the glacier ice decreases as the bed is approached. A knowledge of the true distribution of the vertical ice velocities over the glacier thickness is especially important with regard to the need to determine the age of the ice at great depths, so as to be able to interpret data on the ^{18}O content and to construct models of past climates.

Thus we thought it advisable to reconsider the heat transfer taking into consideration the above-mentioned factors. Because of the complexity of the problem, we have confined our attention to a two-dimensional ice sheet. Any possible divergence or convergence of the flow lines in the vertical was taken to be just an additional source or sink of mass in the continuity equation.

The coordinate origin is taken to be under the ice divide at sea level, axis X is directed horizontally along the lines of movement, and axis Y is vertical. Therefore, the heat transfer in the medium is described by the following equation:

$$\frac{\partial T}{\partial t} = a \frac{\partial^2 T}{\partial y^2} + a \frac{\partial^2 T}{\partial x^2} - v \frac{\partial T}{\partial y} - u \frac{\partial T}{\partial x} + N\Phi. \tag{7.6}$$

The energy dissipation caused by the movement is given by the expression

$$\Phi = \frac{\tau}{\rho c} \frac{\partial u}{\partial y}. \tag{7.7}$$

Here τ is defined as

$$\tau = \rho g(H + y_b - y)\mathrm{tg}\ \alpha, \tag{7.8}$$

where y_b is the height of the glacier bed above sea level.

The velocity components in this glacier are assumed to be related by the continuity equation in the form (3.2), that is, the ice is taken to be incompressible.

When using equation (7.7), we limited ourselves to an ice sheet with a steady-state boundary condition at the upper surface and also with a condition equivalent to a temperaure law modeling the variation of the air temperature in polar regions during the last 100,000 years and allowing for a possible temperature rise in the near future.

For the problem thus stated, Barbash [9] carried out under our direction a numerical modeling of equation (7.6) and of the continuity equation with the following boundary conditions at the upper surface ($y = y_s$):

(a) steady-state case: $\partial H/\partial t = 0$, position of upper surface invariable; $v_s = - A - u_s\ \mathrm{tg}\ \alpha$, the vertical velocity of the ice, is equal to the velocity of ice inflow to (outflow from) the upper surface; $T_s = T\ (y_s)$, the temperature at each point on the glacier surface, is a function of the surface height and is independent of time;

(b) unsteady-state case: surface temperature varies with time. The variation of the surface temperature of the ice sheet during the last 100,000 years, corresponding to a global cooling of the Earth's climate, is for the unsteady-state case taken to be as shown in Figure 7.6.

Inspection of the figure shows the following: the surface temperature of the ice sheet at times more than 100,000 years before the present is taken to be invariable with time and equal to the present temperature; the surface temperature from 100,000 to 20,000 years ago

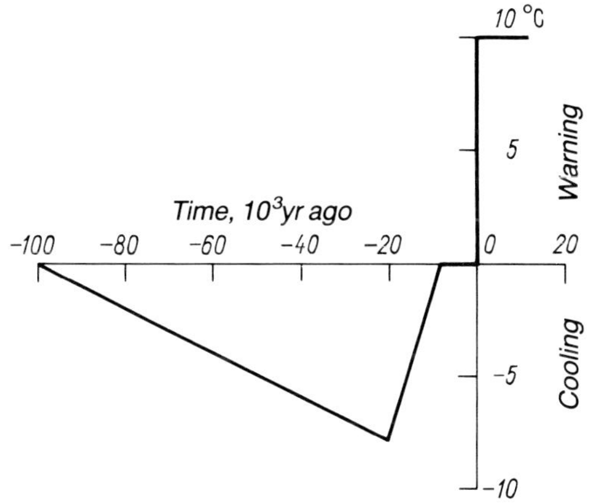

Fig. 7.6. Variation with time of surface temperature of glacier, assumed in order to calculate reaction of Antarctic Ice Sheet to climatic change [9].

is taken to be linearly decreasing in such a way that during the coldest period (about 20,000 years ago) the surface temperature was 8 °C below the present value. This was the developmental period of the Earth's glaciation, corresponding to the Würm, Valdai, and Wisconsin Epochs, according to isotope data for ice cores from deep boreholes as well as an isotopic analysis of deep-sea bottom sediments in the Atlantic [9]. During the last 20,000 years, up until 10,000 years ago, the temperature is assumed to have increased linearly.

In order to predict the effect of future climatic changes, the model assumes an air-temperature increase at polar latitudes of 10 °C, due to the 'greenhouse effect' resulting from the increased retention of solar energy by the atmosphere which will accompany the presumed rise in the CO_2 content of the atmosphere due to anthropogenic factors.

Two possible situations at the lower surface of the ice sheet are considered: the ice temperature is below the melting point or equal to the melting point. In the former case ($T_i < T_m$) the equation of heat balance at the bed is

$$- \lambda \frac{\partial T}{\partial y} = q_g + \frac{1}{K} u_b \rho g H \operatorname{tg} \alpha. \tag{7.9}$$

This case corresponds to conditions whereby the critical glacier thickness is greater than the true thickness.

The first term on the right-hand side of this equation gives the geothermal heat flux to the bed, and the second term gives the heat generated at the bed by the movement. For the case in which the glacier thickness is less than critical, on the right-hand side of Equation (7.9) there will be an additional term corresponding to a heat sink at the bed due to the melting of ice at a rate v_b. The heat-balance equation then has the form

$$- \lambda \frac{\partial T}{\partial y} = q_g + \frac{1}{K} u_b \rho g H \text{ tg } \alpha - v_b \rho g r. \quad (7.10)$$

This, then, is the simplest case of a glacier resting on a water-impermeable bed, under conditions where there is no appreciable overheating of the meltwater at the bottom of the glacier.

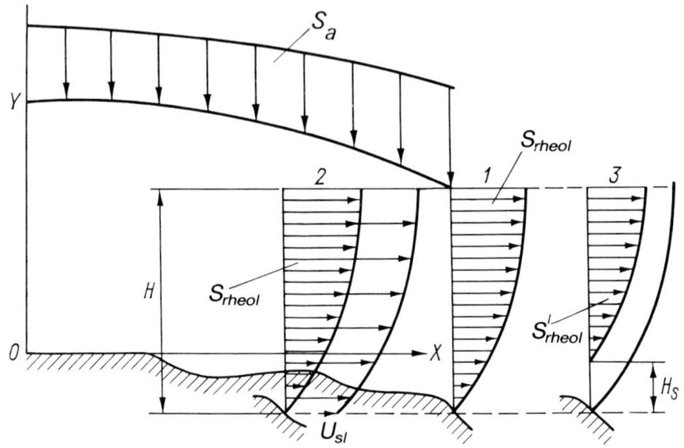

Fig. 7.7. Determination of possible stagnant region at bed H_s or sliding velocity at bed u_{sl}.

The sliding velocity was found from the integral equation of continuity, according to the model depicted in Figure 7.7. It is evident from this model that, provided the glacier surface does not change its position in time, the following cases are possible:

Case 1. The mass flow through Section 1-1, determined by the horizontal outward spreading of ice due to longitudinal rheological deformations (S_{rheol}), is equal to the amount of precipitation S_a accumulated above Section 1-1 right up to the ice divide: $S_a = S_{rheol}$. In this case for a steady-state glacier, the thickness of which does not change, the velocity of horizontal movement of the glacier at the bed is zero.

Case 2. The mass flow of ice through Section 1-1, determined by the horizontal outward spreading due to longitudinal rheological deformations, is less than the amount of precipitation accumulated above Section 1-1 ($S_s > S_{rheol}$). In this case steady-state conditions at the glacier surface will remain only if there is a sliding velocity u_{sl} at the bed. Then

$$S_a = S_{rheol} + u_{sl} H.$$

Case 3. The mass flow through Section 1-1 due to horizontal spreading of the ice S_{rheol} is greater than the rate of accumulation of precipitation above section 1-1 ($S_a < S_{rheol}$). Steady-state conditions at the glacier surface can then exist only if at the bed there is an ice layer which is not involved in the movement, so that $S_a = S'_{rheol}$. Here S'_{rheol} is the mass flow of ice due to the rheological component for the upper, moving part of the glacier of thickness $H-H_s$, where H_s is the thickness of the lower layer of stationary ice.

The boundary condition at the plane of symmetry passing through the ice divide (x=0) is determined by the linear variation of the temperature over the glacier thickness.

As an initial condition, we also assume a linear temperature variation in each vertical section, from the surface temperature T_s down to the temperature at the bed. Since a great many calculation steps are required in order to obtain a steady-state thermal regime with the aid of equation (7.6) and the described initial and boundary conditions, a BESM-6 high-speed computer was used for the calculations. A steady-state regime was defined as a regime in which the temperature variation was close to zero at all points in the glacier for a sufficiently long time.

The dimensionless vertical coordinate y/H, first introduced to analyze the temperature field in a glacier in [76], is generally used to calculate the temperature and velocity in the plane of vertical movement of the glacier. In this case, where the true thickness of the glacier varies with the transport from the central regions to the periphery, the number of points at which the glacier is divided along axis Y does not vary for motion along axis X.

In the model being considered, where the horizontal heat transfer is taken into account and the horizontal ice velocity varies over the glacier thickness, in addition to the dimensionless vertical coordinate, interpolation procedures were also needed to determine the values of T and U off the nodes of the basic grid.

The following procedure was used to determine the horizontal and vertical components of ice motion. The temperature field specified at the initial time and the values of the tangential stress for points on a given vertical section were used to find the tabulated values of the rheological ratio [9], and then to calculate the values of the vertical gradient of the horizontal ice movement caused by the rheology. Next, assuming the velocity at the bed to be zero, we calculated the values of the horizontal velocity at all the grid points. The area of the curve of the rheology-caused horizontal velocity (S_{rheol}) was then calculated and compared with the total mass influx at the glacier surface, from the ice divide to the given section (S_a). The difference between the mass inflow and outflow through the given section, divided by the ice thickness, gives us the possible sliding over the bed, which has to be added to the previously calculated horizontal velocities. In addition, we had to calculate the transverse outflow or inflow of mass produced by the variation in the transverse dimension of the ice flow, as viewed from above. The vertical ice velocity in each section was determined using the continuity equation. Here the boundary condition at the surface was taken to be as indicated above. Numerical integration of the equations with the appropriate boundary conditions was carried out for the Antarctic Ice Sheet (the ice caps of East and

West Antarctica) for lines of ice flow in the vicinity of Vostok and Byrd Stations. These are the only places in central Antarctica for which data from temperature measurements in deep boreholes are available.

The calculations [9] indicated that the true vertical-velocity distribution over the thickness of the Antarctic Ice Sheet at these points differs from linear only in the lower 500-700 m of the glacier. Consequently, to analyze the temperature field in the ice sheet, the linear approximation should be used. This will save a considerable amount of machine time, and it will eliminate the danger that the solution may be unstable.

7.2.2. Modeling the Antarctic Ice Sheet Along Flow Lines Through Vostok Station to Byrd Glacier and Through Byrd Station to the Ross Ice Shelf

Figures 7.8 and 7.9 depict the temperature distribution over the thickness of the ice sheet, along the indicated flow lines. The curves were obtained by numerical modeling on the BESM-6 computer, using the new procedure described in the previous subsection. A geothermal heat flux of 63 mW/m^2 was assumed [9]. Inspection of these figures shows that melting at the bed can be assumed to take place in the entire central region of the Antarctic Ice Sheet.

Figures 7.10 and 7.11 give the distributions of the horizontal ice velocity over the glacier thickness, obtained by numerical modeling of the equations of motion and continuity taking into account the effect of the true temperature field on the rheological properties of the ice [9].

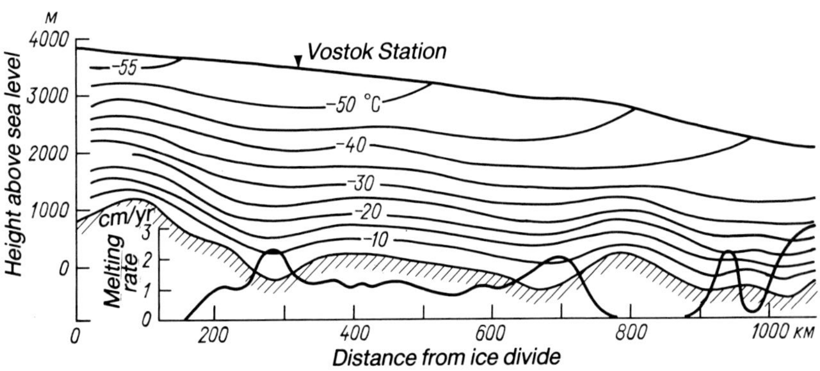

Fig. 7.8. Thermal regime of interior of East Antarctic Ice Sheet along flow line through Vostok Station to Byrd Glacier. Curve at bottom shows rate of melting at bed. Calculation made for q_g = 63 mW/m^2.

As these figures demonstrate, for both profiles in the marginal region there is sliding of the glacier over its bed; for the glacier of West Antarctica the sliding velocity is close to the velocity at the surface (the motion approximates block motion).

With the approach to the central part of the ice sheet, however,

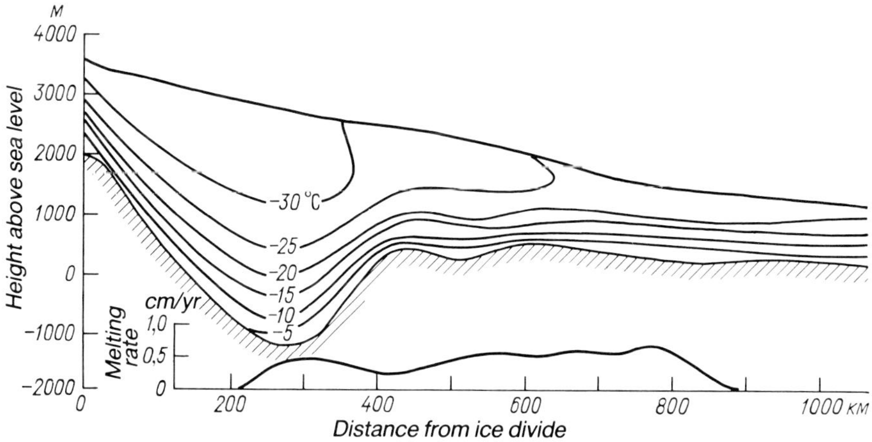

Fig. 7.9. Thermal regime of interior of West Antarctic Ice Sheet along flow line through Byrd Station to Ross Ice Shelf. Curve at bottom shows rate of melting at bed. Calculation made for $q_g = 63\ mW/m^2$.

the portion of the block sliding decreases, and in East Antarctica the steady state of the ice sheet will be preserved (that is, the incoming (S_a in Figure 7.7) and outgoing (S_{rheol}) components of the mass budget of the glacier will still be equal) only if a layer of stationary, that is, 'dead', ice of thickness H_s is assumed to be present at the bed (case 3 in Figure 7.7).

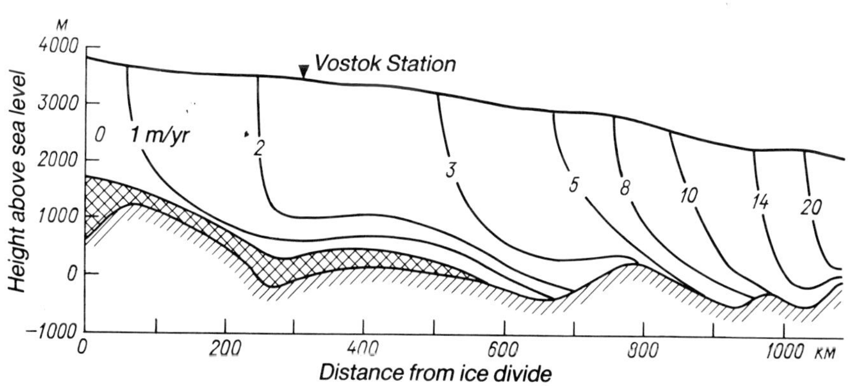

Fig. 7.10. Field of horizontal ice velocities in interior of East Antarctic Ice Sheet. Shaded region indicates stationary ('dead') ice.

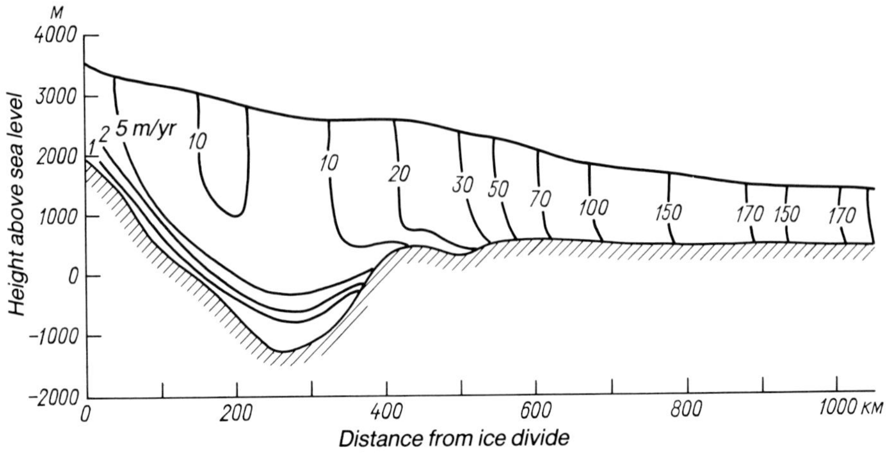

Fig. 7.11. Field of horizontal ice velocities in interior of West Antarctic Ice Sheet. No region of dead ice at bed.

As Figure 7.10 shows, the calculated region of dead ice at the bed of the central part of the East Antarctic Ice Sheet extends for more than 500 km, while the thickness of the dead-ice layer reaches several hundred meters [9].

The presence of a thick layer of stationary ice at the glacier bed in central Antarctica, revealed by these calculations, calls for a re-evaluation of the age of the ice at near-bottom levels and also for a reassessment of the genesis of this ice.

Note, however, that there is an alternative to assuming the presence of dead ice, if we allow the thickness of the ice sheet to vary with time. For example, if we suppose that the ice sheet of East Antarctica is becoming thinner in its central region at a rate of 1 or 2 cm/yr, then the additional mass flow through Section 1-1 of the glacier (Figure 7.7) compensates the mass balance, and we no longer have to assume the presence of a layer of stationary ice at the bed.

7.2.3. The Effect of Climatic Changes on the Thermal State of Large Inland Ice Sheets

This is an important factor with regard to any evaluations of the effect of climatic changes on the behavior of glaciers. Using the Antarctic Ice Sheet as an example, let us consider the effect of climatic changes with the aid of the numerical-modeling method proposed above, taking into account the unsteadiness associated with the variation of the surface temperature of the glacier caused by climatic change. As a first approximation, the accumulation at the glacier surface, as well as the thickness and shape of the glacier, are assumed to be constant in time.

The first of these simplifications was introduced because of the lack of information about how the accumulation of precipitation varied during climatic changes in the past. The assumption of a constant glacier thickness is related to the nature of the numerical model used.

The effect of temperature fluctuations was modeled for the profiles considered above, that is, for sections along flow lines passing through Vostok and Byrd Stations. Figures 7.12 and 7.13 show the temperature distributions within the East Antarctic Ice Sheet at Vostok and within the West Antarctic Ice Sheet at Byrd. These curves were obtained by numerical modeling for the steady state (surface temperature invariable with time) and allowing for the cooling of the climate which took place during the last 100,000 years (Figure 7.6). Inspection of these figures reveals definite differences in the shape of the curves obtained assuming that the temperature of the glacier surface does not change with time and taking into account the temperature variation caused by the cooling of the climate during the last 100 millenia, as approximated by the curve in Fig. 7.6. Consequently, the indicated climatic cooling had a definite effect on the temperature field of the central part of the Antarctic Ice Sheet. Figures 7.12 and 7.13 show the effect of this cooling to be considerably greater for West Antarctica than for East Antarctica. The cooling wave, which went through a maximum about 20,000 years ago, penetrated more deeply into the glacier in the vicinity of

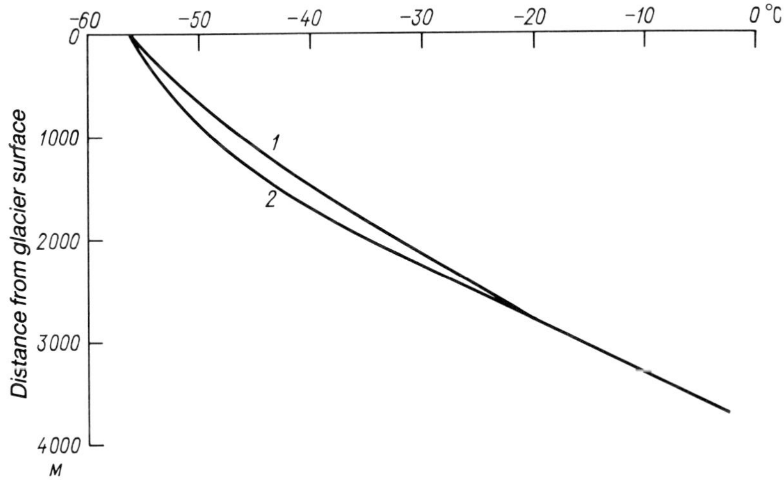

Fig. 7.12. Temperature distribution inside East Antarctic Ice Sheet near Vostok Station.

(1) temperature profile obtained by numerical modeling assuming temperature of glacier surface to be constant in time; (2) present temperature distribution, obtained by numerical modeling assuming surface-temperature variation during last 100,000 yeras according to law in Figure 7.6.

Byrd Station and the attenuation of this wave at great depths was less marked than in the glacier near Vostok Station. This is because the velocity of vertically downward ice flow due to the deposition, spreading, and accumulation of new precipitation at the surface was an order of magnitude higher for Byrd Station (about 20 cm/yr) than for Vostok Station (about 2 cm/yr). Consequently, the rapid, deep

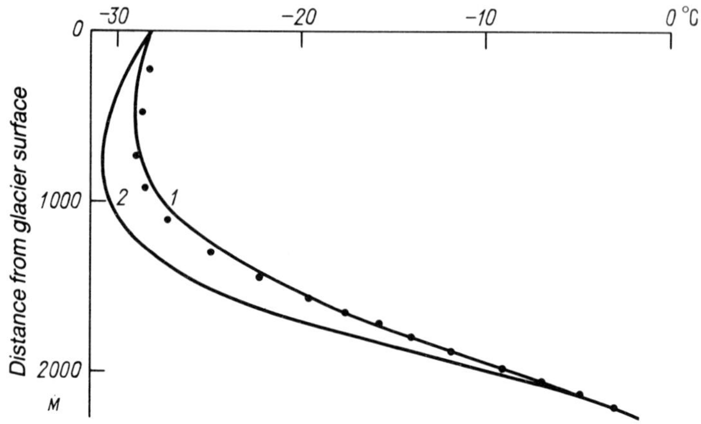

Fig. 7.13. Temperature profiles inside West Antarctic Ice Sheet near Byrd Station.

(1) numerical modeling for constant surface temperature; (2) numerical modeling allowing for past climatic cooling (Figure 7.6). Dots indicate measurements in borehole at Byrd Station.

Fig. 7.14. Temperature distributions inside East Antarctic Ice Sheet at place situated 1000 km below ice divide along flow line through Vostok Station to Byrd Glacier.

(1) present state; (2) 150 years after abrupt warming (Figure 7.6); (3) 1000 years after warming; (4) 5000 years after warming; (5) 20,000 years after warming.

penetration of the cold wave caused by climatic changes in the central part of West Antarctica is explained by the substantially larger role of the convective transfer of cold in the lower layers, due to the vertically downward ice motion [9].

Earlier, when we first considered Figure 7.6, we mentioned that, in the opinion of a number of experts, during the next 100 - 200 years the temperature of the Earth's surface should rise sharply (by 8 - 10 °C at polar latitudes) due to the greenhouse effect caused by the marked anthropogenic increase in the carbon-dioxide content of the atmosphere. Such a heating effect was shown in the right-hand part of Figure 7.6. It would be of interest to employ the method proposed above to estimate the effect of this anthropogenic warming on the thermal regime and other parameters of inland glaciers.

Figure 7.14 depicts the temperature distribution over the interior of the East Antarctic Ice Sheet, at a site located 1000 km from the ice divide (flow line passing through Vostok Station to Byrd Glacier). The temperature distribution corresponding to the present state of the ice sheet (before the anthropogenic surge in the surface temperature) is shown, as well as the (presumed) thermal states at certain times after the beginning of warming. The figure indicates that even 20,000 years after the beginning of the warming the temperature wave produced by the warming will not reach the bed of the glacier [9].

Figure 7.15 shows how the warming of the interior of the ice sheet depicted in Figure 7.14 affects the horizontal flow (spreading) of the ice sheet. According to these curves, 150 years after the start of the warming there is still no perceptible effect on the nature of the outflow of the ice sheet, and only after 20,000 years does the outflow velocity increase somewhat [9].

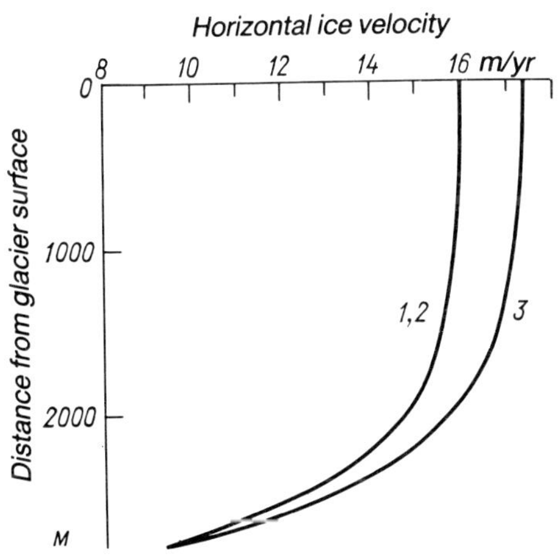

Fig. 7.15. Distribution of horizontal ice velocity in East Antarctic Ice Sheet (for same place as in Figure 7.14).

(1) present state; (2) 150 years after warming; (3) 20,000 years after warming.

Chapter 8

THE THERMAL REGIME OF MOUNTAIN GLACIERS

8.1. The Melting of a Contaminated Glacier Surface

The surface of a glacier is seldom a surface of pure ice or snow. Instead it is usually covered with a larger or smaller layer of dust, fine particles of moraine, etc. If the layer of contamination is not very thick, then the rate of melting of the glacier surface will be high [75].

Let us here consider theoretically just how such surface contamination affects the melting of initially pure ice. This is an important question, both with regard to the general theory and with regard to the theory of the artificial variation of the melting of glaciers by contaminating the surface and the prediction of the possible effects of surface contamination, caused, for instance by volcanic dust or volcanic eruptions, as well as by anthropogenic factors.

The first to investigate on a modern level the dusting of a glacier surface with soot or other dust as an efficient means of speeding up the melting was Avsyuk [2]. Later this topic was considered by many others [20, 21, 46, 83].

The intensification of the surface melting is largely due to an alteration of the surface albedo by the blackening material. Here we will concentrate on methods of determining the albedo of an artificially blackened surface; this is basic to the theory of the modification of glaciers by surface contamination.

The complexity of the problem made it necessary to greatly idealize the process and to devise simplified models which could be verified by comparison with the observations. Some aspects related to the variation of melting by contamination, such as the effect of variations in the turbulent exchange of heat and moisture, and the effect of fresh snowfalls on the albedo of a contaminated surface, will not be dealt with at all. In spite of the obvious importance of evaluating these effects, the author feels that there are not yet sufficient data available to justify the carrying out of such evaluations [47].

Consequently, at present these variations will arbitrarily be neglected, as a first approximation, and later, as further experience is gained, their effects will be gauged and the appropriate corrections introduced, with the aid of empirical or other coefficients. For the time being, the described effects should be evaluated using known methods of determining these components.

8.1.1. Albedo of Blackened Surface. Simplest Case.

If there is melting at a glacier surface, it occurs in a layer known as the melting crust. The melting crust forms a loose porous surface saturated with meltwater. The albedo of a melted surface, even in the absence of contaminating particles, is lower than the albedo of the same surface before melting. The lower albedo of the melting crust is apparently explained by the trapping of some of the incident total radiation inside the layer, due to multiple reflections at the walls of pores, channels, cracks, and ridges in the melting crust. In an optical sense, the layer of the melting crust divided up by these channels can apparently be considered to be a turbid layer into which the entire total incident radiation Q enters and out of which only a part of it AQ exits, where A is the albedo of the melting glacier surface.

If a certain number of particles of some contaminating substance are now deposited onto the surface, then each of the particles falling onto the surface of the melting crust, which is divided up by tiny projections and depressions, will turn up either at the top of a projection or at the bottom of a depression, crack, or channel, or else on the sloping side of a ridge connecting the irregularities. Thus, the contaminating particles will be distributed somehow over the 'turbid layer', as we have called the melting crust. The particles can be either at the surface of the melting crust (at the tops of the projections and ridges) or inside it (at the bottom of the depressions, microchannels, and cracks in the melting crust). Under these conditions the total radiation reaching the particle surface is apparently absorbed by the whole surface of the particle. Note, however, that if a sufficiently large number of particles are present on the surface some of them will have to be very close to one another, so that only a portion of the surface of such a particle can participate in the radiation exchange with the melting crust. Part of the surface will absorb incident solar radiation as well. The particles will thus appear to cover one another.

Consequently, the radiant heat exchange of particles of a contaminating substance with the melting crust has a very complicated mechanism, which depends on the optical properties of the melting crust and on the distribution of particles inside it. At the same time, the characteristics of the melting crust itself become altered by the presence of the particles. This mechanism has not, however, been studied much, either experimentally or theoretically.

At the present level of our knowledge about the structure of the melting crust and about its interaction with particles, it is still difficult to arrive at an accurate method of calculating this mechanism. Therefore, below we will present a very simple model, which, in our opinion, simulates closely the mechanism of the variation of the albedo for contamination of the glacier surface. The semitransparent melting crust prior to this contamination, which in the following will be called 'blackening', is replaced by a smooth surface with an albedo equal to that of the real melting crust before blackening (A_r). The particles of the blackening substance are represented as tiny spheres of identical diameter d.

On the basis of the foregoing, it is assumed that the radiative exchange of the sphere involves its entire surface, provided that the sphere is not close to other spheres. Accordingly, each sphere of the blackening substance in the melting crust is replaced by a disk with an area equal to the surface area of the sphere. The albedo of the surface of each such disk is taken to be equal to the albedo of a hypothetical ground surface, consisting only of particles of the blackening substance. The tiny disks fit snugly onto the contaminated surface [136].

Let F be the area of the region of contaminated surface being considered, and let F_b be the blackened area (that is, the total area of the tiny disks simulating the blackening effect of the tiny spheres). Then, for this mechanism the amount of energy reflected from the part of the surface not covered by disks will be QA $(F - F_b)$; the amount of energy reflected from the area covered by disks will be QA_rF_b, and the amount of energy arriving at the entire surface F will be, correspondingly, QFA_b. According to the definition of the albedo, the albedo of the blackened surface A_b can now be written as

$$A_b = \frac{QA(F - F_b) + QA_rF_b}{QF} = A - (A - A_r)\frac{F_b}{F}. \qquad (8.1)$$

Next let us find F_b. The number of particles, or the number n of disks simulating these, per unit surface is

$$n = \frac{6G}{\pi d^3 \gamma},$$

where G is the rate of dusting of the blackening substance, in mass units per unit area, and γ is the specific weight of the blackening material.

If none of the disks overlap, then the ratio of the area covered by disks to the entire area F will be

$$\frac{F_b}{F} = \frac{\pi d^2 n}{F} = \frac{6G}{d\gamma}. \qquad (8.2)$$

The albedo of the blackened surface is thus given by the expression

$$A_b = A - (A - A_r)\frac{6G}{d\gamma}. \qquad (8.3)$$

Equation (8.3) can be used only for very small numbers of particles per unit area.

As noted above, if particles are dusted onto the surface randomly and individually, then some of them will be bound to influence one another, preventing their interaction with the melting crust and the total radiation; some particles will therefore not participate in varying the total albedo. This effect, whereby particles keep other particles from participating, is reflected in the given model by an overlapping of disks.

8.1.2. Albedo of Blackened Surface. Probability of Particle Overlapping.

Let us first consider the probability that an arbitrarily selected point on surface F will be covered by a disk. If the disk area is πd^2, then the probability p that an arbitrary point will be covered by a disk is equal to $\pi d^2/F$. The probability that a particle will not cover this point is $1 - (\pi d^2)/F$, while the probability that none of the n particles (or disks simulating them) will cover a given point is $[1 - (\pi d^2)/F]^n$. The opposite probability P, that at least one of the n particles will cover the given point, is

$$P = 1 - \left(1 - \frac{\pi d^2}{F}\right)^n. \tag{8.4}$$

Since none of the points on area F differ from the point being considered, expression (8.4) will be valid for the entire surface F. The part F_b of area F covered by at least one of the disks is, accordingly, $F_b = FP$.

Taking (8.4) into account, we now arrive at the following expression for the surface area covered by disks:

$$F_b = F\left[1 - \left(1 - \frac{\pi d^2}{F}\right)^{6GF/(\pi d^3\gamma)}\right].$$

Then the albedo of the blackened surface will be related as follows to the albedo before blackening:

$$A_b = A - (A - A_r)\left[1 - \left(1 - \frac{6G}{d\gamma}\frac{1}{\frac{6G}{\pi d^3\gamma}}\right)^{6GF/(\pi d^3\gamma)}\right]. \tag{8.5}$$

It should be noted that the quantity $6GF/(\pi d^3\gamma)$ determining the number of particles on the given surface is much greater than unity. Suffice it to say that, for a dusting rate of 10 g/m^2, and a density of the dusting substance of about 2 g/cm^3, there will be close to 1000 particles 2 mm in diameter on 1 m^2 of the contaminated surface. For a smaller particle diameter, of course, there will be more particles per m^2. Therefore, expression (8.5) can be simplified, using the extremum formula for a large number of particles, that is, for large $6GF/(\pi d^3\gamma)$:

$$A_b = A - (A - A_r)\left[1 - \exp\left(-\frac{6G}{d\gamma}\right)\right]. \tag{8.6}$$

The relative variation of the albedo of the blackened surface will then be

$$\frac{A_b}{A} = 1 - \left(1 - \frac{A_r}{A}\right)\left[1 - \exp\left(-\frac{6G}{d\gamma}\right)\right]. \tag{8.7}$$

It is evident from Equation (8.7) that the relative variation of the albedo of the blackened surface is a function of the ratio of the albedo of the blackening substance to the albedo of the unblackened surface, and also of the parameter $G/(d\gamma)$. This dimensionless parameter characterizes the degree of blackening of the surface. Formula (8.7) indicates that, as the degree of blackening goes to zero (infinitesimal amount of blackening substance), the surface albedo approaches the albedo before blackening. As the degree of blackening is increased, the albedo of the blackened surface differs more and more from the albedo before blackening. If the degree of blackening is high enough, the surface albedo reaches the albedo of the blackening substance. This value corresponds to conditions where the blackener is so abundant that it covers the surface in a continuous layer. A further increase in the degree of blackening practically does not change the surface albedo at all.

It is evident from the definition of the 'degree of blackening' that for a given blackening substance the same degree of blackening can be achieved either by increasing the dusting rate G per unit surface area or by pulverizing the substance more finely. The smaller the particles, the more effective will be the blackening from the same dusting rate. However, it should be remembered that as the particle diameter is reduced the particle mass decreases in proportion to the diameter cubed, which enhances the effect of the surface-tension forces. Consequently, very small particles can easily be washed away by the meltwater and lost. Experiments [46] have shown that a particle diameter of about 0.2 mm is optimum.

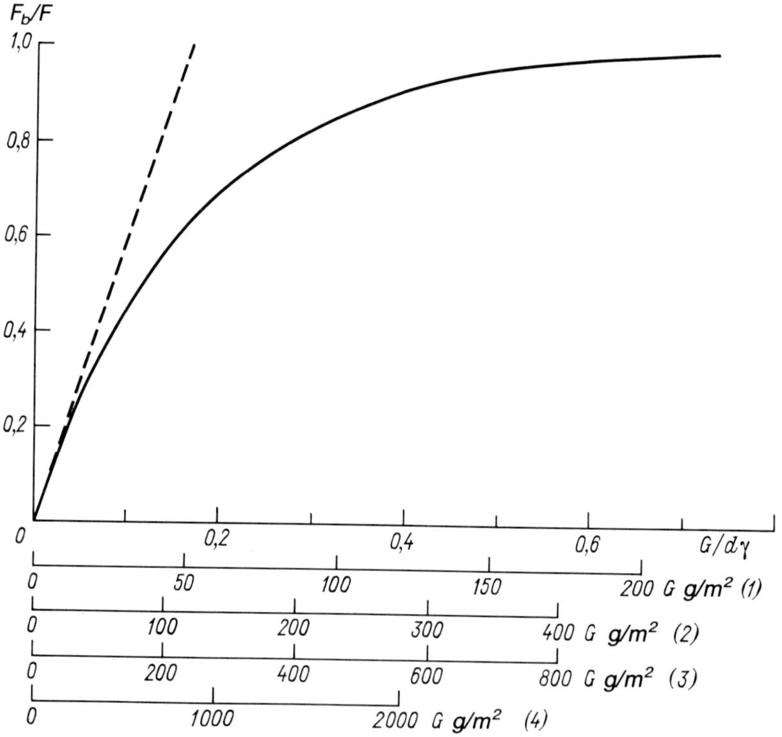

Fig. 8.1. Relative area of blackened surface as function of dimensionless dusting rate, as well as of dusting rate for diameters of 0.2 (scale 1), 0.5 (scale 2), 1 (scale 3), and 2 mm (scale 4).

The factor $\{1 - \exp[-6G/(d\gamma]\}$ in equation (8.7) is the ratio of the area covered by particles to the entire area of blackening. It would be interesting to ascertain how this ratio varies if the overlapping of particles is taken into account. Figure 8.1, based on formula (8.7), shows the ratio F_b/F of the area covered by particles to the entire area as a function of the dimensionless degree of blackening $G/d\gamma$, and also as a function of the dusting rate G for various particle diameters: 0.2, 0.5, 1, and 2 mm. The blackening substance in this case was assumed to have a specific weight of 137.3 kN/m^3, the value for coal.

The dashed line in Figure 8.1 shows the variation of the area ratio F_b/F without allowing for the possibility of particle overlapping (cf., Equation (8.2)). Inspection of the figure shows that, the higher the dusting rate (the degree of blackening), the greater will be the effect of the superposition of particles; beginning at a certain point, practically all the surface is covered with particles and a further increase in the dusting rate has no effect at all.

It also follows from Figure 8.1 that Equation (8.2) can be used only for very low dusting rates, when so few particles are present that it is highly unlikely that one will fall on top of another.

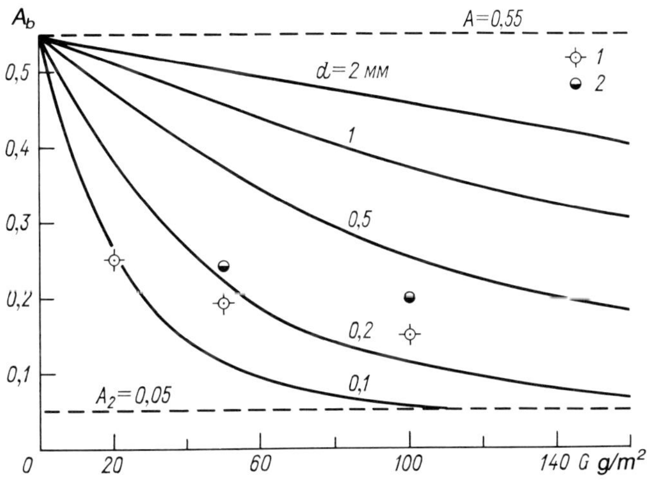

Fig. 8.2. Albedo of blackened surface as function of dusting rate.

Comparison with observations: (1) observations of Kreiter [46] for 1-2 mm particles; (2) observations of Dolgushin [21] for 1-5 mm particles.

In order to evaluate quantitatively the albedo change caused by blackening the glacier surface, we have to know the albedo before blackening and the albedo of the blackening substance. Once again, we will assume the blackening agent to be coal dust (soot), the substance most often used in such experiments. Since no direct data are as yet available on the albedo of a sooty surface, let us as a first approximation assume it to be 0.05, that is, close to the highest albedo of a black painted surface, which according to Mikheev [59] can range from 0.01 to 0.05.

Figure 8.2 shows the albedo of a blackened surface as a function of the dusting rate, for various values of the predusting albedo and for particle sizes of 0.2, 0.5, 1, and 2 mm. The curves were plotted on the basis of formula (8.5). Observational data on the albedo variation are also presented.

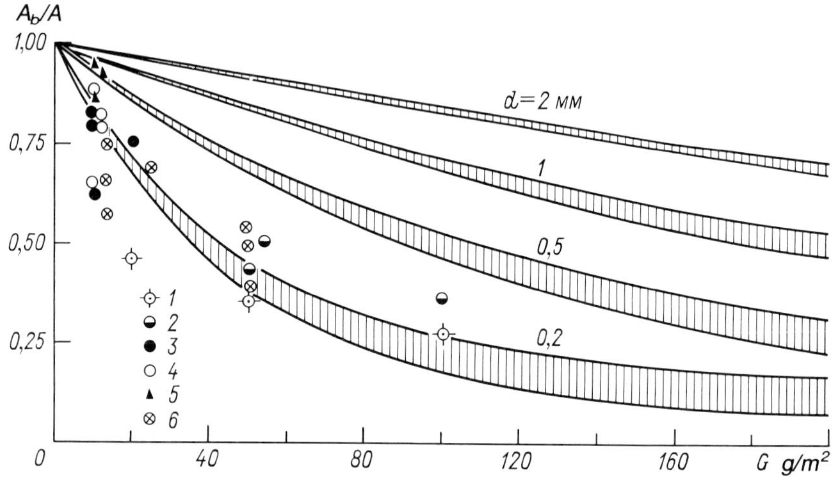

Fig. 8.3. Albedo ratio as function of dusting rate.

(1 and 2) same as in Figure 8.2; (3-5) observations of Kreiter [46] for particle sizes: less than 0.2; 0.2 - 0.5; and 0.5 - 1 mm; (6) observations of Dolgushin [21] for particle sizes less than 1 mm.

According to Figure 8.2, theoretically the surface albedo should at first drop sharply with an increase in blackening, and then level off slowly. Dusting with fine particles should reduce the albedo more than with larger particles. However, a comparison with the experimental points shows these to be quite scattered and in addition that the particle size has considerably less effect than implied by the calculations.

Figure 8.3, plotted using formula (8.7), shows the albedo ratio for blackening as a function of the dusting rate, for the same particle sizes as in Figure 8.2. Note that for each particle size two curves have been plotted: an upper curve for high A (0.7 on the graph) and a lower curve for low A (0.3 on the graph). For all values of A from 0.3 to 0.7 the calculated A_b will lie between these curves (shaded regions). Experimental data of different investigators are also given (notation same as in Figure 8.2).

Just as in Figure 8.2, the scatter of the experimental points is very great, revealing a substantial discrepancy between the albedo values obtained from experiment and from calculations.

8.1.3. Albedo for Blackening with a Mixture of Different-Sized Particles

In order to evaluate the results of the comparison with observational data, let us take a closer look at the data. The experimental points do not reveal significant albedo differences for dusting with particles of different sizes. The observational data for all particle sizes lie, as a rule, between the calculated curves for particles 0.2 and 0.5 mm in size. In order to fathom the reasons for this, we note that the quantity used to characterize the particle size in the various experiments is, as a rule, the mesh gauge of the sifter through which the blackening agent is applied to the surface. However, this size is actually just the maximum diameter of the particles used for the dusting, and it is natural to assume that the blackening substance is a mixture of particles having diameters equal to and smaller than the mesh size of the sifter. The relative number of particles of diameter less than the mesh size is usually unknown.

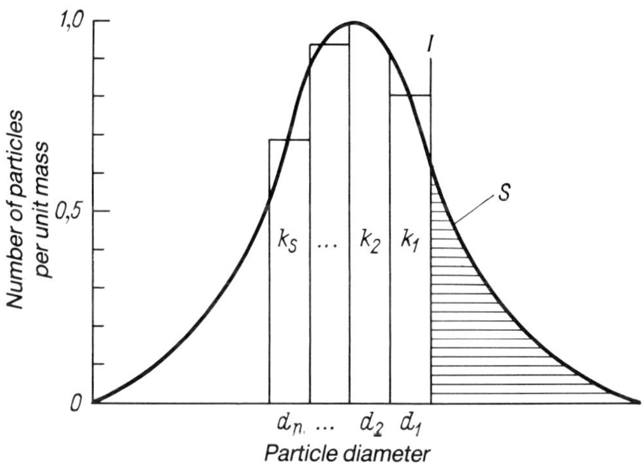

Fig. 8.4. Determination of albedo for blackening with mixture of different-sized particles.

The size distribution of the particle numbers per unit mass depends on the method used to pulverize the blackening substance; this distribution is described by a curve such as the one in Figure 8.4. The portion of the particles S to the right of line 1-1 do not make it through the sifter, and it is the rest of the curve which characterizes the quality of the dusting agent.

In order to demonstrate the effect of the particle-size spectrum on the albedo variation for blackening, let us consider a model mixture of particles with three sizes: d_1, d_2, d_3. The parts by weight of these three particle sizes in the dusting rate are, respectively, k_1, k_2, k_3.

The albedo of a surface dusted with particles having diameters d_1 and d_2 will then be

$$A_{bd_1d_2} = A_{bd_1}\left\{1 - \left(1 - \frac{A_r}{A_{bd_1}}\right)\left[1 - \exp\left(-\frac{6k_2G}{d_2\gamma}\right)\right]\right\}. \tag{8.8}$$

The albedo of a surface dusted with particles of all three sizes: diameter d_1 to an amount k_1G, diameter d_2 to an amount k_2G, and diameter d_3 to an amount k_3G, is

$$A_{bd_1d_2d_3} = A_{bd_1d_2}\left\{1 - \left(1 - \frac{A_r}{A_{bd_1d_2}}\right)\left[1 - \exp\left(-\frac{6k_3G}{d_3\gamma}\right)\right]\right\}.$$

Correspondingly, if the distribution curve for the numbers of particles of different diameters is represented as a sum of s groups of particles having discontinuously varying diameters d_1, d_2, d_3, ..., d_s and the corresponding parts by weight in the dusting rate k_1, k_2, k_3, ..., k_s (Figure 8.4), then the albedo of a surface blackened with this mixture will be

$$A_{b\sum_{i=1}^{s} d_i} = A_{b\sum_{i=1}^{s-1} d_i}\left\{1 - \left(1 - \frac{A_r}{A_{b\sum_{i=1}^{s-1} d_i}}\right) \times \left[1 - \exp\left(-\frac{6k_nG}{d_n\gamma}\right)\right]\right\}. \tag{8.9}$$

Therefore, by successively determining the albedo for blackening by each particle fraction and substituting the values obtained according to formulas (8.7) - (8.9), we can find the albedo for any composition of the particle mixture. Expression (8.9) can be simplified by assuming that the quantity

$$1 - \frac{A_r}{A_b \sum_{i=1}^{\omega} d_i} = \bar{m}$$

is the same for all ω. This is apparently permissible for mixtures with a small number of components. For instance, for a mixture of three components the simplified form of (8.9) is

$$\frac{A_{bd_1d_2d_3}}{A} = \left\{1 - \bar{m}\left[1 - \exp\left(-\frac{6k_1G}{d_1\gamma}\right)\right]\right\}\left\{1 - \bar{m} \times \left[1 - \exp\left(-\frac{6k_2G}{d_2\gamma}\right)\right]\right\}\left\{1 - \bar{m}\left[1 - \exp\left(-\frac{6k_3G}{d_3\gamma}\right)\right]\right\}. \tag{8.10}$$

Figure 8.5 shows the relative surface albedo as a function of the dusting rate for a mixture of three particle groups of equal weight but with diameters of 0.2, 0.5, and 1 mm. The curves were plotted using formula (8.10), with $\bar{m}$ taken to be constant and equal to 0.3. The shading in the figure indicates the limits of error when approximate formula (8.10) is used instead of the accurate formula (8.9) for the various albedos of the unblackened surface. These limits correspond to variations of the albedo prior to blackening from 0.3 to 0.7.

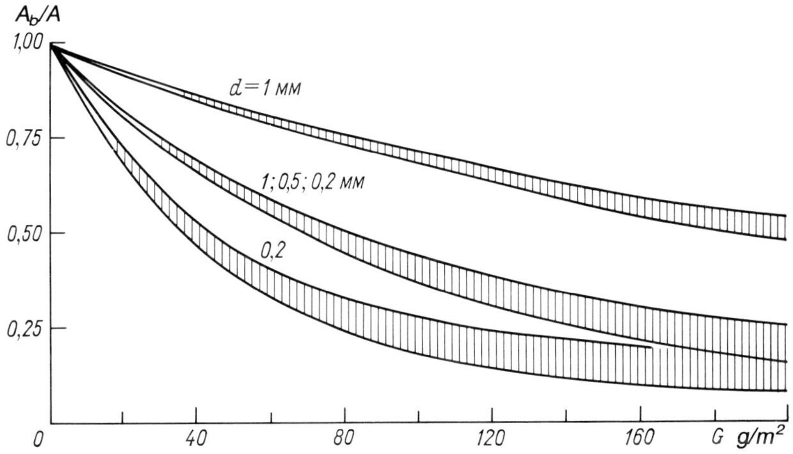

Fig.8.5. Effect of diversity of particle sizes on variation of albedo for blackening of surface.

For comparison, Figure 8.5 also shows once again the curves of the relative albedo as a function of the dusting rate for particles having diameters of 1 mm (upper curves) and 0.2 mm (lower curves). The curves were plotted using formula (8.7) for the rate of dusting of particles of a single size.

A comparison of these curves indicates that, in a mixture of particles of different sizes, the maximum particle size corresponding to the mesh size of the sifter is not a satisfactory criterion of the effectiveness of the blackening properties of the substance. Moreover, at high dusting rates the presence in the mixture of small particles will have more effect than the presence of large particles, so that the blackening effect turns out to be almost the same as for dusting with only small particles.

8.1.4. Effective Particle Size as a Characteristic for Prolonged Blackening Action

The above considerations apparently pertain just to the beginning of melting. Observations indicate that, during a comparatively short period after the start of melting, there is a redistribution of the particles on the surface. This is apparently attended by a change in the reflectivity of the surface. In order to gauge the effect of this redistribution, let us assume that each particle, as it penetrates into the snow or ice, forms a hole with a diameter larger than the characteristic dimension of the particle. In addition, there will exist some distance from the center of the particle at which the thermal effect of this particle on the ice and on other particles is felt.

If two particles lie close enough together that their initial separation is less than this distance, then during the subsequent thawing they will begin to interact with one another. The part of the wall of each hole which is closer to the adjacent hole, and which thus lies within the 'sphere of influence' of the neighboring particle, is bound to melt more rapidly; as a result, the two adjacent particles approach each other, until they appear to join, forming a single common hole in the ice. Then, since there are always slight differences in the sizes, albedos, and other physical properties of particles, one of them will melt through the ice more rapidly, and thus the second particle will move down onto the first, completely or partly covering it.

Consequently, instead of a certain number of particles of diameter d, a smaller number of groups of particles collected together in common holes in the ice will participate in the radiative melting. In view of this, to characterize the prolonged effect of blackening, it is advisable to use, instead of the particle diameter, a quantity D representing the mean diameter of a tiny 'sphere', the volume of which equals the combined volume of the particles which are joined together during the radiative melting of the blackened surface.

This volume, other conditions being the same, is a function of the number of particles z present in the mutual 'sphere of influence'. The region within which each particle interacts with the sphere of influence of an adjacent particle is defined by a circle of diameter 2z. Then the number of particles whose spheres of influence interact with the sphere of action of the central particle will be

$$\frac{6G}{\pi d^3 \gamma} \frac{\pi(4z^2 - d^2)}{4} .$$

The total number of particles present in a zone of diameter 2z including the central particle is

$$1 + \frac{6G}{\pi d^3 \gamma} \frac{\pi(4z^2 - d^2)}{4}.$$

If we now assume that the square of the diameter of the sphere of influence is considerably larger than the square of the diameter of the particle, and all the more so that $4 z^2 \gg d^2$, then it follows that for a maximum packing density of the tiny 'spheres' in a sphere of diameter D we have

$$\frac{\pi D^3}{6} = \frac{\pi d^3}{6} \left(1 + \frac{6G}{d^3 \gamma} z^2\right).$$

After certain transformations, this becomes

$$D = \sqrt[3]{d^3 + \frac{6z^2}{\gamma} G}. \tag{8.11}$$

The diameter 2z of the sphere of influence of the particle is apparently a function of time, of the mechanism of interaction of the particle with the ice, and of other factors. This diameter is found either with the aid of a special theoretical study or experimentally. Apparently, after some time z reaches a constant value and becomes independent of time.

8.1.5. Determination of Effective Diameter and Sphere of Influence

It is evident from Equation (8.11) that for intensive blackening of the surface, when $6\ Gz^2/\gamma >> d^3$, the effective diameter D is independent of the initial particle size. For low dusting rates, on the other hand, the effective diameter will be equal to the particle diameter. This will be the case when there are so few particles that the distance between them precludes their interference.

If (8.11) is substituted into formula (8.6), we arrive at the following expression for the surface albedo at the end of a certain time after the dusting:

$$A_b = A - \left(1 - \frac{A_r}{A}\right)\left(1 - \frac{6G}{\gamma \sqrt[3]{d^3 + \frac{6z^2}{\gamma} G}}\right). \tag{8.12}$$

Let us compare this calculated formula with the observational data of Peschanskii [64], obtained in the Arctic. Coal was used as the blackening agent. The experiments were carried out for dusting rates from 0.4 to 4 N/m^2.

The dots in Figure 8.6 show Peschanskii's results, for comparison with the calculated curves for three values of z. The value of z was assumed to be independent of the dusting rate. The middle curve (for z = 1.6 mm) gives the best fit with the experimental data.

Figure 8.7 depicts the effective blackening diameter D as a function of the dusting rate G, for different particle diameters. Note that for dusting rates of about 3 or 4 N/m^2 the effective blackening diameter lies between 1.4 and 1.2 mm.

According to the observations of Glazovskaya (cited in [136]), the dust constituting the natural contamination of a glacier surface does not form a more or less uniform layer, but rather is grouped into small spheres from 1 to 3 mm in size. This is apparently the effective diameter of the natural blackening to which the glacier surface is subjected.

Observations [21] indicate that the natural contamination of glacier tongues ranges from 3 to 4 N/m^2. Consequently the effective diameter for a naturally contaminated glacier surface is very close to the value obtained by us by comparing the calculated curves with Peschanskii's observations.

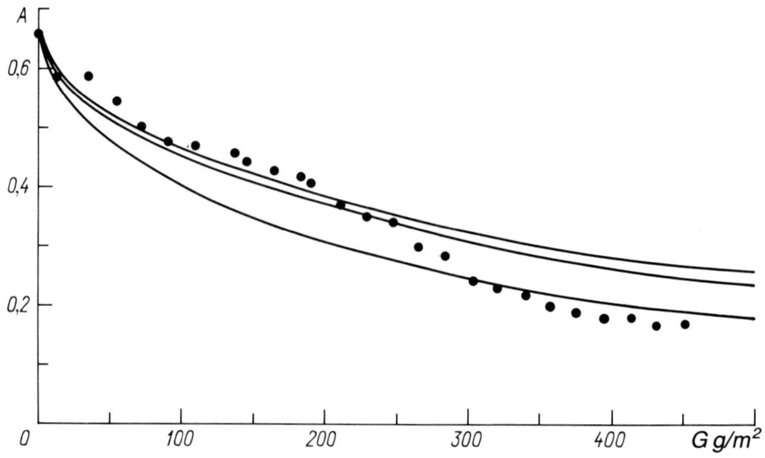

Fig. 8.6. Albedo plotted as function of dusting rate for prolonged blackening action.

Curves give calculated results, dots give data of Peschanskii [64].

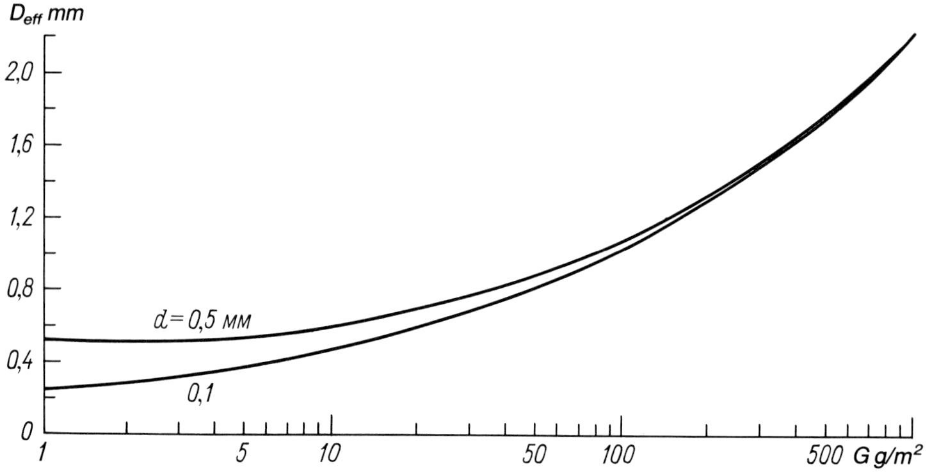

Fig. 8.7. Effective size as function of dusting rate for different particle diameters.

8.1.6. Intensity of Artificial Melting

The heat-balance equation for the melting surface of a glacier has the form

$$B_B + P_B + x_B T_{li} = M_B + L_B E_B + C_B, \tag{8.13}$$

where B_B is the radiative heat flux, determined by the radiation balance: $B_B = Q(1 - A_b) - E_B$ (here E_B is the effective emission); $x_b T_{li}$ is the supply of heat to the melting surface by liquid precipitation, falling with a temperature T_{li} (this component is very small and will be neglected in the following); P_B is the turbulent flux caused by heat exchange between the air and the glacier surface; M_B is the amount of heat consumed to melt the ice; L_B is the turbulent heat flux caused by evaporation or condensation; and C_B is the amount of heat spent to heat the layers underlying the melting surface. If all the components of the heat balance are assumed to pertain to the same time interval Δt_B, then in view of (8.12) we can write the amount by weight m_b of meltwater formed during this time as

$$m_b = \frac{Q}{r}\left\{\left\{1 - A + A\left(1 - \frac{A_r}{A}\right)\left[1 - \exp\left(- \frac{6G}{\gamma\sqrt{d^3 + \frac{6z^2}{\gamma} G}}\right)\right]\right\} + \frac{P_{Bb}}{Q} - \frac{L_{Bb}}{Q} - \frac{C_{Bb}}{Q}\right\}, \tag{8.14}$$

where r is the latent heat of fusion.

One characteristic of the acceleration of the melting caused by artificial blackening of the surface is the ratio m_b/m, where m_b is the amount of water formed by melting of the blackened surface, and m is the amount of meltwater which would be formed under the same conditions without blackening of the surface. This ratio is defined as

$$\frac{m_b}{m} = \frac{1 - A + A\left(1 - \frac{A_r}{A}\right)\left[1 - \exp\left(- \frac{6G}{\gamma\sqrt[3]{d^3 + \frac{6z^2}{\gamma} G}}\right)\right] - \frac{L_{Bb}}{Q} - \frac{C_{Bb}}{Q} + \frac{P_{Bb}}{Q}}{1 - A - \frac{L_B E_B}{Q} - \frac{C_B}{Q} + \frac{P_B}{Q}} \tag{8.15}$$

Here the subscript 'b' refers to the conditions for an articially blackened surface. The absence of any subscript indicates that natural conditions are referred to (no blackening).

At present not enough data are available to enable a reliable estimate of the variations of components P_B, L_B, E_B, and C_B under conditions of blackening. Consequently, we will not attempt to analyze these variations here, and in the following the quantity $(L_B E_B/Q)$ +

$(C_B/Q) - (P_B/Q)$, to be designated as N_B, will be assumed to be the same for the natural surface as for an artificially blackened surface. Then the relative variation of the melting rate due to blackening will be

$$\frac{m_b}{m} = -\frac{1 - A + A\left(1 - \frac{A_r}{A}\right)\left[1 + \exp\left(- \frac{6G}{\gamma \sqrt{d^3 + \frac{6z^2}{\gamma} G}}\right)\right] - N_B}{1 - A - N_B} . \tag{8.16}$$

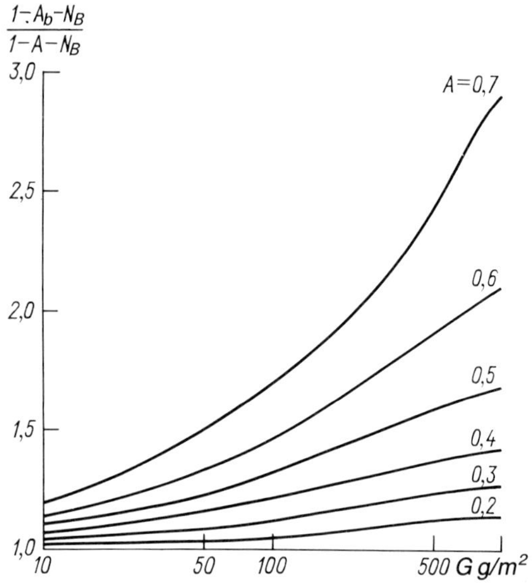

Fig. 8.8. Calculated curves of relative acceleration of melting for various albedos. $N_B = 0$.

Figures 8.8 and 8.9 show how the relative acceleration of the surface melting m_b/m varies as a function of the dusting rate G, for various albedos of an unblackened surface and for $N_B = 0$ (Figure 8.8) and $N_B = 0.327$ (Figure 8.9).

It would be interesting to compare the expressions obtained here with some observational results, for instance, with data [70] on the increase in the melting rate of ice samples subjected to various degrees of blackening by dusting with pulverized coal and slag under laboratory conditions. The air temperature over the ice samples was 0°C, and the heat flux of the short-wave radiation to the ice surface was, according to the data of [70], 1.39 kW/m^2. The observations were made using samples dusted with coal and slag dust at a rate of 0.01 to 6 N/m^2.

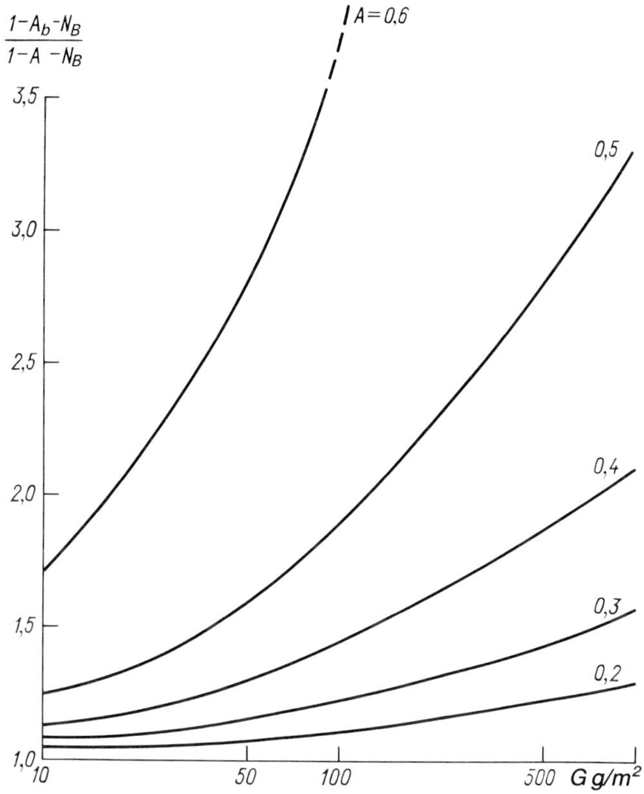

Fig. 8.9. Calculated curves of relative acceleration of melting for various albedos. $N_B = 0.327$.

The dots and circles in Figure 8.10 depict the results of these experiments. For the black dots in the figure the ordinate is the ratio m_b/m of the melting rate for blackening with various amounts of coal dust to the melting rate for samples dusted with only 1 g/m^2 of coal. Thus samples dusted with the minimum amount of coal, 1 g/m^2, are arbitrarily taken to be unblackened. For the circles in the figure the ordinate is the analogous ratio for blackening with slag dust. The solid and dashed curves in Figure 8.10 give the values of m_b/m calculated with formula (8.16) for $N_B = 0$ (dashed) and $N_B = 0.38$ (solid). The albedo of the 'unblackened' (dusted with only 1 g/m^2 of the coal or slag) ice surface was assumed to be 0.4, while that of the blackening agent was assumed to be 0.05. The value $N_B = 0.38$ reflects the effect of removal of heat from the melting surface via evaporation; it was obtained by calculation, for $N_B = L_B E_B/Q$ with $Q = 1.39$ kW/m^2.

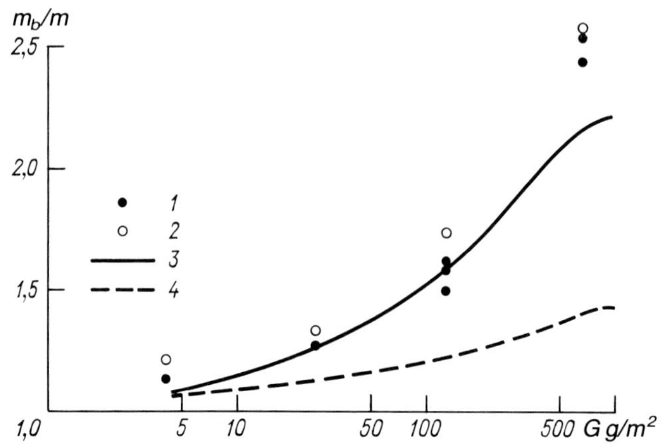

Fig. 8.10. Relative acceleration of melting for prolonged blackening. Curves represent calculations, points represent observations [70].

(1) coal dust; (2) slag dust; (3) calculation for conditions of experiment; (4) calculation for $N_b = 0$.

The quantity $L_B E_B$ was calculated as the difference between the heat accumulated by an 'unblackened' surface with an albedo of 0.4 and $Q = 1.39$ kW/m^2 and the heat required to melt this surface at the rate obtained at the time of the experiments. A comparison of the calculated curve for $N_B = 0.38$ with the observations shows that this calculation method reflects quite well the melting of a blackened surface. At the same time, it is seen that an increase in N_B causes a sharp rise in the value of ratio m_b/m. For sufficiently high N_B, for instance when the evaporation and the inward heat removal are both sizable, the value of m_b/m can be many times higher than the ratio of the shortwave-radiation fluxes incident upon a blackened surface and a natural surface. At the same time, an increase in the heat input to the melting surface caused by turbulent heat exchange with the ground-adjacent air layer will reduce m_b/m and it will diminish the relative effect of the blackening.

Under the conditions of mountain glaciers, N_B apparently increases with elevation, so that the ratio m_b/m will be larger at greater absolute heights of the surface, even if the albedo prior to blackening remains the same. The same will be true with regard to the effect of the time of year. In springtime, when the air temperature is still low, N_B will be higher than in summer, all other conditions being the same, and the relative enhancement of the melting upon blackening m_b/m will be greater than in summer.

It should be kept in mind, however, that the indicated increase in ratio m_b/m is caused not by an increase in the actual amount of meltwater from the surface due to its blackening, but rather by the fact that the amount of meltwater from a natural unblackened surface is less. Therefore, the ratio of the meltwater from a blackened surface to the meltwater from a natural surface can be used as a characteristic of the enhancement of the melting only if the quantity N_B is known.

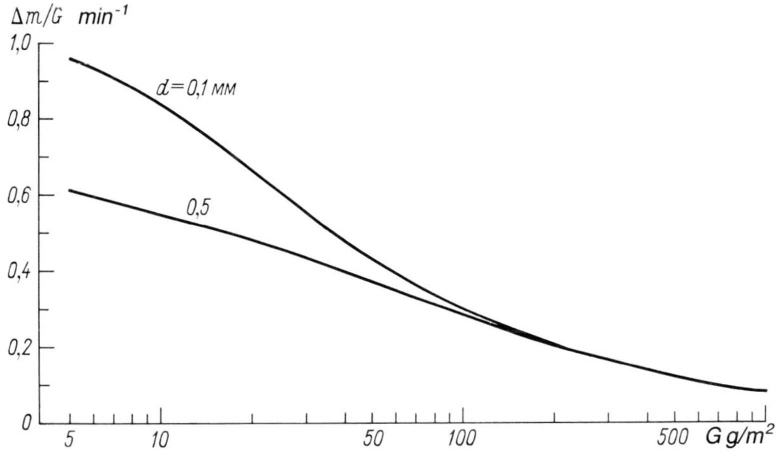

Fig. 8.11. Effectiveness of blackening as function of dusting rate.

Another very important gauge of the effect which artificial blackening has on the melting process is the additional amount of meltwater formed because of the blackening. If this is designated as Δm, then we have

$$\Delta m = \frac{Q}{r} \Lambda\left(1 - \frac{A_r}{A}\right)\left[1 - \exp\left(- \frac{6G}{\gamma \sqrt[3]{d^3 + \frac{6z^2}{\gamma} G}}\right)\right]. \qquad (8.17)$$

The ratio $\Delta m/G$ of this additional meltwater to the dusting rate G is a characteristic of the effectiveness of the artificial blackening. This ratio, which we will call η_b, represents the efficiency of a unit mass of blackening agent. It is a measure of the economic feasibility of the blackening, and it is described by the expression

$$\eta_b = \frac{\Delta m}{C} = \frac{Q}{rC} A\left(1 - \frac{A_r}{\Lambda}\right)\left[1 - \exp\left(- \frac{6G}{\gamma \sqrt[3]{d^3 + \frac{6z^2}{\gamma} G}}\right)\right]. \qquad (8.18)$$

Figure 8.11 shows calculated curves for ratio $\Delta m/G$ as a function of the dusting rate, calculated using formula (8.18). The curves were plotted for an ice albedo of 0.4, an albedo of the blackening agent of 0.05, and a Q value of 1.39 kW/m^2 (corresponding to the experiments of Tsykin and Tsykina). It is seen from the figure that the effectiveness of a unit of blackener decreases as the dusting rate is increased. This lowered effectiveness is apparently due to the screening of one particle by another, an effect which is enhanced at higher dusting rates, but which practically does not show up at all at very low dusting rates. Consequently, as $G \to 0$ the curves in Fig. 8.11 become straight lines parallel to the G axis, that is,

$$\left.\frac{d\eta_b}{dG}\right|_{G \to 0} \to 0.$$

An analysis indicates that $d\eta_b/dG$ actually goes to zero as G goes to zero. Therefore, for the model being considered, the blackening effect relative to a unit mass of blackening agent theoretically reaches a maximum only for a zero dusting rate.

8.2. Water in Mountain Glaciers

8.2.1. The Thermics of the Water Under Mountain Glaciers

We have already made a study of subglacial water, using the antarctic ice caps as an example [32]. The basic assumptions relating to, for example, the thickness of the mean water layer at the bed can be applied as well to mountain glaciers, provided the special features of mountain glaciation are taken into account (the small sizes of the glaciers, the steeply sloping beds, etc.).

A more careful consideration has shown, however, that the thermal regime of mountain glaciers has, in view of the thermics of the glacier water, some important peculiarities which were previously ignored. One of these, which will be considered here, can be summarized as follows: water flowing downward along the glacier surface, through cracks and cavities in the interior, and also along the bed of the glacier, will produce an amount of heat due to friction which is much greater than the heat generated by the movement of the glacier itself and by the geothermal heat flux.

In order to verify this statement, let us consider a typical mountain glacier. Its length L, as viewed from above, will be a few kilometers (2 km, say, to be specific). The mean angle of the bed tg α_b of this glacier is assumed to be around 0.25. The height differential ΔH_L from the upper boundary of the nourishment region to the outlet grotto of the tongue, for a glacier 2000 m long, will be about 500 m. The melting of ice during the ablation period will take place, as a rule, over the entire surface of the glacier; on the average, for the whole surface, the melting rate will be a few meters a year (4 m/yr, say, to be specific).

As a first approximation, we will assume that the annual melting increases linearly from zero at the upper boundary of the firn fields to a maximum in the lower part of the glacier tongue. Then the mean height Δh_{av} of runoff of all the meltwater, reckoned from the height of the outlet grotto, will be close to one third of ΔH_L:

$$\Delta h_{av} \approx 1/3 \Delta H = 200 \text{ m}.$$

Without deviating much from reality, we can take the width S of the glacier to be about 1/3 of its length, that is, for our glacier $S \approx 600$ m. For a specific weight γ of the ice close to 9.6 kN/m^3, the annual water runoff by weight G_w from the glacier will be $G_w = LSA\gamma = 50 \times 10^6$ kN/yr. The work performed during the descent of this water over a vertical distance Δh_{av} m is then $P - G_w \Delta h \approx 400$ kW.

Some of this work goes to increase the water velocity, that is, it is converted into kinetic energy of the water outflow from under the glacier. The rest of the energy is converted into heat. If the rate of water outflow from under the glacier $u_w = 3$ m/s, then the kinetic energy P_K of the water outflow will be

$$\frac{G}{g} \frac{(u_w)^2}{2} = 0.78 \text{ kW},$$

or only about 1/1000 of P. Thus practically all of the work of descent of the water in the gravitational field is converted into heat.

The mean global value of the geothermal heat flux q_g is about 50 mW/m^2. Consequently, the entire lower surface of the glacier is supplied with an annual amount of geothermal heat $Q_g = g_g LS = 59.7$ kW.

The mean amount of heat generated in a year by the motion of the outflowing water, averaged over the entire glacier, is thus seen to be several times greater than the heat supplied to the glacier bed from the interior of the Earth.

The amount of heat produced by the movement of the ice Q_m is calculated as the thermal effect of the work done as the ice descends in the gravitational field. In order to estimate this heat, we note that the mean thickness of the glacier II is usually less than 100 m. Then the ice mass of the glacier $M = LSH \rho g = 1200 \times 10^6$ t. The velocity of ice movement over the bed u, averaged over the thickness and length of the glacier, can be assumed to be about 10 m/yr. The glacier will have such an ice velocity if, for example, its nourishment region occupies half the area of the glacier, while the rate of accumulation in the nourishment region and the rate of ablation at the glacier tongue, both averaged over the area, are close to 2 m/yr. In this case the work of the glacier movement in the gravitational field P_m will be $P_m \approx Mu \sin \alpha_b = 93.25$ kW.

Consequently, the heat generated by the ice movement is, as a whole over the year, also substantially less than the thermal effect of the motion of the meltwater.

It should be noted, however, that the production of heat by the meltwater flow takes place only during the four months of summer and early autumn, so that the hodograph of the discharge during these months is extremely irregular. Therefore, during the summer (when thawing occurs at the glacier surface) the intensity of the heat sources associated with water flow in the glacier can be more than ten times greater than the intensities of the sources associated with geothermicity and the movement of the glacier ice.

Heat is produced by the meltwater stream both at the walls of the water channels and in the water itself. This means that the release of frictional heat in the water as the meltwater moves from higher parts of the glacier to lower parts may be accompanied by heating of the water. For instance, as the water descends 100 m, about 0.837 kJ of heat per kilogram of water is generated, an amount sufficient to heat this water by 0.2 °C (assuming no heat exchange with the ice and with the glacier bed).

It should be noted that the conditions of the water flow under a glacier are by no means adiabatic. This being so, the heat exchange between the meltwater and the ice is such as to enhance the melting of the ice at the bottom of the glacier.

Experimental studies of the thermal regime of the water under mountain glaciers were carried out by us for Dzhankuat Glacier in the Central Caucasus and for Medvezhii Glacier and Russian Geographical Society Glacier in the Pamirs. The water temperature was measured with TM-1 metastatic mercury thermometers (Beckmann thermometers), which provided temperature readings with accuracies of up to 0.01 °C.

Our most detailed observations were those on Dzhankuat Glacier. The measurements were made in late July 1970 on sunny as well as overcast days. The temperature of the water in the streams emerging from beneath the glacier was always found to be 0.15 to 0.2 °C higher than the melting point of the ice. Measurements of the water temperature in rain channels on the glacier surface, carried out at 20 points in the ablation zone, showed that the water temperature in the rain channels ranged from 0.04 to 0.06 °C. Consequently, the water was heated not at the glacier surface, but rather as it made its way down through cracks and holes in the ice and at the bed itself. On the average, the water discharge via streamlets emerging from under the glacier was close to 2 m^3/s, and this water carried off with it about 1700 kW of excess heat, relative to the water at the melting temperature.

The amount of heat of geothermal origin incident upon the entire lower surface of a glacier can be estimated if we assume that the geothermal flux equals the mean value for the Earth's surface. This gives a value of about 36 kW for the geothermal heat incident upon the entire bed surface of a glacier about 0.7 km^2 in area.

We estimated the heat produced by the glacier movement, assuming Blockschollen movement of the ice. The glacier thickness was taken to be 60 m, for a mean speed of ice movement of 10 m/yr over the whole glacier and a bed slope of about 0.3 (according to hypsometer data). Even for these excessively high values, the thermal effect of the ice movement did not exceed the geothermal heat flux into the ice, also being close to 34 kW.

Consequently, the heat generated in the water by friction as it moved from higher parts of the glacier to lower parts was found to be 50 times as great as each of the two components usually taken into account when analyzing the thermal regime of a glacier.

These results have the following corollary: for long glaciers with a sufficient height differential, the temperature of the water emerging from under the ice grottos in the lower part of the tongue should be elevated considerably. To ascertain this, we carried out temperature measurements at the outlet grottos of Medvezhii Glacier and Russian

Geographical Society Glacier.

Several years ago Medvezhii Glacier executed a rapid movement, and it is now in a state preceding a new movement. The active part of this glacier, which contains the region of active summer melting, is about 3500 m long; the height differential in this zone reaches 600 m. Therefore, by analogy with Dzhankuat Glacier, the temperature increase in the water emerging from under the glacier can be up to 0.7 °C. An even higher temperature (more than 1 °C) is to be expected for the water discharged from the grottos in the lower part of Russian Geographical Society Glacier. This glacier is 10 km long, and the height differential in the region of possible melting is 1000 m.

Measurements of water temperatures at the outlet grottos of Medvezhii Glacier and Russian Geographical Society Glacier were carried out in September 1970. They indicated that, despite the large sizes of these glaciers and the steep slopes, the temperature of the emerging water is only 0.02-0.04 °C above the melting point of ice, that is, it is practically the same as the water temperature at the glacier surface.

The obtained temperature difference for the water at the outlet of the glacier grottos is due, in our opinion, to differences in the nature of the water flow along the glacier-bed boundary; accordingly, the data on the temperature of the water emerging from under a glacier should give us information about the interaction of the glacier with its bed. For Dzhankuat Glacier, for example, the water temperature was quite high, indicating that there is little heat exchange of the water under the glacier with either the glacier ice or the bed. This is possible if the subglacial water flows through relatively large grotto-channels and does not contact the ice directly. For the water flowing along the beds of Medvezhii Glacier and Russian Geographical Society Glacier, on the other hand, the heat exchange with the ice is apparently great; this water probably flows along the bed via numerous small channels, distributed over a considerable portion of the glacier bed.

8.2.2. The Temperature of the Water on a Melting Glacier Surface

Measurements of the temperature of the water on the melting surface of a glacier, contaminated with moraine particles, eolian dust, or a blackening substance, show that this temperature is somewhat higher than the melting point of ice (sometimes some tenths of a degree higher). That such overheating does occur is borne out by many indirect data as well, especially by the erosive effect of the water streams on the glacier surface over which they flow.

Although the problem of the overheating of the water (let us call it ΔT_r) is a complex one [17], let us try to evaluate the amount of overheat with the aid of simple models. We choose a model of a melting ice surface covered with particles of a blackening substance, that is, a material with an albedo lower than the albedo of the ice. In this case the temperature of the possible overheating of the water located right over particles lying in their tiny meltholes will be close to, but not higher than, the temperature of the upper sunward-pointing surface of the particles. In order to estimate the possible temperature of the upper surface of the particles, we assume that the solar energy needed to maintain the melting only reaches the upper particle surface. The problem will be taken to be one-dimensional, that is, we will neglect the loss of heat from the upper 'surface of energy accumulation' to the 'side' walls of the particle. If λ_r is the thermal conductivity of the particle material, and m_r is the rate of melting of the ice under the particle, then under steady-state conditions, since the thermal resistivity of the water layer between the lower surface of the particle (or group of particles) and the melting ice is negligible, we have

$$m_r r\gamma = \lambda_r \frac{\Delta T_r}{D}.$$

Consequently, the temperature of the upper particle surface needed to ensure that an amount of heat corresponding to a melting rate $m_r = m_b$ reaches the lower surface will be

$$\Delta T_r = \frac{m_r r\gamma}{\lambda_r} D. \tag{8.19}$$

For example, if the thermal conductivity of the particle material is 0.59 J/(cm min °C) (the value for damp soil), while the particle thickness is 1.6 mm, then for a melting rate of about 0.6 cm/hr the water temperature at the upper particle surface will be 0.8 °C.

If conditions are such that the melting rate of the ice under the particle (or group of particles) is equal to the melting rate of the entire surface, then this rate is given by expression (8.14). Substituting the result obtained into formula (8.19), we then get the following expression for the prolonged effect of blackening the surface with particles of the same size:

$$\Delta T_r = \frac{D}{\lambda_r} Q \left\{1 - A - (A - A_r)\left[1 - \exp\left(- \frac{6G}{4\gamma \sqrt[3]{d^3 + \frac{6z^2}{\gamma} G}}\right)\right]\right\}. \tag{8.20}$$

Equation (8.20) can be used to find the temperature of overheating of the water, assuming this temperature to be equal to the temperature of the upper surface of the particles of blackening substance lying at the bottom of the meltholes. For D = 1.6 mm, A_r = 0.05, Q = 0.98 kW/m^2, λ_r = 0.65 W/m^2, and N_B = 0.32, calculations give the following overheating temperatures for the indicated dusting rates: ΔT_r = 0.54 °C for G = 1.7 N/m^2; ΔT_r = 0.42 °C for G = 0.5 N/m^2; ΔT_r = 0.38 °C for G = 0.3 N/m^2; and ΔT_r = 0.32 °C for G = 0.15 N/m^2.

Are the amounts of overheating given by these estimates large or small? In order to answer this question, we must take a closer look at the mechanism of the melting at a surface covered with particles of foreign matter having a low albedo.

8.2.3. Melting Mechanism of Blackened Surface

Let us assume that each particle of the blackening agent descends into the ice with a melting rate m_r corresponding to the amount of energy supplied to the particle. A surface situated between the blackening particles, on the other hand, melts at a rate m corresponding to the energy influx to the unblackened surface. As in Section 8.1, we will assume here as a first approximation that each particle forms, as the melting proceeds, a hole in the ice having an area larger than that of the midsection of the particle itself. The dusted area will be designated as F, and the part of F covered with particles as F_b. The area occupied by meltholes caused by the additional heat arriving directly from the particles must be larger than F_b; let us call this area F'.

It will be assumed that within area F' the heat supplied to the surface is consumed by the phase transition (melting) and by the overheating of the water forming in the meltholes from the melting temperature to some temperature ΔT_r. We will ignore the conductive heat loss from zone F', since the temperature of the ice interior in the melting-crust region is apparently close to 0 °C. We also assume that area F' and the overheating temperature both remain constant in time. Then the amount of heat consumed in melting under the particles at a rate m_r and in maintaining a corresponding invariable area F' will be $m_r F'\gamma(r + c\Delta T_r)$.

The quantity of heat arriving at the surface of area F' is equal to the sum of the heat incident upon the area F_b covered with particles (and hence having an albedo A_r) plus the heat incident upon the area $F' - F_b$ without particles but still within the melthole region. This part of the surface is assumed to have the albedo of pure ice A. If Q is the intensity of the incident short-wave radiation, then the amount of heat arriving at surface F' will be

$$Q(1 - A_r)F_b + Q(1 - A)(F' - F_b).$$

Some of this heat is inevitably lost as long-wave radiation, even in the absence of turbulent exchange, evaporation-condensation, and heating up of the ice. If the intensity of the long-wave radiation, which depends on the temperature of the surface, is E(T), then the corresponding amount of heat leaving surface F' will be E(T)F'. The heat balance at surface F' can then be expressed as

$$m_r F'\gamma(r + \Delta T_r c) = (1 - A_r)F_b Q + (1 - A)(F' - F_b)Q - E(T)F'. \qquad (8.21)$$

According to the Stefan-Boltzmann law $E(T) = \varepsilon_n \sigma (T + 273)^4$, where ε_n is the degree of blackness (equal to about unity for snow, ice, and water), and σ is a radiation constant, equal to 3470 J/(cm^2 min K). For a melting surface (T = 0 °C), intensity $E(T) \approx E(0)$ and equals 1.9 J/(cm^2 min).

Now let us return to the melting conditions in the part of the surface F - F' where the blackening particles have no direct effect. The melting in this zone is assumed to be caused, firstly, by the radiation component (1 - A) (F - F') - E(0) (F - F)' and, secondly, by the fact that the overheated water forming in zone F' gives up its heat to zone F - F' and cools down again to the melting temperature. The amount of heat given up is $m_r \Delta T_r c\gamma F'$. According to this scenario, the redistribution of energy between zone F' (direct effect of blackening) and zone F - F' is effected by the outflow of overheated water from the meltholes onto the remaining part of surface F. Naturally, it is assumed here that the melting in zone F - F' takes place without overheating of the water, so that the amount of heat consumed for this melting at a rate m is $m\gamma r(F - F')$. The heat balance of the part of the surface F - F' then has the form

$$m(F - F')\gamma r = (1 - A)\ (F - F')Q - E(0)\ (F - F') + m_r \Delta T_r c\gamma F'. \tag{8.22}$$

Now, by adding together both sides of balance equations (8.21) and (8.22), we obtain the following equation for the heat balance at the entire surface F:

$$\gamma[m_r F' + m(F - F')] = \frac{QF}{r}\left\{1 - A\left\{1 - A\left[1 - \left(1 - \frac{A_r}{A}\right)\frac{F_b}{F}\right]\right\}\right\} - E(0)F. \tag{8.23}$$

The left-hand side of this equation is seen to be none other than the amount of water m_b formed per unit time by the melting of the blackened surface.

8.2.4. The Stable Prolonged Melting of a Blackened Surface (Simplest Case)

Heat-balance equations (8.22) and (8.23) for zones with areas F' and F - F' were obtained while describing the simplest mechanism of melting from a blackened surface. It was also demonstrated that, for any (not in general identical) rates of melting from the part of the surface F' not directly affected by the particles and from the part F - F' affected only indirectly by the particles, via the influx of heat with the overheated water, the amount of water formed under conditions of radiation melting is determined by the albedo variation caused by the blackening.

However, if the melting rates in zones F' and F - F' are not the same, then this conclusion will be valid only during the initial period of melting. If, for example, the melting rate under the blackening particles is higher than the melting rate in region F - F', then as they become submerged in the ice the particles move downward more rapidly than the slower surface of zone F - F'. But does this 'out-distancing' process continue on and on? Observations [70] indicate that they do not. Over a certain time the rate of submergence of the particles decreases, while the rate of descent of the surface in zone F - F' increases. But this process is complicated: sometimes rates m_r and m increase and sometimes they decrease. And these increases and decreases are nonsynchronous, which causes a rise or a drop in the mean value (over the entire surface) of the melting rate m_b and the thickness of the melting crust, relative to the mean values of these quantities considered by us.

According to Equations (8.21) and (8.22), the melting rates are

$$m_r = \frac{(A - A_r)\frac{F_b}{F}\frac{F}{F'}Q + (1 - A)Q - E(0)}{\gamma(r + \Delta T_r c)}; \tag{8.24}$$

$$m = \frac{(1 - A)Q}{\gamma r} + m_r \frac{\frac{F'}{F}}{1 - \frac{F'}{F}} \frac{c\gamma\Delta T_r}{\gamma r} - \frac{E(0)}{\gamma r}. \tag{8.25}$$

However, stable prolonged melting can take place only if the melting under the particles and the melting at the sides of the particles proceed in the same way.

From the condition of equality of the melting rates, we have

$$m_r = m = m_b. \tag{8.26}$$

We now substitute (8.24) and (8.25) into (8.26). For blackening with particles of the same size under conditions of prolonged melting, ratio F_b/F will, according to (8.16) and (8.11), be given by the expression:

$$\frac{F_b}{F} = 1 - \exp\left(-\frac{6G}{4\gamma\sqrt{d^3 + \frac{6z^2}{\gamma}G}}\right). \tag{8.27}$$

Then the amount of overheating ΔT_r will be related as follows to ratio F'/F and the dusting rate G, under conditions of possible equality of the melting rates under the blackening particles and outside of the zone of their direct influence:

$$\Delta T_r = \frac{r}{c}\,\frac{\left(1 - \frac{F'}{F}\right)\frac{F}{F'}(A - A_r)\left[1 - \exp\left(-\frac{6G}{4\gamma\sqrt[3]{d^3 + \frac{6z^2}{\gamma}G}}\right)\right]}{1 - A + \left[1 - \exp\left(-\frac{6G}{4\gamma\sqrt[3]{d^3 + \frac{6z^2}{\gamma}G}}\right)\right](A - A_r) - \frac{E(0)}{Q}} \qquad (8.28)$$

The equation also states the condition for possible prolonged melting of a blackened glacier surface for the very simple model being considered.

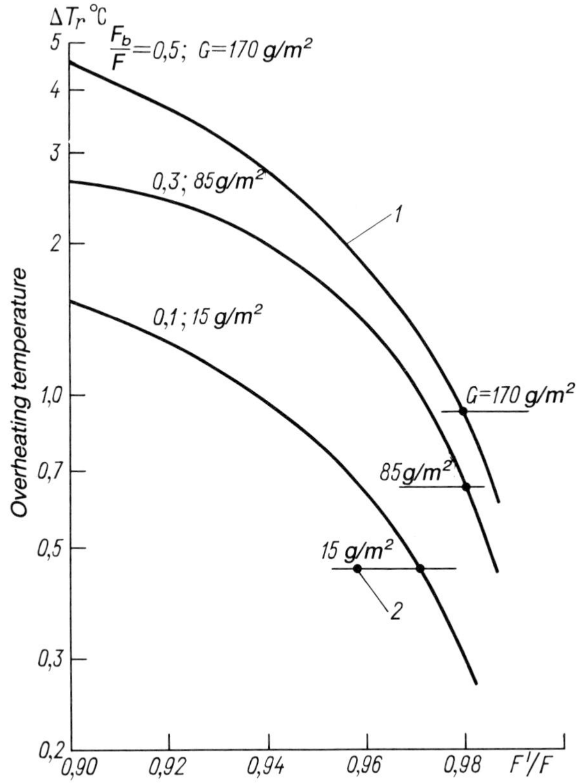

Fig. 8.12. Determination of overheating temperature of water and melting conditions for blackened surface.

In Figure 8.12 curves 1 show the overheating temperature as a function of ratio F'/f for different dusting rates. The curves were plotted on the basis of calculations for $A = 0.5$ and $A_r = 0.05$. In accordance with our previous discussion, z was taken to be 1.6 mm. The horizontal segments 2 for these same dusting rates indicate the maximum values of the overheating temperature for this same z value. The conditions of possible prolonged melting of the blackened ice surface in the coordinates shown in the figure correspond to the points of intersection between curves 1 and horizontal segments 2. This graph shows that the overheating temperature of the water, so as to ensure that this water will impart enough heat to the sides to melt pure ice, depends considerably on the relative area of the ice situated outside of the zone of direct thermal effect of the contaminating particles. An increase in the relative area, that is, a decrease in F'/F, makes necessary a higher overheating temperature.

This temperature will be higher, the greater the dusting rate. Note that an increase in the dusting rate is attended by a rise in the rate of melting of the blackened surface, and thus a higher enthalpy of the overheated water is required in order to ensure this enhancement of the melting. In the model considered, by the way, the necessary overheating turns out to be very great (2 to 4 °C), even when 90% of the entire surface lies within the zone of direct thermal effect of the particles.

Chapter 9

THE THERMOPHYSICS OF ICE SHELVES

The seaward extensions of land glaciers are known as ice shelves. Such large attached floating glaciers are now commonly found only in the Antarctic, where they comprise more than 11,700 km (or 49%) of the coastline. During the Quarternary glacial period ice shelves were much more widespread both in Antarctica and in the Northern Hemisphere.

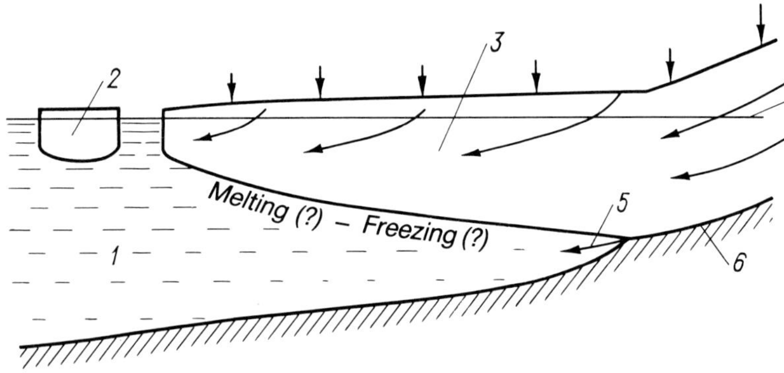

Fig. 9.1. Schematic representation of ice shelf.

(1) subglacial sea; (2) iceberg which has broken away from ice shelf; (3) ice shelf; (4) inland ice sheet, nourishing this shelf; (5) fresh meltwater flowing out to lower surface of shelf in its seaward part; (6) bottom of subglacial sea and bed of land glacier.

Figure 9.1 gives a good representation of an ice shelf and the thermophysical problems associated with it. An important feature of these problems is the presence of a large amount of water under the glacier, with the attendant question of whether melting or freezing takes place at the bottom of the ice shelf, and at what intensity. The thermophysical problems associated with icebergs, which are for the most part pieces of ice shelves which have broken off and floated away, are essentially the same as the problems involving the ice shelves themselves.

9.1. Melting at the Bottom of an Ice Shelf

9.1.1. The Englacial Temperature Field as an Indicator of Melting

It has always been difficult to decide whether melting or freezing occurs at the bottom of an ice shelf, since until recently it was impossible to observe directly the processes going on at the lower surface of a shelf. Thus these processes could be evaluated only on the basis of indirect characteristics or by analyzing the components of the mass balance of the glacier ice. Unfortunately, however, neither of these approaches, especially during the initial stages of study, was accurate. It was suggested that freezing is to be expected at the bottom of an ice shelf, since the seawater temperature in places where ice shelves occur was always found to be lower than 0 °C, the melting temperature of the fresh-water ice making up the bulk of the ice shelf.

There were also other, indirect, signs of bottom freezing. For instance, in his early studies of the western extremity of the Ross Ice Shelf, Debenham [94] detected in that part of the shelf which is a floating extension of Koettlits Glacier, close to the Dailey Islands, remains of fish, corals, sponges, and other marine organisms. Debenham assumed the only possible explanation of the presence of these remains to be that the marine organisms had been frozen into the bottom of the glacier, had risen to the glacier surface, and then had melted out. There were clear indications of intensive melting at the glacier surface.

For a long time the main method, in our view, of studying the intensity and direction of the processes at the bed has been the method which we have called [32] the thermophysical method of determining the mass exchange at the glacier surface. This method consists in determining the melting-freezing, that is, the rate of vertical ice motion, on the basis of the temperature field over the glacier thickness. The first measurements of the temperature field in an ice shelf were carried out by a Norwegian-Swedish-British antarctic expedition in a borehole drilled through Maudheim Glacier. These measurements revealed considerable nonlinearity of the temperature distribution over the glacier thickness (Figure 9.2). Robin [115] explained this distribution as an effect of the downward vertical movement of the ice, which under the conditions at Maudheim Glacier could correspond to melting beneath the glacier (Figure 9.2). Later calculations indicated that the melting rate at the lower surface should be close to 40 cm/yr.

Actually, back in Chapter 3 we demonstrated that there is a definite close relationship between the nature of the englacial temperature field and the direction and intensity of the processes of accumulation or melting at the upper and lower surfaces of the glacier. Therefore, if the temperature field and the conditions of accumulation or melting at the upper surface of the glacier are known, the mass exchange at the bottom of the glacier can be assessed. Accordingly, the temperature distribution over the thickness of Maudheim Glacier would be associated by us with a regime of cooling by the vertical downward movement of the ice, corresponding to accumulation at the upper surface of the glacier and melting at the bottom (see Figures 4.2 - 4.4).

Systematic studies of processes at the lower surfaces of ice shelves thus date from the time of the first deep boreholes in the shelves and the first studies of the temperature fields.

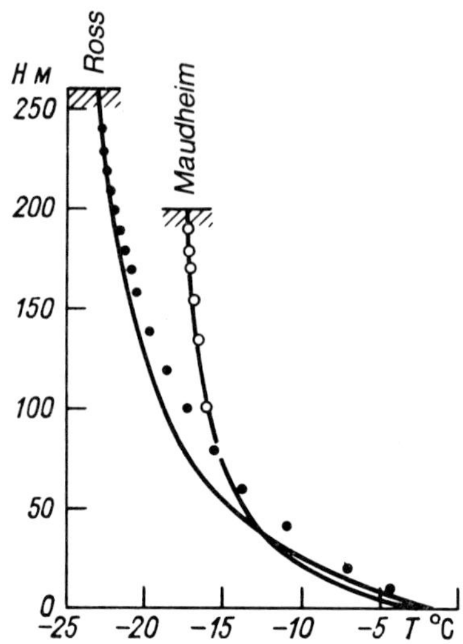

Fig. 9.2. Temperature distributions in marginal regions of Maudheim and Ross Ice Shelves.

Black or white circles indicate measurements, and curves show theoretical distribution for conditions of bottom melting.

A deep borehole was drilled almost to the bottom of the Ross Ice Shelf near Little America V Station, which was situated at the glacier margin to the west of Roosevelt Island. Temperature measurements over the entire borehole depth verified the conclusion that there is melting in the part of the ice shelf close to the margin. The temperature distribution measured in this glacier also corresponded to a regime of cooling due to downward ice movement over the whole glacier thickness (see Figure 9.2).

Crary [93] used the temperature distribution in the borehole and data on the accumulation rate at the glacier surface to determine the rate of bottom melting. He obtained a value of about 60 cm/yr.

Oceanological observations have shown, however, that the temperature of the seawater near the ice shelf, at the depths corresponding to its lower boundary, are around -1.89 or -1.87 °C; this is appreciably lower than the melting temperature of the fresh-water ice making up the glacier. On the other hand, this temperature is about 0.2 - 0.3 °C higher than the freezing point of seawater with the corresponding salinity and pressure at the depth of the bottom of the glacier.

The question arises as to whether it is possible for the fresh-water ice to melt upon contact with seawater having a temperature above the freezing point of this saltwater. Such a melting process can indeed occur if it is a process of heat and mass exchange. Here the mass exchange is a transfer of salt from the water to the surface of the ice. The higher the salt concentration of the surface layers of the initially fresh-water ice, the lower will be the melting point of these layers. The melting of the salt-saturated surface layers of what was originally fresh-water ice can start only when the melting temperature of these layers drops below the temperature of the water surrounding the ice, which gives rise to a heat flux from the water that can melt the ice. Once this melting has begun, a gradient of the salt concentration is set up in the water layer next to the ice, thereby saturating with salt the next layers of the initially fresh-water ice.

If the melting rate is steady, then the rate of transfer of salt from the water to the ice surface will be such as to keep the salinity of the deep ice layers (relative to the surface) within limits holding the melting point of the surface layer of ice below the water temperature.

A salinity gradient can be set up in the water layer next to the ice, in order to maintain the transfer of salt to the ice, provided that the salinity of the surface layer of melting ice is lower than the salinity of the water.

Consequently, the freezing point of this surface layer of initially fresh-water ice will be below the temperature of the water but higher than the freezing point of the seawater. The higher the rate of melting of the fresh-water ice in the seawater, the greater will be the difference between the temperature at the melting ice surface and the freezing point of the surrounding water.

This melting process may in some cases be governed by the rate of transfer of salt, that is, by the 'salt exchange', and in other cases by the ordinary heat exchange.

In view of the difficulty of arriving at a general theoretical solution of the problem of the heat exchange and salt exchange at the bottom of an ice shelf, it was advisable, as a first approximation, to assume that the temperature at an ice surface (initially fresh-water ice) melting in seawater is essentially equal to the freezing temperature of the seawater. This will be the case if the salt concentration in the surface layer of ice is essentially equal to the salt concentration in the water. Thus, this assumption is equivalent to stating that the 'salt transfer' has a high intensity and that the melting rate is determined only by the heat exchange.

This statement has been verified, to some degree, by experiments on the melting of cubes of fresh-water ice in seawater [106]. Treshnikov [69], analyzing the results obtained during the fifth and sixth cruises of the diesel-electric vessel Ob', noted the emergence of supercooled (relative to the freezing temperature at the surface) water from under ice shelves in very different parts of the antarctic coast. However, as the author notes, because the freezing point of the water drops with increasing depth in the sea, water lying at depths corresponding to the location of the bottom of an ice shelf (170-250 m) will have a temperature higher than its freezing point at these depths.

Treshnikov's considerations can be interpreted in two ways: either bottom melting under the ice shelves can occur everywhere or melting is confined to those parts of the glacier not too far away from the margin. Apparently, the bulk of the water which goes under an ice shelf does not penetrate too far under the ice, so that it does not cool down to the freezing point at the pressure of the shelf base. Data [54] indicate that the water emerging from under an ice shelf has a temperature about 0.1 °C higher than the freezing temperature under the conditions obtaining beneath the glacier.

9.1.2. Modeling the Heat Exchange Under a Glacier as Flow around Plates and Slabs

The problem of the heat exchange at the lower surface of an ice shelf floating on seawater is a very complicated one. As mentioned above, this process is a combined exchange of heat and salt under conditions where the shelf bottom has a wide spectrum of roughness, with irregularities ranging from several centimeters to meters in size. In addition, the turbulence characteristics in the water near the shelf bottom are affected considerably not only by the interaction between the water and the solid boundary, but also by external factors, some of which enhance the convective mixing and some of which diminish it.

Experiments [130] on the melting of ice cylinders with water flowing around them did not reveal any differences in the dimensionless characteristics of the heat-exchange intensity, as compared with the heat exchange for cylinders made of nonmelting material. Thus we hypothesized that the heat exchange at the surface of an ice shelf or iceberg around which seawater flows is determined by the same dimensionless parameters as the heat exchange of immersed plates and slabs, which have been thoroughly studied for engineering purposes. The experimental data on the heat exchange when liquids or gases flow around a solid body are usually represented in engineering practice as a relationship of the type

$$Nu = \bar{K} Re^{\bar{n}} Pr^{\bar{m}},$$

where Nu is a dimensionless heat-transfer coefficient called the Nusselt number; Re is the dimensionless Reynolds number, characterizing the ratio between the inertial and viscosity forces in the fluid; and Pr is the dimensionless Prandtl number, characterizing the physical properties of the fluid flowing around the solid surface (the ratio between the kinematic viscosity and the thermal diffusivity). For water at a temperature close to the freezing point, Pr is about 13.7. The quantities $\bar{K}$, $\bar{n}$, and $\bar{m}$ in the formula are experimental coefficients. Let us introduce the following quantities for the water: L_w is a characteristic linear dimension defining the flow process and the heat exchange: λ_w and κ_w are the thermal conductivity and kinematic viscosity of the water; and $\alpha_w = q_w/(T_w - T_o)$ is the heat-transfer coefficient, that is, the coefficient of proportionality between the heat flux q_w for the liquid and the solid surface and the temperature differential between the liquid and the solid surface $(T_w - T_o)$. Now we can write

$$Nu = \frac{\alpha_w L_w}{\lambda_w} \quad \text{and} \quad Re = \frac{u_w L_w}{\kappa_w},$$

where u_w is the rate of flow of the water around the ice.

On the basis of an anlaysis of different experimental data on the heat exchange at the surface of a plate or slab, we arrived at the following dimensionless expression for evaluating the heat exchange at the lower surface of an ice shelf, at a point located a distance L_w from the place where the water first came in contact with the glacier (along a flow line of the water, see Figure 9.1), for a velocity u_w of water flow along the glacier surface [76]:

$$Nu = 0.037Re^{0.8}Pr^{0.33}. \qquad (9.1)$$

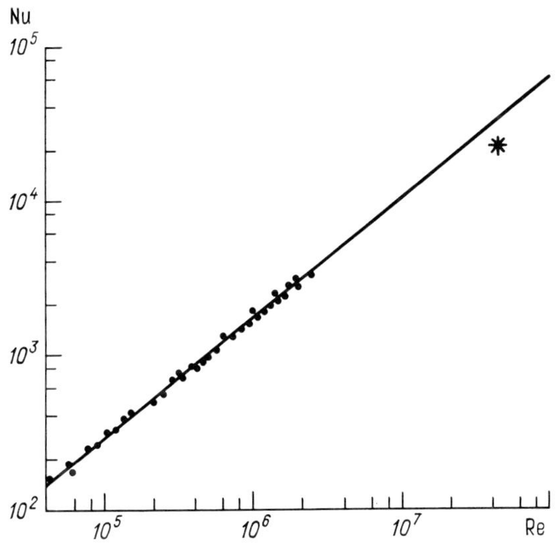

Fig. 9.3. Dimensionless heat-transfer coefficient Nu as function of Re, according to measurements of heat exchange for fluid flow around plates or slabs [59].

Asterisk indicates heat transfer at bottom of Ross Ice Shelf.

This expression is valid, strictly speaking, as a description of the turbulent heat exchange for flow around flat plates or slabs with characteristic lengths of the order of a few meters. Although it holds true (see Figure 9.3) over a wide range of Reynolds numbers, including the values typical of ice shelves (of the order of 10^8 to 10^9), it should be noted that the high Re in these experiments was due to the high (hundreds or thousands of times higher than for glaciers) velocities of fluid motion for characteristic linear dimensions that were low by this same factor [59].

The applicability of the proposed formula to calculate the heat exchange at the lower surface of an ice shelf was verified by us on the basis of data on the melting rate at the bottom of the Ross Ice Shelf in the vicinity of Little America Station. The temperature of the sea-water was assumed to be 0.2 °C above its freezing point, and the velocity at which the water flowed under the shelf and along its lower surface was taken to be from 10 to 15 cm/s. The point in the upper right-hand corner of Figure 9.3, which corresponds to these values, shows a good fit with the other experimental data and with the straight line (described by formula (9.1)) generalizing these data.

9.1.3. Melting at the Bottom of an Ice Shelf

The above approach makes it possible to plot the melting rate at the bottom of an ice shelf as a function of the distance from the shelf margin, the mean elevation of the water temperature above the freezing point, and the rate of water inflow under the ice shelf. Actually, let us assume that in a certain place on the ice shelf water flows under the shelf as a result of the currents there. Prior to going under the ice shelf, the water had a velocity u_w and a temperature T_w which were the same over the entire water layer. We further assume that as the water goes under the ice shelf the heat exchange between the water and the melting or freezing shelf bottom only affects the boundary water layer right next to the ice, which grows in thickness with an increase in the distance L_w traversed by the water under the ice shelf. It will also be assumed that this boundary layer is much thinner than the entire water layer between the lower surface of the ice shelf and the sea bottom, so that the effect of the sea bottom on the heat exchange between the water and the ice can be neglected.

In this case the heat exchange under the ice shelf conforms to formula (9.1), and the melting can be found from the equation describing the heat balance at the lower surface. By substituting into (9.1) the definitions of Nu and Re, we arrive at the following expression for the heat flux q_w from the water to the bottom of the ice shelf:

$$q_w = \alpha_w(T_w - T_0) = \bar{\bar{K}}(T_w - T_0)u_w^{0.8}L_w^{0.2}, \tag{9.2}$$

where $\bar{\bar{K}} = 0.037\ \lambda_w/\kappa^{0.8}\ Pr^{0.33}$ is a constant depending only on the physical parameters of the water and on the experimental coefficient $\bar{K} = 0.037$.

The rate of melting under the ice shelf is given by the equation describing the heat balance at the bottom of the shelf, which can be written as

$$q_w = \lambda\frac{T_0 - T_H}{H}\bar{\zeta}_{y=0} + v_0\rho g r; \tag{9.3}$$

where $\bar{\zeta}_{y=0}$ is the dimensionless temperature gradient for y=0, v_0 is the melting rate at the bottom of the shelf, and q_w is the heat flux from the water to the melting surface, as defined by equation (9.2).

By substituting (9.2) into (9.3), we get the following expression for the melting rate at the bottom of the ice shelf:

$$v_0 = \frac{1}{\rho g r}\left[\bar{\bar{K}}(T_w - T_0)u_w^{0.8}L_w^{0.2} - \lambda\frac{T_0 - T_H}{H}\bar{\zeta}_{y=0}\right]. \tag{9.4}$$

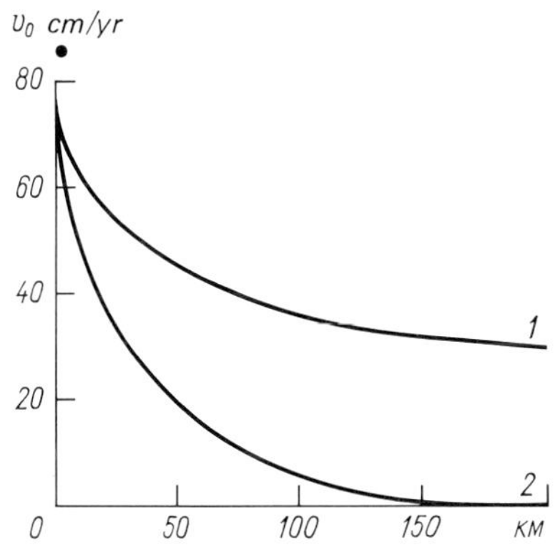

Fig. 9.4. Calculated melting rate at bottom of ice shelf as function of distance traversed by water under shelf. Calculation carried out (after Yu. V. Raikovskii) for water temperature elevation above freezing point ΔT_w of 0.3 °C and for rate of water inflow under shelf of 0.1 cm/s.

(1) ΔT_w constant over entire subglacial surface; (2) ΔT_w decreases with cooling of water layer 100 m thick.

Figure 9.4 shows a typical plot of the melting rate at the lower surface of an ice shelf as a function of the distance from the shelf margin. Here the temperature of the water inflow under the ice is assumed to be 0.3 °C above the freezing point at all points beneath the shelf, for a rate of inflow of about 0.1 cm/s. This figure shows that, given these assumptions, the entire lower surface of the ice shelf should be in a state of melting.

The idea proposed in [76] that the methods of the hydrodynamic theory of heat exchange be used to analyze melting under ice shelves and icebergs has subsequently been developed by various scientists. Lebedev [53], for instance, used expression (9.2) to analyze the heat exchange at the bottom of an iceberg.

Yin-Chao Jen and Chi-Tien [131] evaluated the heat exchange at the lower surface of a floating glacier with the aid of a similar dimensionless equation relating the Nusselt number to the Prandtl number and the Reynolds number. However, the equation they took was the one recommended by Schlichting [73] to calculate the heat transfer at a laminar boundary layer:

$$Nu = 0.332 Re^{0.5} Pr^{1/3}. \tag{9.5}$$

But the use of formula (9.5) to analyze the processes at the bottom of an ice shelf is hardly justified, since for the high Reynolds numbers typical of the process being considered expression (9.5) is not borne out by the experimental data.

On the other hand, it was correctly pointed out in [131] that the use of dimensionless expressions like (9.4) and (9.5) to calculate the hydrodynamic heat exchange at the bottom of a glacier is in need of further analysis. The problem is that, while the melting or freezing occurs, the velocity of the water along the water-ice interface has a vertical (normal) component which is not zero. For melting the vertical velocity is directed away from the interface and its magnitude is equal to the melting rate. In the case of freezing, the vertical velocity has an upward direction and a magnitude equal to the freezing rate. The effect of this motion is to weaken the heat exchange for melting and to enhance the heat exchange in the case of freezing.

Calculations carried out by us using the data of [73] indicate, however, that the intensity variation for heat exchange under real conditions of melting or freezing at the bottom of a glacier does not exceed 1 or 2% and is thus negligible. Such a variation would become significant only for a controlling temperature differential between the water and the melting surface of the order of 10 °C, that is, only if an iceberg manages to drift into warm waters.

For comparison, the dot in Figure 9.4 indicates the value of the melting rate at the bottom of the Ross Ice Shelf near Little America Station obtained by Crary on the basis of experimental data [93]. It is seen that the use of formula (9.2) gives a rate of bottom melting which is lower than the measured value. However, the difference between the calculated and measured values is surprisingly small, considering the complexity of the heat-exchange processes at the bottom of a glacier and the approximate nature of the model used.

For smaller differentials between the melting point of the ice and the temperature of the seawater the rate of bottom melting is correspondingly lower. As can be seen from Equation (9.3), the rate of bottom melting decreases linearly with a reduction of this temperature difference.

Studies based on an analysis of the velocities of motion of the Ross Ice Shelf indicate that the rate of bottom melting drops with increasing distance from the shelf margin much more markedly than shown by curve 1 in Figure 9.4. This sharp drop in the melting rate with increasing distance from the margin of the Ross Ice Shelf is ascribed in [32] to a decrease in the above-mentioned temperature differential further away from the shelf margin. Actually, many investigators have noted a cooling of the water emerging from under an ice shelf. However, this emerging water usually has a temperature about 0.1 °C higher than its freezing point at the depths corresponding to the lower surface of an ice shelf.

These findings were interpreted as follows: (1) it was assumed that bottom melting takes place almost everywhere under the floating part of the ice shelf; (2) melting was assumed to take place only in parts of the shelf not too far away from its margin. Apparently, the water flowing into the region under the ice shelf does not penetrate too far under the shelf and thus does not become cooled down to the freezing point at the pressure of the level of the shelf bottom. Later measurements of the temperature of the water under the rear (seaward) part of the Ross Ice Shelf (see Section 9.3 below) indicate, on the other hand, that there may well be a layer of relatively warm water under the water layer that has been cooled to the freezing point by the heat exchange with the melting shelf. The thickness of the water layer which was cooled down to a temperature close to the freezing point was found to be less than 100 m.

According to the data of [32], for a mean rate of inflow of water under the Ross Ice Shelf of 20 cm/s and an initial 0.6 °C elevation of the temperature of this water above its freezing point, the amount of heat which can be given up by a layer of this water 100 m thick as it is cooled down to the freezing point was found to be 1.99 kW per cm of the margin of the ice shelf. This value is close to the value of 1.99-3.98 kW per cm of shoreline given by Rubin [120] in his calculations.

This amount of heat is equal to that needed to melt the bottom of the ice shelf in a coastal strip 50 km wide, if the melting rate varies with the distance from the margin as shown by curve 1 in Figure 9.4.

Consequently, the heat lost by the water as it moves under an ice shelf under conditions of bottom melting is commensurate with the total heat reserves of the water flowing into the region under the ice. This causes a sizable reduction of the temperature differential in the upper, cooling layer of water as this water moves under the shelf. Thus, the melting of the bottom of the shelf will slow down more rapidly than indicated by formulas (9.3) and (9.4) for a different $T_w - T_0$ which does not vary with the distance from the margin. For example, if the melting rate decreases linearly with increasing distance from the margin, then the melting rate will already be close to zero 100 or 150 km from the shelf margin.

In view of the foregoing, we calculated the variation in melting rate at various distances from the margin of an ice shelf using formula (9.3), assuming that the water temperature in a layer 100 m thick decreases as the water penetrates beneath the ice. This temperaure drop was determined by the change in the enthalpy of the water as its heat was used for melting. Curve 2 in Figure 9.4 depicts the variation of the melting rate in this case. According to this curve, at a distance of about 200 km the rate of bottom melting is already close to zero.

After a study of currents and temperatures in McMurdo Sound near Ross Island, Gilmour [96, 98] concluded that the heat arriving under the ice shelf is distributed nonuniformly over the year. The bulk of the heat, out of the total annual input of heat to the region under the shelf, arrives during the first four months of the year. It can thus be assumed that the bottom melting proceeds nonuniformly over the year as well, being confined mainly to the first quarter.

In [48] it was concluded that in the northeastern part of the Lazarev Ice Shelf the melting rates should, judging by the turbulent mixing, be of the same order of magnitude as those in the marginal part of the Ross Ice Shelf. This result agrees well with the data of [54], which give water temperatures more than 0.5 °C above the freezing point near the eastern edge of the Lazarev Ice Shelf at depths close to the shelf bottom.

The situation is apparently similar near the margin of the Shackleton Ice Shelf, where the temperature of the water entering the region under the ice was found to be substantially higher than the freezing point [32].

Now let us estimate the amount of ice melting from the lower surface for all the ice shelves of Antarctica. To do this, it will be convenient to assume that the melting rate varies with the distance traversed by the water under the ice approximately as shown by curve 2 in Figure 9.4.

As previously [32], it must be kept in mind that only rarely does the water enter the region under the ice along the normal to the seaward edge of the shelf. If the approach of the water to the region beneath the shelf is independent and random, then the angle between the water inflow and the coastline (shelf margin) should be close to 45°, but this also seems somewhat high to us. The action of the Western Coastal Current is such as to make this angle lower, and, in our opinion, we will not be too far off if we assume that it is close to 30°.

In this case the bottom melting of the ice shelves, averaged over Antarctica, takes place essentially in a 'coastal' strip about 150 km wide. Within this region the melting decreases from around 50-60 cm/yr to values close to zero, giving an average melting rate of about 30 cm/yr for the coastal strip.

Given these assumptions, we obtain a figure of about 300 km^3 per year for the amount of ice melting from the lower surfaces of the antarctic ice shelves [32].

9.2. Freezing at the Bottom of an Ice Shelf

9.2.1. The Englacial Temperature Field as an Indicator of Freezing

As mentioned earlier, the temperature field in the interior of a glacier constitutes a sensitive instrument for determining the speed and direction of the vertical ice movement, and thus, for some glacier models, the direction and intensity of melting (ablation) or freezing (accumulation) at the surface of the glacier. The foundation for the use of this method was actually already laid in [29], which presented some possible temperature distributions in steady-state glaciers.

Previously, with regard to ice caps and ice shelves, we considered only the regime of motion-caused cooling of the ice due to its vertical downward movement, that is, we considered a scenario of accumulation at the top and melting at the bottom. If, on the other hand, there is freezing at the bottom, we should expect to have an S-shaped temperature field (see Figure 4.6) for accumulation at the top and a distribution corresponding to warming of the ice (see Figures 4.4 and 4.5). Such temperature distributions have actually been measured in the interior of glaciers.

For instance, Figure 9.5 shows the temperature distribution in the interior of the Amery Ice Shelf, obtained from measurements in a borehole drilled at the point shown on the map in Figure 9.6 [107]. The plot in Figure 9.5 shows the measured temperature field to be S-shaped, corresponding to accumulation at the upper surface and freezing at the shelf base. Examination of a core extracted from the borehole in which the temperature was measured verified this conclusion. The ratio of oxygen isotopes ^{16}O and ^{18}O for the upper 270 m of the ice shelf was found to correspond to ice formed by accumulation of precipitation at the shelf surface and also to glacier ice brought in from Lambert Glacier and from the central regions of the Antarctic Ice Sheet. The lower part of the ice shelf, however, down to a depth of 410 m beneath the surface, is made up of ice formed by the freezing of seawater at the shelf bottom. The content of the ^{18}O isotope in this ice is close to that of seawater. These findings were also verified by measurements

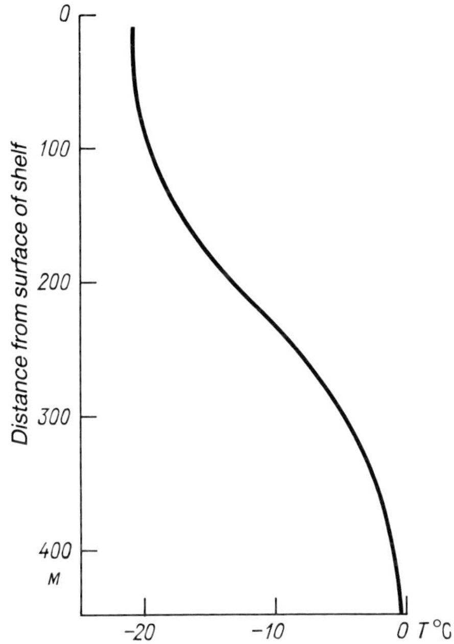

Fig. 9.5. Temperature distribution in interior of Amery Ice Shelf (S-shaped distribution corresponding to freezing at shelf base).

of the electrical conductivity of the core ice [107].

The freezing rate at the bottom of the Amery Ice Shelf, calculated for the best fit of the S-shaped temperature curve with the observed distribution (Figure 9.5), was about 0.4 m/yr. This value is close to the rate (0.3 m/yr) obtained in [107] for a layer of bottom-frozen ice 158 m thick and a time of bottom freezing which was also known (determined by the distance from the borehole to the place where Lambert Glacier becomes the floating Amery Ice Shelf and by the rate of horizontal movement of this glacier).

Figure 9.7 shows the temperature distribution measured in the interior of the floating tongue of Koettlits Glacier [135]. As noted in [135], this glacier (Fig. 9.8) has attracted the attention of many investigators, ever since Debenham published his account of vestiges of marine fauna at its surface. The tongue, which is an extension of Koettlits Glacier, is bounded on the west by the coast of Victoria Land and on the east by Brown Peninsula, while its northern part abuts upon a number of islands (the Dailey Islands). It was precisely here, among the Dailey Islands and just to the south of them, that Debenham and others found the remains of fish, corals, and sponges. The largest quantity of remains of marine organisms was discovered near the eastern island of the Dailey group. The depth of the borehole at this site was 17.5 m, for a total ice thickness in the vicinity of about 40 m.

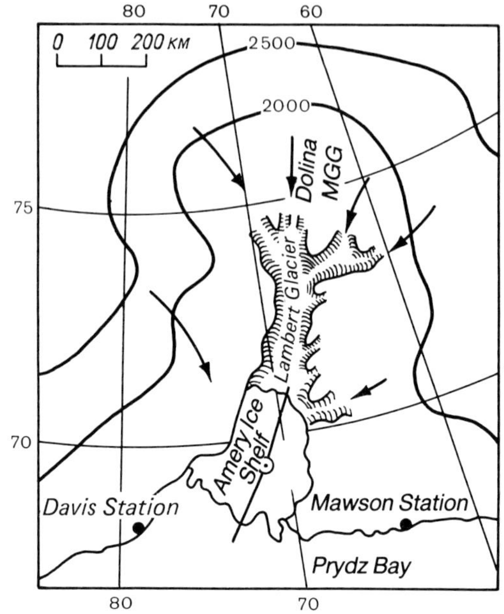

Fig. 9.6. Amery Ice Shelf (Antarctica).

Dot indicates site of borehole in which temperature measurements were made (Figure 9.5).

The white circles in Figure 9.7 show temperatures measured in the upper 12 m of ice, which are affected by the annual temperature fluctuations, while the black circles show the temperatures below the active layer. The mean values, over many years, of the surface temperature and the temperature at the 'bottom' of the glacier are given here.

The straight lines joining points corresponding to the mean many-year values of the temperature at the upper and lower surfaces of the glacier define the temperature distribution in a glacier without any vertical movement of the ice and with temperatures at its surfaces which are constant in time.

It is seen from Figure 9.7 that temperatures measured at depths of more than 12 m beneath the surface, that is, at depths where there are no seasonal temperature fluctuations, are about 4 °C higher than they would be if there were no vertical movement of the ice in the glacier.

There is only one possible explanation of the discrepancy noted in [135]: there must be a vertical upward movement of the ice in the glacier, which causes warming of the glacier interior. If the glacier thickness is assumed to be constant in time, as indicated by a comparison of descriptions of the glacier made 40 years ago and at present, then a necessary condition for such a vertical movement of the ice is ablation at the surface of the glacier and freezing at the bottom.

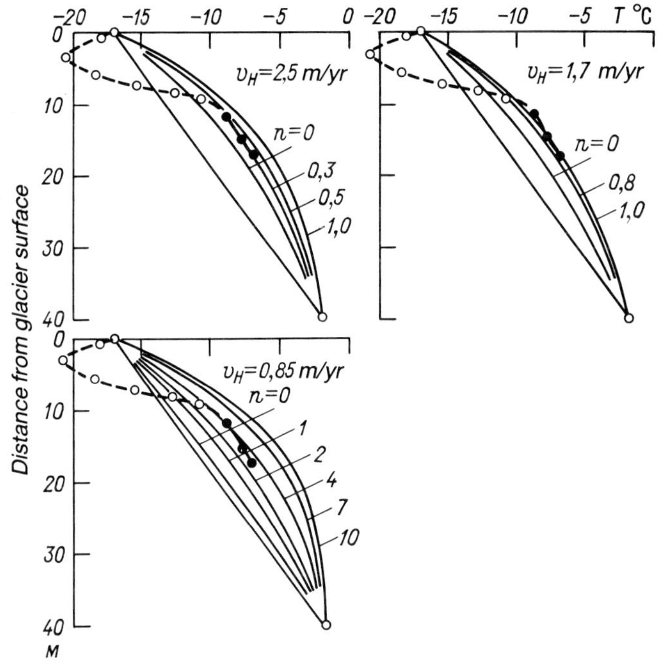

Fig. 9.7. Measured temperature distribution and theoretical distribution for bottom freezing. Koettlits Ice Tongue.

The series of curves shown in Figure 9.7 were calculated using the method described in Chapter 4 for the case of vertical upward movement of the ice in the glacier (warming of the ice due to its movement). The curves were plotted for annual rates of ablation at the surface of 0.85, 1.7, and 2.5 m/yr. Several curves determining different rates of bottom freezing pertain to each of these ablation rates. The value of $\bar{v}_0$ for each curve gives the ratio of the bottom-freezing rate to the surface-ablation rate. The value $\bar{v}_0 = 0$ indicates the absence of freezing at the base, and $\bar{v}_0 = 1$ means that the bottom freezing and the surface ablation have the same rate.

These graphs show that, depending on the ratio between the rates of surface ablation and bottom freezing, the observed temperature values can be explained as different values of the ablation and freezing rates. An analysis showed a surface-ablation rate of 1.7 m/yr and a bottom-freezing rate of 1.0 m/yr to be the most probable values, in view of the data for the Koettlits Ice Tongue given above.

9.2.2. Freezing Accompanying a Reduced Salinity at the Ice-Seawater Interface.

Consideration of the heat-balance equation in form (9.3) reveals a possible mechanism of freezing at the bottom of an ice shelf. Freezing will occur when the left-hand side of (9.3), which gives the heat influx

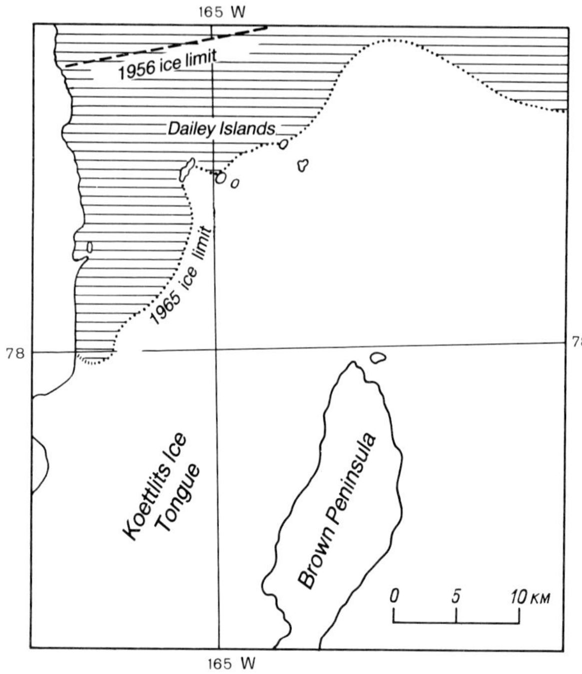

Fig. 9.8. Floating tongue of Koettlits Glacier (Antarctica).

to the ice-seawater interface from below, is either equal to zero or is less than the heat flow into the ice via thermal conduction. In order to satisfy this condition, the water temperature has to be very close to the freezing point. Then the freezing rate will be

$$v_0 = \frac{\lambda}{\rho g r} \frac{T_0 - T_H}{H} \bar{\zeta}_{y=0}. \tag{9.6}$$

An analysis of the temperature gradients at the lower surfaces of real ice shelves indicates, however, that the freezing rate according to (9.6) cannot under actual conditions be higher than 3 to 5 cm/yr. Therefore, we concluded [32] that the main mechanism behind the increase in freezing rate from several cm/yr to some m/yr is the arrival of fresh water at the lower surface of the shelf, which is washed by the very cold, salty seawater. A sharp enhancement of the growth of the ice can occur under such conditions because 'new ice' forms, due to the supercooling of the freshened ice-adjacent water layer beneath the ice shelf. This possibility was pointed out by Zubov [38], Crary [90,91], and Untersteiner [125].

The mechanism of the formation of such ice is, however, complicated and has not been studied much. For our purposes, following [32], we simplified the problem by assuming that the layer of fresh or slightly saline water does not mix with the saltier seawater. As a first approximation, we also assume that the heat exchange at the water-ice interface takes place as at a solid wall.

A calculation of the heat transfer at the boundary between a water layer of reduced salinity and seawater, based on Equation (9.1) and assuming that the water layers do not mix, indicated the following: even if the salinity of the less salty layer is about 15 ‰ (freezing temperature -0.7 °C) and the seawater salinity is about 35 ‰ (freezing temperature around -1.8 °C), the less saline water can freeze at a rate as high as 1 m/yr, provided the flow velocity of the seawater under the ice tongue is of the order of 10 cm/s.

The layer of less saline water at the bottom of an ice shelf forms because fresh water reaches the seaward part of the shelf from the inner continental regions, as a result of the above-described melting at the bottom of an ice sheet, together with the arrival of meltwater from the upper surface of the glacier during the ablation period.

Fresh water was detected under the tongue of Koettlits Glacier during observations at the western edge of the Dailey Islands. The layer of fresh or slightly saline water apparently formed in this case due to the penetration of meltwater from the surface via tiny tidally formed fissures or other cracks in the glacier.

On the basis of the model used, it also became clear why intensive freezing was detected under the Amery Ice Shelf in Prydz Bay and under the seaward extension of the Lambert Outlet Glacier (see Figure 9.6). Lambert Glacier constitutes the main outlet channel not only of ice, but also of subglacial fresh water, from the vast continental glacial basin, centered on Dolina MGG. This basin is located, to a considerable extent, in a region of melting at the glacier bed. Lambert Outlet Glacier, at the bottom of which melting takes place, extends inland to the bottom-melting zone of the central part of the glacier basin, and thus it forms a bridge between the basin of subglacial runoff from the ice-sheet center and the sea. The water thus flows seaward along the bed of Lambert Glacier and under the Amery Ice Shelf.

Taking into account the area of the region of bottom melting in the central part of this basin, it is to be expected that about 3 to 3.5 km^3 of water per year will run off along the bed of Lambert Glacier. An additional 1 km^3 of water per year should come from the bottom melting at the bed of Lambert Outlet Glacier. This value was obtained by assuming an area of 24,000 km^2 for Lambert Glacier and its tributaries.

Consequently, about 4 km^3 of fresh water per year runs off along the bed of Lambert Glacier and under the ice shelf. Thus the total amount of ice which can form at the bottom of the glacier by the freezing of fresh water is close to 7 km^3 per year. This is about one third of the quantity of ice formed from precipitation onto the surface of the Amery Ice Shelf in its seaward and central regions.

The growth of young shelf ice, formed from the old (perennial) shore ice, presents a somewhat different picture. Following [32], we note that the conditions for a prolonged existence of shore ice are satisfied in bays, inlets, and straits, wherever there is protection from the destructive action of the sea. In places where such old shore ice comes in contact with the large ice shelf, rapid freezing at the bottom of the shelf may occur. These conditions will be produced where from beneath a massive ice shelf there is an outflow of water which has cooled down considerably because of the bottom melting under the shelf.

Treshnikov, Kruchinin, and others have shown that this water can be very close to its freezing point at the depths of about 200 m corresponding to the bottom of a large ice shelf. This water runoff from under the ice shelf turns out to be supercooled when it arrives at the bottom surface of the old shore ice. The amount of supercooling, according to Treshnikov, is around 0.1 °C and it leads to an intensive formation of new ice, which freezes onto the bottom of the shore ice. Such bottom freezing, stimulating the growth of the young shore ice, is apparently quite widespread [48].

In view of the foregoing, it would be interesting to employ thermophysical methods to trace the evolution of young ice shelves, contiguous to large 'old' ice shelves.

It should be kept in mind that the growth process at the lower boundary of a young ice shelf is intrinsically an unsteady-state process. As the young ice shelf becomes thicker, the effect of the supercooling of the water emerging from under the old ice shelf is diminished, until, when the 'young' shelf becomes as thick as the 'old' shelf, this process, and thus the formation of ice at the bottom, must cease. This process is apparently quite common.

9.2.3. Bottom Freezing in the Seaward Part of a Large Ice Shelf

It was mentioned above that the melting at the lower surface of a large ice shelf can extend very far, and in [32] it was shown that the melting region for the Ross Ice Shelf is about 200 km wide (Figure 9.9). The situation is apparently similar in the seaward parts of the Ronne and Filchner Ice Shelves; the same can be said of the massive ice shelves which existed in Europe and North America during the Quaternary.

The possibility of bottom freezing in the seaward parts of large ice shelves, and the Ross Ice Shelf in particular, has been mooted and discussed for years. Zumberge [138] assumed that bottom freezing occurs in the seaward parts of the Ross Ice Shelf, on the basis of an analysis of the glacier movement between Ross Island and Roosevelt Island; he obtained a mean rate of bottom freezing of 12 cm/yr, at points located south of a grounding line drawn 150 km from the shelf barrier. Gordon (see [102]) assumed that the bottom-adjacent water of high salinity observed in the Ross Sea originated beneath the Ross Ice Shelf, where brine is produced as a result of the freezing of seawater onto the shelf bottom.

New data on the salinity and temperature of the seawater under the ice shelf at Camp J-9 [34, 97, 102] reveal the presence of a cold isothermal water layer (about 80 m thick) of high salinity in the upper part of the subglacial sea close to the water-ice interface. The temperature of this water layer is close to the freezing point.

The results of a thermophysical study of the Ross Ice Shelf, to be presented in the next section, indicate that in the region studied (near Camp J-9, see Figure 9.9) heat exchange between the shelf bottom and the sea is completely absent, and bottom freezing occurs at a rate governed by the upward transport of heat through the shelf via thermal conduction. These results imply the absence of any significant heat exchange in the sea in the horizontal direction, and they also suggest that the temperature profile in the subglacial water is highly invariable in the direction of the current near Camp J-9.

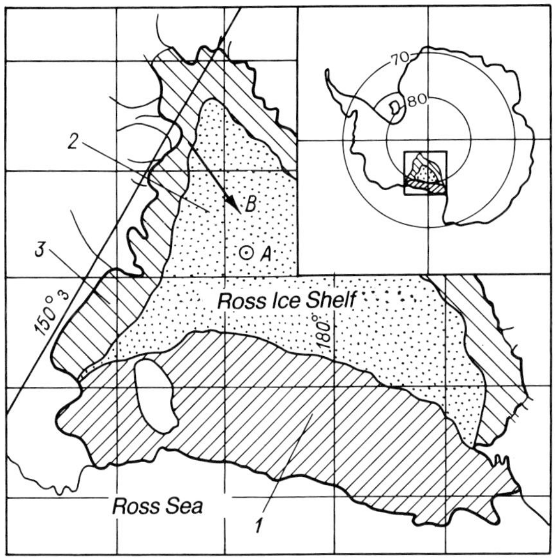

Fig. 9.9. Map of Antarctica and Ross Ice Shelf.

(A) Site of drilling (Camp J-9); (B) direction of glacier movement; (1) melting region; (2) region of bottom freezing caused by upward heat transport through shelf; (3) region of bottom freezing enhanced by arrival of fresh water under shelf.

An important result of the studies in the vicinity of Camp J-9 was verification of the enhancement of the bottom freezing due to the arrival of fresh water at the bottom of the shelf (see Section 9.3 below). By verifying the model of bottom freezing of the Amery Ice Shelf and the Koettlits Ice Tongue, this experiment once again raises the question of the role of the arrival of fresh water formed by the bottom melting at the beds of ice sheets adjacent to ice shelves or else nourishing these shelves via bottom freezing.

In [32] the role of the bottom freezing of ice shelves was analyzed by considering the conditions at the western seaward extremity of the Ross Ice Shelf. The region of runoff from the middle of the Antarctic Ice Sheet to the western part of the Ross Ice Shelf was found to have an area of about 650,000 km^2. Bottom melting is typical of this region, and the total amount of fresh water produced in the drainage system has been estimated to be of the order of 3.3 km^3/yr. This is just the water formed in the central part of the ice sheet, where the variations in thickness are small and the glacial runoff is undifferentiated.

This part of the antarctic ice plateau is cut off from the Ross Ice Shelf by mountains, cut through by massive outlet glaciers. Along with the drainage, these glaciers also produce large quantities of fresh water, due to the friction at their beds as they move rapidly along. Such outlet glaciers as Scott, Amundsen, Liv, Nimrod, Milton, and

Shackleton produce an additional 0.4 to 0.5 km^3/yr of water. About 0.8 km^3/yr of water forms under Byrd Glacier, and about 0.6 km^3/yr under Beardmore Glacier.

Thus we can draw the following conclusions concerning freezing and melting at the bottom of the Ross Ice Shelf, conclusions which are in general applicable to other large glaciers as well. There is a region of melting in the northern part of the glacier which extends approximately 200 km from the barrier (region 1 in Figure 9.9). Active heat exchange takes place in the northern part of this region. The heat exchange becomes less and less as we progress southward in the region, until the heat flux is just sufficient to compensate for the heat losses from the water through the ice. Further south, bottom freezing occurs, at a rate of about 2 cm/yr (region 2 in Figure 9.9). Here there is practically no heat exchange at the ice-water interface, and the bottom freezing is governed by the conductive transport of heat through the ice shelf.

Still further south is a zone of intensive bottom freezing, initiated by the arrival of fresh water at the ice-seawater interface, the temperature of this incoming water being close to its freezing point.

9.3. Thermophysical Studies of Melting and Freezing at the Bottom of the Ross Ice Shelf

As an example of the complex thermophysical approach to solving a specific problem, let us consider the studies of melting and freezing at the bottom of the Ross Ice Shelf, carried out as a part of the international 'Ross Ice Shelf Project' [88], financed by the National Science Foundation (U.S.A.). This work was based at Camp J-9, which is located in the central part of the Ross Ice Shelf, about 450 km from the northern margin (Figure 9.9).

9.3.1. The Thermal Regime of the Ross Sea Under the Ice Shelf

The thermal regime of the sea under the Ross Ice Shelf was investigated on the basis of the temperature distributions. The latter were measured with the aid of a Hewlett-Packard quartz thermometer, lowered into the subglacial sea through a borehole drilled 420 m through the ice at Camp J-9 [34, 88].

The temperature sensor, suspended on a coaxial cable, was moved up and down, slowly but continuously, through the water under the ice shelf at a velocity of about 0.1 m/s. A frequency meter provided a frequency (and temperature) comparison every 10 seconds, to an accuracy of 0.001 °C. Thus, the measured temperature was the mean temperature of the water layer through which the sensor moved in 10 s. The thickness of this layer did not exceed 1 m. The thermometer readings were displayed on the frequency meter readout and recorded in tables, which were then corrected on the basis of a check of the zero setting of the meter.

Temperature measurements were carried out in the water under the ice shelf on 17 and 22 December 1978. The results have been plotted in Figure 9.10. Note that 230 points were used to obtain the temperature distribution for 17 December, and more than 100 points for 22 December. As the graph shows, the temperature distribution in the sea under the ice shelf is very complicated and it varies with time. However, there are certain details which remain invariable and are of great importance;

these have also been discovered on other distributions [34]. From Figure 9.10 it is possible to discern the following important regularity: there are two definite water layers under the ice shelf, one with a temperature close to the freezing point and one, situated lower down, which is relatively warm. The layer of cold water is about 70 m thick, and the lower, warmer, layer is about 80 m thick. The temperature of the lower layer is approximately 0.5 °C above the freezing point of sea-water at the corresponding pressure. The cold and warmer water layers are separated by a transition layer, in which the temperature variation is nearly linear.

As already mentioned, a comparison of the curves in Figure 9.10 suggests that the water temperature beneath the ice shelf fluctuates with time. In order to verify this, the thermometer was left at one of the levels (43 m from the sea bottom) for 5 hr, from 0400 to 0900 on 22 December. The temperature was measured every 10 seconds. These measurements showed that fluctuations actually do exist, but their amplitude does not exceed 0.01 °C (Figure 9.11).

The discovery of a warmer layer of water under the ice shelf was quite unexpected, and it disproved the assumption that the bottom-adjacent water under an ice shelf is supercooled. The existence of two almost isothermal layers with an intermediate transition layer having a high temperature gradient indicates that the conditions in the transition layer are such as to prevent the vertical natural convection; for this to be the case, the salinity of the water must increase sharply, from the shelf bottom down to the sea bottom [34].

The existence of these clearly defined water layers in the sea under the ice shelf served as an incentive for studying the currents under the shelf. However, when conventional current meters were lowered beneath the ice shelf, it was found that the current speeds in the subglacial sea were at the lower sensitivity limit of the instruments.

To aid in the detection of these low current velocities, we devised a special instrument which was to be attached to a television camera lowered under the ice. A horizontal disk with a compass was placed in the field of view of the TV camera, about 50 cm from its objective. At the center of the disk was a small nozzle, which emitted a stream of colored liquid from a special container behind the disk. The liquid stream and its velocity of motion along the disk were easily viewed on the telecamera screen, against the background of the disk and compass. Video recordings of the behavior of the liquid relative to the disk and compass were obtained for some tens of points at verious levels between the bottom of the ice shelf and the sea bed.

Currents were indeed detected under the ice shelf, but their velocities were low (several centimeters per second) and considerably variable in time (perhaps because of tidal effects). At the same time, the subglacial basin of the Ross Sea in the vicinity of Camp J-9 is by no means a stagnant zone, and there is apparently a lively water exchange with the open sea to the north. There seems to be a general tendency for the currents to become aligned with valleys on the sea floor. Thus, in the vicinity of Camp J-9, the prevailing currents should have directions of 290-320° and 120-160°, which is close to the preliminary observational data.

If there were no heat exchange due to horizontal transfer, as these studies seemed to indicate, then in the central regions of the ice shelf the entire difference between the geothermal flux and the heat flow up

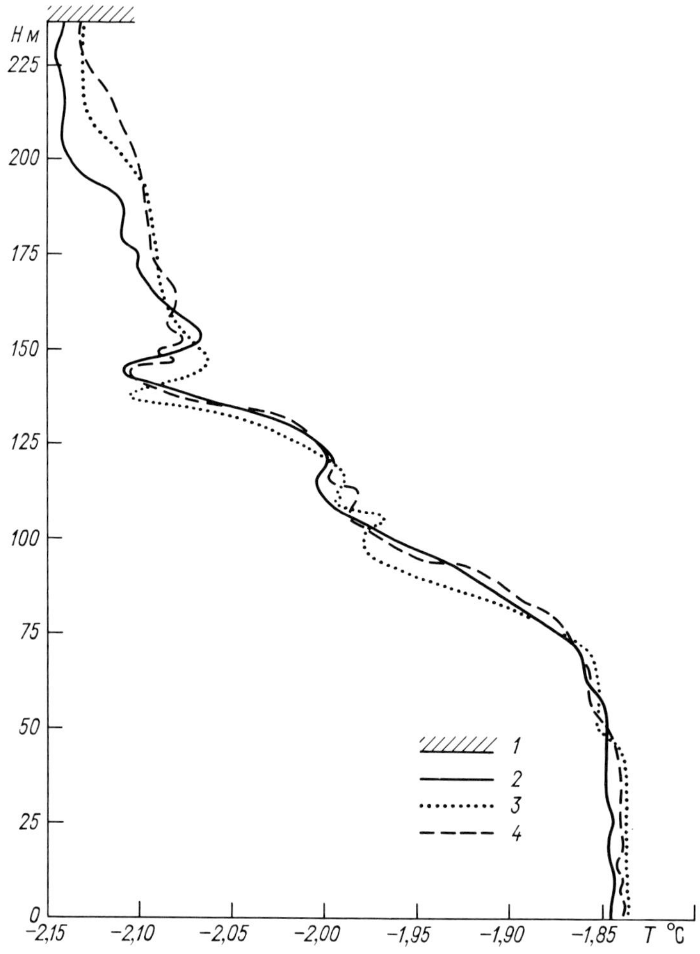

Fig. 9.10. Temperature distribution in seawater beneath Ross Ice Shelf [34].

(1) shelf bottom; (2) 16 December 1978 (measurement interval 1.55 m); (3) 17 December 1978 (interval 1.20 m); (4) 21 December 1978 (interval 1.50 m).

into the ice would go for either melting or freezing at the shelf bottom. The geothermal heat flux in the vicinity of Camp J-9 is, according to the data of T. Foster (private communication), close to 46 mW/m^2. The heat flux up into the ice, for a shelf thickness of about 400 m (as near Camp J-9), is close to 170 mW/m^2. Consequently, at the bottom of the ice shelf the flux of heat into the ice exceeds the incident geothermal flux and bottom freezing should take place at a rate of about

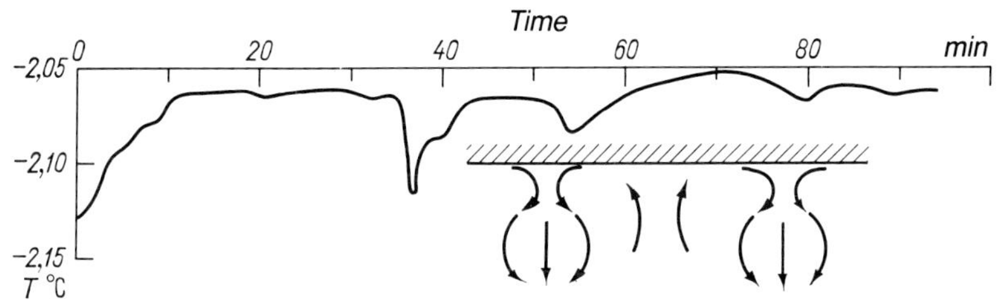

Fig. 9.11. Temperature fluctuations of seawater under ice shelf and a possible model of this phenomenon.

1.5 cm/yr. However, such a conclusion cannot be drawn just from the data presented, since the absence of horizontal heat exchange in the seawater has not been proven.

Back in Figure 2.14 (1) we gave a plot of the temperature distribution over the thickness of the Ross Ice Shelf, measured in a borehole near Camp J-9 by Clough and Hansen [88]. The nature of this distribution implies that conditions of cooling exist over most of the ice, that is, a downward vertical movement of ice, and thus melting at the bottom of the shelf. Estimates indicate, however, that even in the presence of bottom freezing at a rate of the order of 1.5 cm/yr the temperature curve, shaped mainly by the vertically downward ice movement due to longitudinal outflow or due to accumulation at the shelf surface to the tune of tens of centimeters per year, remains close to the curve shown in Figure 2.14. Consequently, the question of whether melting or freezing occurs at the bottom of the shelf had to be answered on the basis of direct observations.

9.3.2. Detection of Bottom Freezing by Thermal Core Drilling

Core drilling through the entire thickness of the ice shelf, under conditions of bottom freezing, enables us to find the boundary between the ice of atmospheric origin (formed by the transformation of precipitation falling onto the surface of the ice shelf) and the layer of ice formed by the freezing of seawater. The first such ice layer formed by bottom freezing on an ice shelf was discovered by core drilling of the Amery Ice Shelf [107]. In this case, if the time during which the bottom part of the ice shelf near the borehole was in contact with the sea is known, then it becomes possible to estimate the mean rate of bottom freezing during this time.

In view of the foregoing, in December 1978 we employed the anti-freeze-thermal method [134] with an annular electrical heater to drill a borehole through the entire 416-meter thickenss of the ice shelf, and then extracted ice samples (cores) right down to the shelf bottom. The drilling was carried out at the American Camp J-9, located in the central part of the Ross Ice Shelf, about 500 km from the open sea (82°22'S, 168°37'W) (Figure 9.9).

The ice shelf was found to be made up of two layers. The upper layer is composed of the usual fresh-water glacial ice, originating from the antarctic precipitation; it is 410 m thick. The lower layer, only

6 m thick, is very different from the upper layer; it extends to the bottom of the ice shelf. Unlike the upper layer, this layer is composed of somewhat saline ice, with a salinity varying from 2.2 $^{o}/oo$ at the boundary with the fresh-water ice to 4.5 $^{o}/oo$ at a distance of 0.5 m from the shelf bottom. The saline ice, which has in it clearly evident cells of seawater brine, has all the earmarks of ice formed by the freezing of seawater and thus can be called sea ice. This ice is made up of vertical crystals about 5 mm in diameter, arranged not randomly but rather in definite parallel rows of vertical platelets, oriented precisely in one direction along the horizontal.

The ice at the upper boundary of the sea ice is porous, slushy, and water-saturated. The layer of this transition ice at the boundary between the upper and lower layers is 5 to 7 cm thick. When the drill reached this transition layer, the level of the water in the borehole suddenly rose sharply by more than 20 m, reaching sea level. This showed that the transition layer was in good hydraulic contact with the sea under the layer of sea ice, and thus that the six-meter sea-ice layer has in it vertical cracks ensuring this contact.

9.3.3. Structure of Frozen-on Base of Ice Shelf and Freezing Rate

In Chapter 1 we described some of the difficulties involved in extracting undamaged cores from the very bottom of a glacier. In view of these difficulties, cores from the base regions of glaciers have not as yet been described in the literature, and thus the nature of the lower surfaces of glaciers has remained unknown. It would be especially interesting and important to extract a core containing a portion of the 'bottom' of an ice shelf. The processes governing the heat and mass exchange at the undersurface of an ice shelf can be very diverse and very intensive, making it almost impossible to predict the nature of the glacier undersurface, which may turn out to be smooth, solid, and monolithic, or loose, spongy, and brine-saturated, or even fluffy and covered with ice needles growing in different directions (crystals of new ice).

As mentioned in Chapter 1, we succeeded in bringing to the surface an undamaged core sample, the lower surface of which was a piece of the bottom of the Ross Ice Shelf. Examination of this part of the core led to the first description of the base of the ice shelf. The undersurface of the Ross Ice Shelf was found to be in a state of bottom freezing, which manifested itself as a pronounced 'pimply' wafflelike surface. The tiny protrusions, the 'pimples', on the undersurface are the rounded ends of vertical ice crystals about 5 mm in diamter: these were discussed previously. The entire lower 6-m layer of frozen-on ice is made up of such crystals. Figure 9.12 shows the generalized structure of the base of the ice shelf, deduced from a study of the core.

It is seen that the tiny protrusions, the 'pimples', on the shelf base are by no means randomly arranged; rather they form horizontally parallel rows. This same surprising orientation of the crystals, the ends of which pimple the shelf base, extends over the entire thickness of the frozen-on layer.

The origin of this unexpected orientation of the crystals of the undersurface has not yet been explained satisfactorily. Apparently, the forming ice is influenced by sea currents that have a definite prevailing direction. If this is indeed the case, then the observed

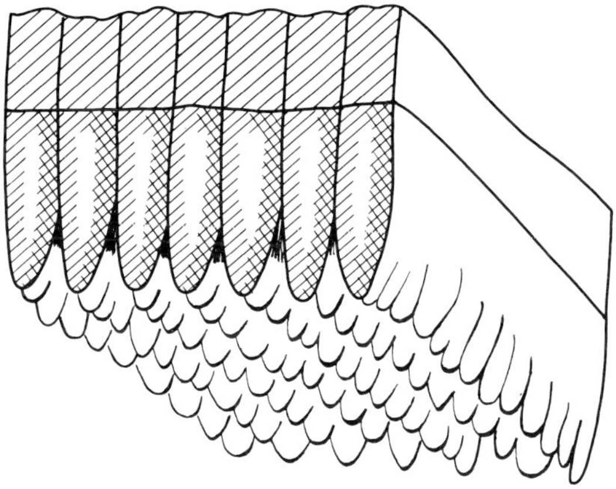

Fig. 9.12. Generalized diagram of undersurface of Ross Ice Shelf and structure of bottom-adjacent ice layer.

orientation indicates the presence of prevailing sea currents beneath the ice shelf, both in the vicinity of the borehole and along the line of movement of the shelf above the borehole. This crystal orientation could be used for a remote (from the shelf surface) survey of the direction of the currents under the Ross Ice Shelf.

The sizes of the bottom crystals and the distances between the vertical platelets making up the sea ice of the Ross Ice Shelf are nearly 10 times larger than the analogous characteristics of the frozen-on sea ice of the northern seas. This implies that the rate of growth of the ice on the base of the Ross Ice Shelf is many times less than the rate of growth of ordinary sea ice.

As Figure 9.12 shows, some of the bottom sea ice differs markedly from the rest of the sea-ice layer. A definite horizontal boundary sets the lower two centimeters apart from the rest of the sea ice; it is the only horizontal boundary observable in the six-meter layer of sea ice. The presence of this boundary indicates sudden significant changes in the conditions of ice formation in this part of the shelf. These changes may well have been due to a previous drilling through the ice shelf with a jet drill, carried out about 200 m from the borehole where the bottom core was extracted. If the boundary indeed originated from the jet drilling, which was in December 1977, then the rate of bottom freezing was 2 cm/yr.

The rate of bottom freezing over a long period of time can be estimated from the thickness of the sea-ice layer and the rate of movement of the Ross Ice Shelf. Measurements of the velocity and direction of ice movement along a flow line from the floating line to Camp J-9 give the distance traversed by the ice column from the time it started floating to the time it reached Camp J-9. This distance was found to be about 200 km, for a mean velocity of ice movement of 300 to 400 m/yr. Consequently, the bottom part of the ice column (core) at Camp J-9 was in seawater for about 400 to 600 yr and the mean annual rate of bottom freezing was 1 to 1.3 cm/yr. This value is close to the present annual rate of freezing at Camp J-9, deduced above from the core analysis.

9.3.4. Direct Monitoring of Melting-Freezing Processes at the Base of the Ross Ice Shelf

These processes can be monitored directly by measuring the velocity of motion of the melting-freezing front relative to reference marks securely attached to the ice mass close to its undersurface. This velocity can be measured by repeated determinations of the position of the front relative to the fixed reference marks, which may be, for example, frozen into the lower portion of the ice shelf.

(1) A vertical string of electrical thermometers can be so frozen into the lower part of the ice shelf that some of the thermometers are at various distances inside the shelf ice (are frozen into it), while others are in the seawater beneath the melting-freezing front, also at various distances from the front. Remote temperature measurements, made at the surface of the shelf, indicate the moment of passage of the front past each of the sensors, and thus provide a determination of the direction and mean velocity of the front between two sensors. A disadvantage of this method, however, is that the front velocity can be found only as a mean value over the time the front travels between the two sensors. Note, too, that for an ice shelf in which the temperature of the near-bottom ice layers is close to the temperature of the seawater it is difficult to ascertain the exact moment the front goes past the thermometer.

We had used this method successfully for five months back in 1959 to investigate freezing at the undersurface of sea ice in the vicinity of Mirny Station. ETP-310 platinum resistance thermometers were used for this, the units being frozen into the ice or suspended in the water in a horizontal arrangement, that is, along the presumed isotherm, which made the instruments more sensitive for determining the passage of the freezing front through the level of the thermometer.

The indicated method was proposed as one of the alternatives for measuring the melting or freezing beneath the Ross Ice Shelf (U. Urheim, private communication).

(2) A vertical rod with scale divisions on it can be so frozen into the lower part of the ice shelf that its upper part is in the ice, while its lower part sticks out into the seawater. Repeated photographing of the rod, once it has become frozen into the ice and does not move relative to it, then provides the necessary information about the melting or freezing at the shelf base, in the manner of ordinary snow-gauge observations of ablation or accumulation at the upper surface of a glacier. This method, proposed by Urheim (private communication) as an alternative for studying melting at the base of the Ross Ice Shelf, is apparently somewhat complicated to put into practice.

(3) Sensitive instruments, securely attached to the shelf ice, can be placed in the interior of the ice shelf or in the seawater under it. These devices measure the distance away from the melting-freezing front, that is, the distance from the ice-water interface. Remote methods are used for this (radar, ultrasound, etc.), whereby the location of the front is tracked as an interface between two media having very different physical properties. This was the method used by us successfully to monitor directly the melting or freezing at the base of the Ross Ice Shelf. Sources of ultrasound oscillations, connected securely to the

shelf ice, were placed in the seawater beneath the shelf, not far from its undersurface. The distance from the fixed transmitter to the melting-freezing front was determined on the basis of the time needed for the ultrasound to go from the transmitter to the ice-water interface and back. Repeated measurements of this time were used to estimate the direction and velocity of the front motion.

The ultrasound sources were lowered beneath the ice shelf through a specially drilled access hole more than 400 m deep. They were mounted on a long tube (about 6 m long), the lower part of which stuck out into the water and the upper part of which remained in the borehole (was frozen into the ice).

Figure 9.13 shows schematically the measurement setup, and the arrangement of the transmitters under the ice shelf. The ultrasound sources (tiny antennas) are so arranged in the subice water that they face upward toward places in the shelf which have not been disturbed by the drilling, that is, which are located to the side of the borehole opening. On the whole, the apparatus suspended beneath the ice shelf is reminiscent, in both its external appearance and its principle of action, of an umbrella (see Figure 9.13). When folded up, the device is enclosed in a protective case, which is filled with alcohol, to keep ice from forming on the movable parts of the assembly. After the apparatus has been lowered through the borehole, on command from the surface the protective case is jettisoned and the 'umbrella' is opened. The exact position of the assembly relative to the borehole and the base of the ice shelf is determined with the aid of sensors pointing toward the shelf base and the borehole. The sensors which remained in the borehole were to be used later in observations of freezing inside the hole.

In these studies we were the first to employ the active sonar method to monitor melting-freezing processes at relatively high frequencies. These measurements were preceded by laboratory experiments to determine the reflectivity of ice surfaces of various roughnesses, taking into account the attenuation of an ultrasound signal as it passes through water and the attenuation of an electrical signal as it traverses a cable 500 to 800 m long. This enabled us to select the working frequencies of the acoustic transducers. The assembly under the ice shelf contained 23 ultrasound antennas with frequencies of 0.45, 0.6, 1.25, and 2.5 MHz. As a result, miniature components could be used, suitable for lowering through an access hole 110 m in diameter. At the lowest frequency an ultrasound antenna 30 mm in diameter and 5 mm thick, made of barium titanate, had a width of the directional diagram at the 0.5 level equal to about 10°. An antenna with a working frequency of 2.5 MHz had a width of the directional diagram at the 0.5 level of about 2° [34].

On 4 December 1978 the device was deployed through an access hole drilled with the aid of hot water. The assembly made it possible to carry out measurements of the distance in the water from the base of the shelf with an accuracy of 0.3 to 0.5 mm.

In addition to the above-described access hole, a second hole was drilled through the ice shelf so as to deploy a second set of measuring devices, to investigate the melting-freezing processes. The second borehole, at the bottom of which part of the measuring tube was located, was 40 m away from the first hole. Since both access holes were drilled with the aid of a jet of hot water, at the moment the drilling stopped

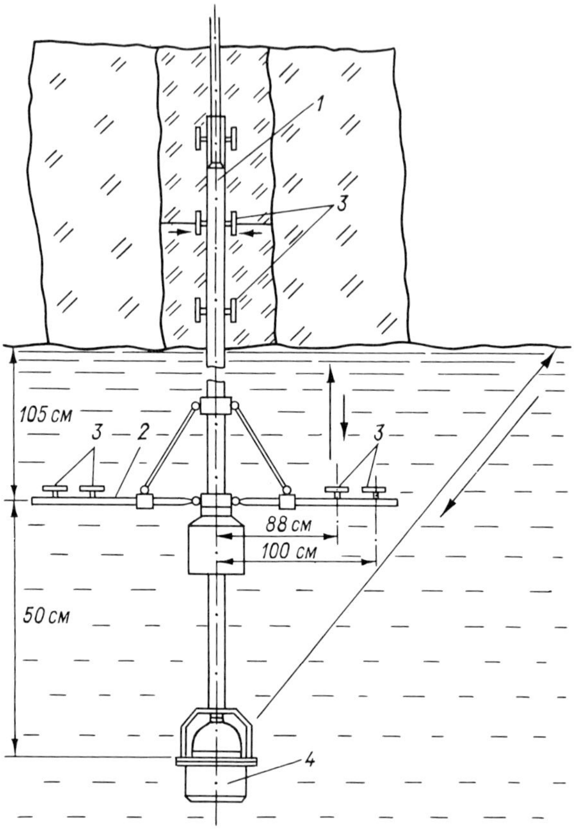

Fig. 9.13. Schematic diagram of ultrasound system, partly lowered beneath Ross Ice Shelf and partly frozen into it. Ribs of 'umbrella' open and their sensors facing upward.

(1) main tube, partly frozen into ice; (2) ribs of 'umbrella'; (3) ultrasound transmitters; (4) scanning device.

both holes were filled with fresh water. The borehole through which the first assembly had been deployed was plugged from above with snow and left to freeze. The second borehole, however, first had all its fresh water pumped out and only then was allowed to freeze up. Both holes apparently started freezing at the upper, colder end, and only then did they freeze further down the line. Because of the difference in the densities of water and ice, the new ice had a greater volume than the water, so that the freezing was accompanied by the forcing out of excess water at the bottom of the borehole. Consequently, at the time of this freezing, along the undersurface of the ice shelf close to the first access hole fresh water was emerging under the shelf. At the undersurface of the shelf near the second borehole water was also emerging as the hole froze up, but this water was salt water.

Readings were taken twice with the instruments deployed through the borehole from which fresh water had emerged (first borehole): in January 1979, and almost a year later, in late December 1979. It was found that during this period the distance from the sensor to the undersurface of the ice shelf had diminished by 3 cm, giving a bottom freezing rate of 3 cm per year. This value is close to the rate obtained by us earlier from an analysis of other data.

A completely different picture emerged from the readings of an instrument which measured the melting-freezing processes under conditions whereby fresh water enters the region under the shelf. Such observations were carried out for 24 days in December 1978.

During the observation period the reduction of the distance from the sensor to the ice reached 19 cm; it was attributed by us to freezing onto the base of the ice shelf. The freezing rate in the vicinity of this sensor was estimated to be 2.40 m/yr, which is two orders of magnitude higher than the rate determined from the core analysis and from measurements with the same apparatus at the second borehole.

The difference between the measured freezing rates may, in our opinion, be due to an enhancement of the freezing when fresh water enters the region under an ice shelf in contact with cold water (this was discussed in the previous section). Thus, the results being considered constitute experimental proof that freezing onto the base of the Ross Ice Shelf is enhanced by the entry of fresh water under the shelf. They verify both the possibility and the effectiveness of this process, and hence they indicate that there is a region of intensified freezing in the seaward part of the ice shelf.

These data on the freezing onto the base of the Ross Ice Shelf in the vicinity of Camp J-9, together with the data on the sudden enhancement of this basal freezing when fresh water enters the region under the shelf, enable us to re-evaluate the reaction of the Ross Ice Shelf to a warming of the Earth's climate, and in particular to the warming which may occur in 50 to 100 years due to an anthropogenic increase in the CO_2 content of the atmosphere, with its attendant enhancement of the greenhouse effect.

Such a warming of the climate would raise the water temperature of the World Ocean, so that warmer seawater would enter the region under the shelf, right up to the floating line, and would aid the melting of the shelf base. This would thin down the ice shelf and make it more buoyant, and thus would move the floating line 'upstream'. The ice shelf would thus be more exposed to destructive effects.

The conclusion arrived at in this study, that the region of basal melting of the Ross Ice Shelf is at present limited, and in any case does not extend to Camp J-9, implies that the ice shelf is not being thinned down. Actually, the enthalpy of the water entering the region under the shelf, determined with respect to the water cooled to its freezing point, increases with the warming of the seawater. However, there will be a like increase in the hydrodynamic heat exchange between the shelf surface and the water (it also is a linear function of the difference between the true temperature and the freezing point of the water). Consequently, the length of the path over which the water cools down to the freezing point does not vary in comparison with existing conditions; thus, in region 2 of Fig. 9.9 the water will once again be cooled down to its freezing point, in spite of the warming of the water entering the region beneath the ice shelf.

If the polar climate warmed up by 5 to 10 °C, as a result of the above-mentioned greenhouse effect, then the melting at the surface of the Ross Ice Shelf would increase sharply. This melting would be particularly great in the seaward parts of the shelf adjacent to mountains, and in the vicinity of outlet glaciers intersecting these mountains. The albedo of the shelf surface would be lowest there, due to contaminating substances borne in by the wind. The formation of meltwater at the shelf surface in summer, and the winter freezing of this water in pores in the firn and in cracks in the ice, would raise the mean temperature in the interior of the ice shelf, and thus would increase the plasticity of the ice. It is thought, too, that this would speed up the movement of the outlet glaciers and of the main mass of the inland ice sheet, in places close to the seaward parts of the Ross Ice Shelf, and thus would create a negative balance of the ice of West Antarctica.

However, data on the bottom freezing in region 2 of Figure 9.9, together with our data on the marked enhancement of freezing under the Ross Ice Shelf in its seaward regions when fresh water enters there, appear to cast some doubt on this conclusion. An accelerated movement of the outlet glaciers and of the ice sheet of West Antarctica would result in the formation of more meltwater at the beds of the outlet glaciers and ice sheet. This additional meltwater, plus some of the meltwater from the surface, should enter the region under the ice shelf in its seaward parts near the floating line, and this subice water should then refreeze. The fresh water thereby becomes mixed with the cold seawater, forming new ice with a volume greater than that of the fresh water entering the region under the shelf. Consequently, the Ross Ice Shelf close to the floating line and throughout region 3 in Figure 9.9 should become thicker, causing the floating line to shift 'downstream', increasing the resistance to the shelf movement, and causing the ice shelf to slow its movement and start to grow.

These circumstances call for a reconsideration of the reaction of the Ross Ice Shelf to a change in the Earth's climate, especially to the expected warming due to the increased CO_2 content in the atmosphere. Such a warming should by no means break up the Ross Ice Shelf; rather it should strengthen its base and move its floating line northward.

Chapter 10

APPLICATIONS OF THE THERMOPHYSICS OF GLACIERS

10.1. The Thermal Pollution of Ice Caps with Heat-Emitting Industrial Wastes

One of the main problems now confronting nations making wide use of atomic energy is the problem of the prolonged storage of the ever increasing amounts of radioactive industrial wastes. A substantial part of these waste materials includes long-lived isotopes, that is, elements heavier than uranium, with a high degree of radiation toxicity. When stored, these wastes must be carefully kept out of contact with the biosphere, which means that they must be kept out of the atmosphere, the surface waters, and the soil for as long as 250,000 years [132]. Accordingly, the storage sites must be protected from possible accidents, terrorist acts, natural calamities, etc., for this amount of time.

However, the amounts of radioactive wastes are piling up so rapidly throughout the entire world that it is becoming more and more imperative to find reliable inexpensive means of storage. It appears that this problem will have to be solved sometime during the next 10 to 20 years, and that all our engineering expertise will have to be marshaled for this.

In addition to purely engineering solutions, the possibility of using the sea bottom, exhausted salt mines, and other sites is also being examined [132]. In 1972 a group of American scientists [132] proposed to the International Atomic Energy Agency that the antarctic ice cap be considered as a possible site for the prolonged storage (burial) of radioactive industrial wastes [129]. This proposal was considered by a number of scientists, and it was decided that burial at the bottom of the Antarctic Ice Sheet would not ensure the required isolation of the waste materials from the biosphere, since at the bed of the ice cover melting takes place, with runoff of the meltwater into the World Ocean. This opinion was supported at a meeting of the Scientific Committee on Antarctic Research (SCAR). The Soviet delegation at this meeting also came out against the use of Antarctica for such storage.

However, in view of the growing urgency of the problem, SCAR decided to carry out further studies of this proposal. In the following, the author will present some of his thermophysical evaluations of the situation.

The disposal of heat-emitting industrial wastes in the Antarctic has a purely scientific aspect, related to the possible methods to be used for the disposal and its possible consequences, and it also has an ethical and legal aspect. We will consider here only the purely scientific aspect of the problem.

Radioactive wastes to be buried in the ice sheet will apparently have to be either placed in separate containers or else formed into individual solid blocks. There are several possible ways of disposing of wastes in the ice.

10.1.1. Disposal at the Glacier Bed

The individual waste deposits (blocks, containers) to be buried must have parameters ensuring their descent to the glacier bed (Figure 10.1 I). For instance, the mean density of the container of waste must

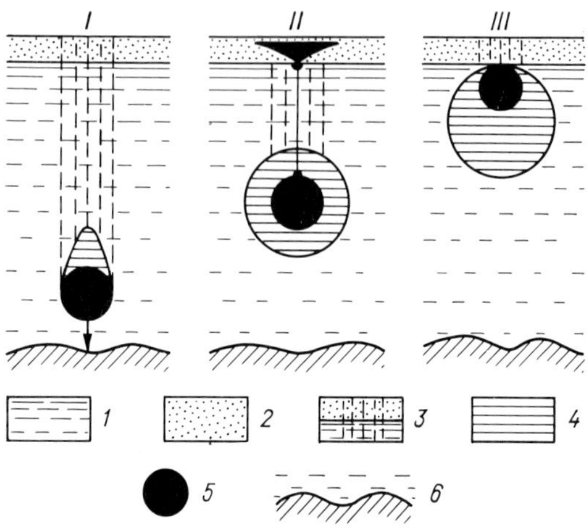

Fig. 10.1. Various possible ways of disposing of wastes in an ice sheet.

(I) descent of heat-emitting waste container to glacier bed; (II) anchoring of container in ice; (III) container floating in water pocket at bottom of water-permeable layer; (1) ice interior; (2) water-permeable layer; (3) refreezing of 'path' of container which has melted its way through ice; (4) water pocket around container; (5) heat-emitting container; (6) glacier bed.

be higher than the density of water, while the thermal power (heat output) per unit container volume must be sufficient to melt through the ice. According to the data of [132], a container of waste in the form of a cylinder 30 cm in diameter and 3 m long has a thermal power of about 5400 W, and it will melt its way through ice with a temperature of -50 °C and a thickness of about 1.5 km in 3 or 4 years [129]. Consequently, the waste containers may descend to the bed very rapidly and become isolated from the glacier surface by the water which has frozen in the channels melted through by the containers. This means of waste disposal is, however, a hazardous one. As we have demonstrated above, the temperature at the bed of a considerable portion of the Antarctic Ice Sheet is equal to the melting temperature, so that melting takes place at the bed. The meltwater so formed is forced out via run-off channels into peripheral regions, and some of it goes into the surrounding ocean. In this case the radioactive wastes placed at the bed of the ice sheet will come in direct contact with the water of the World Ocean.

If, on the other hand, the temperature at the bed of the ice sheet is lower than the melting point of the ice, then the thermal state of the ice sheet is extremely sensitive to any additional influx of heat to the ice at the bed. For example, if the heat flux at the bed in central Antarctica were to increase from 42 to 83 mW/m^2, then the ice at the base in the central regions of the ice sheet, which was initially at about -10 °C, would instead be in a state of melting [32]. An increase in the liberation of heat at the bed even by 0.1 q_g (where q_g is the geothermal flux) considerably alters the thermal state of the ice at the glacier bed. Accordingly, such an increase can change the conditions of movement of the ice at the bed, with an attendant abrupt outflow of the ice sheet, a discharge of ice into the sea, and a possible catastrophic raising of the level of the World Ocean [129]. The ice temperature at the bed over vast areas could also rise briefly to the melting point, because of the frictional heat of the rapid movement, once again making possible direct contact between the buried wastes and the water of the World Ocean, via the meltwater formed. Here the possibility of mechanical breakage of the waste units (containers or blocks) should also be kept in mind, since they will be located at the boundary between the moving ice and the stationary bed of the glacier.

The possibility of, and the danger of, a thermal effect of the waste materials on the temperature conditions at the glacier bed will depend directly on the specific (relative to a unit surface of the ice sheet) thermal power of the buried radioactive wastes. Figure 10.2 shows the specific thermal power of the buried wastes q_{bur} as a function of the radius of a circular area within which the wastes are uniformly distributed. The total thermal power q_Σ of the wastes is taken to be the sum of the individual thermal powers of all the waste which will apparently accumulate in the world up to the year 2000. For simplicity, this amount, which is close to 630 MW [132], will in the following be taken to be constant in time. In other words, we will assume that the burial of new wastes, which will undoubtedly follow the dumping before the year 2000, will compensate for the lowering of the activity and thermal power of the previously buried wastes.

The plot in Figure 10.2 shows that the burial of wastes in an area about 20 km in radius results in a specific thermal power of the buried wastes q_{bur} which is equal to the geothermal flux. To reduce this heat to 0.1 q_g, we must increase the radius of the disposal area to approximately 70 km. To reduce the heat output of the wastes to 1/100 of the geothermal flux (in this case q_{bur} will apparently have little effect on the temperature regime of the ice sheet), the radius of the disposal area has to be increased to 200 km. However, increasing the size of the disposal region involves other hazards, since one of the conditions of waste burial is that the radioactive materials be kept out of contact with the environment for 250,000 years. At the same time, the waste containers will move horizontally with the ambient ice. Curve 2 in Figure 10.2 shows the time of undifferentiated motion of the wastes toward the ice-sheet margin in East Antarctica, as a function of the distance of the disposal site from the center of outflow of the sheet. The rate of movement of the ice sheet is taken from the conditions of steady-state axisymmetric outflow for Blockschollen movement (block flow) of the ice. The distance between the outflow center and the margin of the ice sheet is taken to be 1200 km. The time needed for containers buried close to the outflow center to migrate to the margin

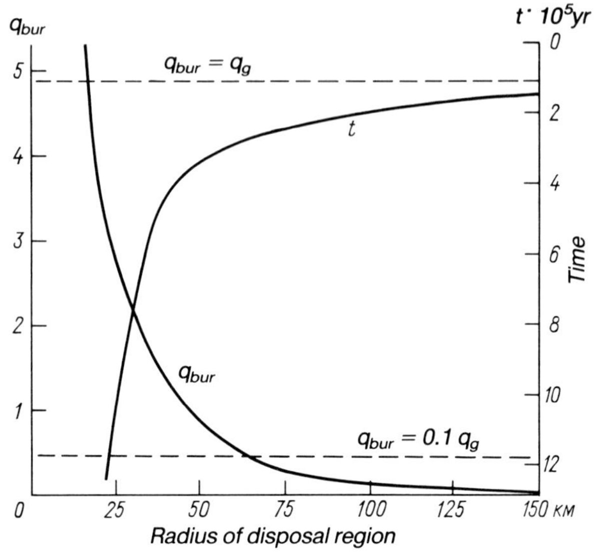

Fig. 10.2. Specific thermal power of buried wastes q_{bur} (in 10 mW/m^2) and time of movement of wastes toward margin of Antarctic Ice Sheet t, as functions of radius of waste-disposal area in center of ice sheet.

of the ice sheet is seen to be more than 1 million years, but this time decreases rapidly with increasing distance from the center. For containers buried at sites about 100 km from the center, the time for migration to the margin is less than 250,000 years.

It should be kept in mind that this calculation pertains to an axisymmetric model, that is, the possibility of a local, more rapid discharge via outlet glaciers is neglected. If deposits of buried wastes are able to enter such discharge channels, the time needed for migration to the margin of the ice sheet may be reduced significantly. In view of this, using regions of the ice sheet located more than 50 km from the ice divide is apparently also dangerous, making it very difficult indeed to reduce the specific thermal power of the buried waste products to 1/100 of the geothermal heat flux.

10.1.2. Slowing the Descent of Heat-Emitting Wastes to the Glacier Bed

Individual waste containers can be slowed down in their descent through the interior of the ice sheet. Then they will spend a quite long time in regions of the ice far above the bed, that is, regions where the temperature is definitely below the melting point. There are, in principle, two ways of slowing the descent of a waste container with a density higher than the density of water.

(1) The first means is reducing the thermal activity of the container of waste, while keeping its size the same. This can be accomplished by 'diluting' the radioactive wastes with a nonactive filler. The rate at which the container melts through the ice is thereby so

reduced that, because its heat output diminishes with time, the container will remain above the bed for a long period. However, in order to reduce the melting rate substantially, an amount of 'diluting additive' hundreds of times greater than the mass of the wastes themselves must be added. Moreover, since the total mass of the wastes themselves will by the year 2000 be more than 30,000 tons [132], millions of tons of diluting additive would be needed. Transporting such vast quantities of material to the middle of Antarctica does not appear to be very feasible.

(2) The second means of slowing the descent of the waste is to stop it completely by hanging the container on an 'anchor', frozen into the ice some distance away from the container (for instance, at the surface, as in Figure 10.1 II). The waste container is connected to the anchor by a flexible line [113]. However, in this case the waste deposit will nevertheless descend at the velocity of vertical movement of the glacier ice, because of the sinking that accompanies the continual accumulation of precipitation at the surface of the ice sheet. Calculations carried out for the vicinity of Vostok Station in the middle of the ice sheet of East Antarctica show that a waste deposit suspended from an anchor on a cable 100 m long will traverse the distance of more than 3000 m from the surface to the bed in 450,000 years, if the vertical velocity of the glacier ice varies linearly from 3 cm/yr at the surface to zero at the bed (no melting at bed). If, on the other hand, at the bed of the ice sheet melting takes place at the rate of, say, about 0.8 cm/yr, then this time will be only 250,000 years. The use of such anchors eliminates the need to introduce a neutral diluting substance that lowers the volume heat output of the deposit. However, the presence of an anchor and a connecting line makes this method dangerous, too, since the connecting device will almost inevitably become damaged during the course of 250,000 long years.

10.1.3. Burial of Heat-Emitting Wastes Close to the Glacier Surface

The deep and bottom-adjacent regions of an ice sheet will be subjected to the least thermal disturbance when the heat-emitting wastes buried in the ice are always kept close to the upper surface of the ice sheet. This can be accomplished if the mean density of each heat-emitting deposit is considerably lower than the density of water (so that a container will float on water; see Figure 10.1 III). Such a heat-emitting container placed on the surface of the Antarctic Ice Sheet will, via gravity melting, descend through the upper 40 or 50 m of the ice sheet, which is usually made up of firn having a porosity such that all the meltwater formed can run off. The author has verified this by melting boreholes in the central part of Antarctica, close to Vostok and Komsomol'skaya Stations. At depths greater than 50 m bubbles in the ice are apparently no longer able to combine, and the meltwater, which is no longer able to get away from the heating source, surrounds it. As a consequence, the heat-emitting deposit will float in a water-filled pocket in unporous ice. The waste container will not descend further into the depths of the ice as long it remains inside the glacier, since it heats up and melts mainly the upper layers of the ice surrounding the pocket, while seemingly floating in the ice relative to the layers moving downward due to sinking. The boundary of this region of floating will be the level where the ice becomes firn, capable of diverting to the side the water forming around the container of waste (this level is about 50 m from the surface). Let us consider the effect of such disposals on the ice sheet.

10.1.4. The Temperature Disturbance of the Interior of the Ice Sheet (Antarctica) by Heat-Emitting Wastes

The disposal of heat-emitting objects near the surface of the ice sheet will undoubtedly raise the temperature of the surface and of part of the interior of the glacier. In order to estimate the thermal disturbance caused by such wastes, let us consider a simplified model of the phenomenon. The heat emission is assumed to be spread uniformly over the entire disposal area (the ice-firn interface), rather than to come discretely from each individual waste container. Let us also assume that the initial ice temperature is constant over the thickness of the glacier (which is nearly the case for the upper thousand meters of the 3-km-thick Antarctic Ice Sheet in its central regions), and that the vertical ice movement has a negligible effect. Finally, we assume that the radiative and convective heat exchange at the surface do not change when additional radioactive heat sources are placed near the glacier surface. In this case the additional heat release at the surface, and the accompanying rise in surface temperature, are assumed to take place under conditions whereby all the heat is dissipated only in the lower-lying layers of ice. If the process is examined over a time interval less than the time needed for the temperature wave to reach a depth equal to the glacier thickness, then the problem reduces to a study of heat transfer in a semibounded body with an initial condition of the second kind. If the Y' axis is directed downward and the origin is at the glacier surface, then this condition can be written as follows:

$$-\lambda \left.\frac{\partial T}{\partial y'}\right|_{y'=0} = \text{const} = q_{bur}; \quad \left.\frac{\partial T}{\partial y'}\right|_{y'=\infty} = 0; \quad T_{y'=\infty} = T_0; \tag{10.1}$$

where λ is the thermal conductivity of the ice, T_0 is the initial ice temperature, constant over the thickness, and q_{bur} is the mean specific heat flux due to radioactive decay of the waste deposits.

Then the initial condition has the form

$$T\Big|_{t_{bur}=0} = T_0,$$

where t_{bur} is the time since the wastes were buried.

If the ice is assumed to be isotropic (the thermal diffusivity a is constant over the thickness and in time), then the equation of heat transfer in the glacier ice will be the usual heat-conduction equation

$$\frac{\partial T}{\partial t} = a \frac{\partial^2 T}{\partial (y')^2},$$

which must be solved with boundary conditions (10.1). The solution of this equation [65] gives the following expression for the variation in time of the surface temperature of the glacier:

$$T_{y'=0}(t_{bur}) = T_0 + \frac{2}{\sqrt{\pi}} \frac{q_{bur}\sqrt{t_{bur}}}{\sqrt{\lambda c \rho}}, \qquad (10.2)$$

where t_{bur} is the time since the first burial of wastes; λ, c, and ρ are the thermal conductivity, specific heat, and density of the ice; and $T_{y'=0}$ is the temperature of the glacier surface.

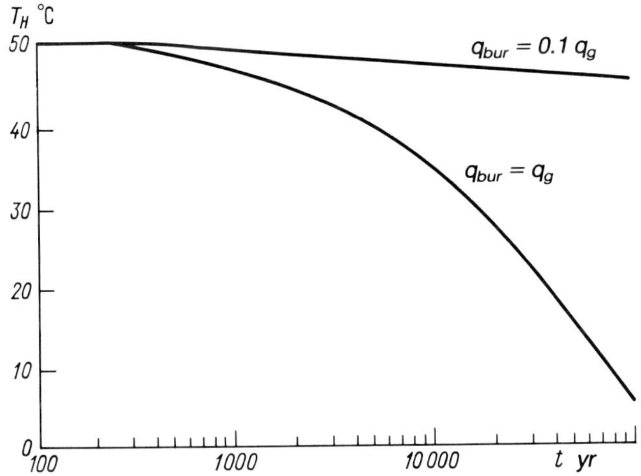

Fig. 10.3. Variation in time of temperature of glacier surface beneath which heat-emitting wastes have been placed, for different specific thermal powers.

It is seen from formula (10.2) that, other conditions being the same, the rate of change of the surface temperature is directly proportional to the specific intensity of the emitted heat. This rate decreases with increasing time from the burial of the wastes according to the expression

$$\frac{\partial T}{\partial t} = \frac{1}{\pi} \frac{q_{bur}}{\sqrt{\lambda c \rho}} . \qquad (10.3)$$

Figure 10.3 shows the variation in the temperature of the glacier surface for q_{bur} values of 0.1 q_g and q_g. Inspection of the figure shows that disposal of the heat-emitting wastes in a very small area (q_{bur} = q_g) leads to a sharp rise in the temperature of the glacier surface during a very short time. An enlargement of the disposal area to a size corresponding to q_{bur} = 0.1 q_g, on the other hand, reduces considerably this rise in the surface temperature.

Figure 10.4 shows the distribution of the temperature over the glacier thickness for different times since the burial of the wastes. These plots indicate that, if q_{bur} is equal to the geothermal heat flux

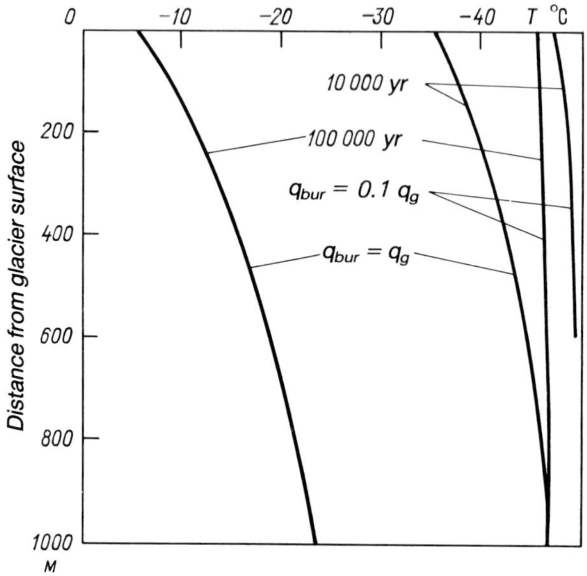

Fig. 10.4. Distribution of temperature disturbance due to waste burial, for different times since burial.

for the first 10,000 years, then the heating wave practically does not reach a depth of 1000 m, but later there is a sudden heating of the ice at this level. This means, firstly, that beginning at this time the effect of the heating due to the radioactive wastes becomes appreciable throughout the interior of the ice sheet and, secondly, that the model being used is no longer suitable for making estimates. A reduction of q_{bur} to 0.1 q_g will naturally reduce considerably the heating of the ice, but even so the ice temperature over the entire glacier interior after 100,000 years will be quite different from (about 5°C higher than) the temperature before burial of the wastes.

10.1.5. Formation of a Pocket. The Thermal Disturbance at the Surface Due to Heat-Emitting Waste Deposits.

The velocity of the waste containers with respect to the descending ice particles is equal to the rate of accumulation of precipitation, that is, it is close to 3 cm/yr for the central part of Antarctica and about 5 times as great as this for Greenland. Each waste container melts the ice around it and forms a pocket, close to spherical in shape, which is filled with meltwater. We introduce the following notation: q_c is the thermal power of each container, and R_p is the radius of the meltwater pocket forming around it. As a first approximation, the walls of the water pocket and the isolines of the temperature in the ice interior are taken to be spherical. Then, if the thermal state is steady, the heat flux through the ice from the container at any given distance

$R > R_p$ will be the same and equal to

$$q_c = -4\pi\lambda R^2 \frac{\partial T}{\partial R},$$

which gives

$$T = \frac{q_c}{4\pi\lambda}\frac{1}{R} + \text{const.} \tag{10.4}$$

Since for $R = R_p$ we know that $T = 0$ °C, while as $R \to \infty$ we have $T = T_0$, the radius of the meltwater pocket will be

$$R_p = \frac{q_c}{4\pi\lambda T_0}. \tag{10.5}$$

Thus the size of the pocket will be directly proportional to the heat output of the waste deposit. For the container with an output of about 5.5 kW in ice at -50 °C, which was considered above, the water pocket will have a radius of about 4 m. Note, however, that (10.5) cannot be used for very high values of q_c, since a steady state is reached only after a very long time. For example, for our value of q_c this time will be 1 or 2 years, whereas for $q_c = 63$ kW (R_p about 40 m) it is already around 200 years.

The pocket radius is a function of the heat output of the waste deposit, and, to ensure that the 'ceiling' over the pocket is strong enough, the pocket diameter has to be, as a first approximation, about equal to the thickness of the 'ceiling', defined as the distance ΔH_{bur} from the waste deposit to the glacier surface. Thus there will be some maximum allowed thermal power of the container $q_{c\ max}$:

$$q_{c\ max} \leqslant -\frac{\Delta H_{bur}}{2} \cdot 4\pi\lambda T_0.$$

For ice at -50 °C and for ΔH_{bur} of about 50 m, $q_{c\ max}$ should be no more than 30 kW.

It should be noted that the radius of the thermal effect of each waste deposit is appreciably larger than the radius of the water pocket. To estimate the radius of the effect, we have to consider the temperature field in the ice interior close to the pocket. On the basis of (10.5), we can write this field as

$$T = T_0\left(1 - \frac{R_p}{R}\right). \tag{10.6}$$

Figure 10.5 shows the temperature in the ice interior as a function of the dimensionless distance from the pocket center, for an initial ice temperature of -50 °C. It is seen from formula (10.6) and Figure 10.5 that, even at a distance ten times greater than the pocket radius, the ice temperature will still differ by 10% (5 °C) from the temperature of the undisturbed ice.

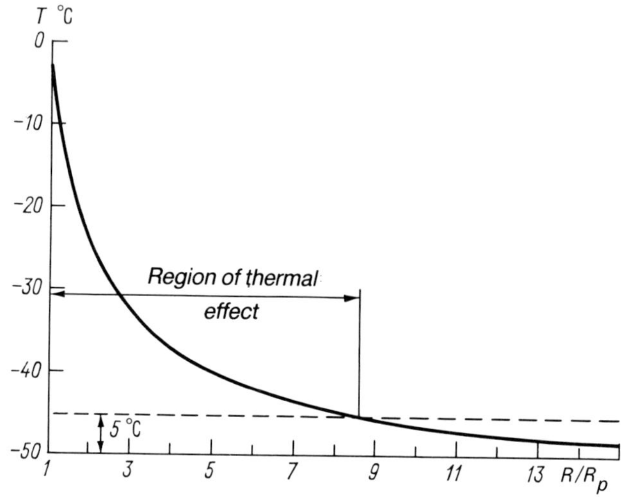

Fig. 10.5. Temperature disturbance in glacier interior with ice temperature of -50 °C, as function of dimensionless distance from center of water pocket.

10.1.6. Interference of Heat-Emitting Waste Deposits

One of the hazards connected with the burial of radioactive wastes close to the surface of an ice sheet is their interference. According to the data in Figure 10.2, the area of the circle in which wastes can be buried until the year 2000 is about 300 km^2 for $q_{bur} = q_g$ and about 3000 km^2 for $q_{bur} = 0.1\ q_g$. At the same time, despite the smallness of the total projected area of the water-filled pockets (about 6 km^2) for the example being considered (q_c = 5.5 kW), the circular regions within which the temperature is more than 10% higher than the undisturbed ice temperature have an area as high as 600 km^2, that is, almost 20% of the entire waste-disposal area, even for $q_c = 0.1\ q_g$.

Every few tens of years a like number of new waste deposits will probably be buried in the same place. Thus the total area of the regions where the temperature differs from its undisturbed value will very rapidly become equal to the area of the entire waste-disposal region. Consequently, the zones of thermal influence of several deposits will overlap. A redistribution of the deposits on the surface will inevitably take place as well.

Deposits with overlapping zones of influence will approach one another as the ice is melted through, and in time a considerable number of these may be in dangerous proximity.

To conclude this discussion of the various possibilities for disposing of heat-emitting toxic wastes, we would like to state that none of the alternatives considered can be recommended as yet, in view of the hazards involved — hazards which cannot be thoroughly evaluated at the present level of our knowledge about the behavior of ice sheets.

Therefore, we concur with the conclusion reached in [129] that it is not advisable to bury active wastes in the Antarctic, because of the dangers associated with such burial.

10.2. Reconstruction of the Quaternary Inland Ice Sheet of Europe

Thermophysical studies of glaciers have also been important for the reconstruction of Quaternary glaciations, since they provided new material needed to formulate the theory of these phenomena, leading to new approaches to the search for certain minerals. A primary example is the discovery of subglacial melting in the central region of the ancient ice sheets, as well as a peripheral ring of permafrost beneath the glacier. Moreover, these findings implied the existence of a pressure gradient in the deep ground water under the ice sheet, so directed that these waters were forced out of a large area toward the margins of the ice sheet. The reconstruction of these ancient ice sheets on the basis of a thermophysical approach to the analysis of the phenomenon has some important practical applications.

Ancient ice sheets are now reconstructed using the 'principle of actualism'. This principle assumes that all the main features of the large present-day ice sheets, of the type of the Antarctic Ice Sheet, were also typical of ancient ice sheets, and in particular the Quarternary ice sheets that covered parts of Europe and North America. For instance, the ice sheets of the past are assumed to have had a form similar to that of the present-day antarctic ice cap.

However, the accumulated geomorphological data on traces of the Quarternary glaciation of the Northern Hemisphere do not verify this, largely hypothetical, conclusion. For instance, numerous geological and geomorphological facts about the thickness of the Quaternary continental ice sheet close to the limit of its maximum expansion, together with data on its thickness during the period of retreat, consistently indicate that in the marginal (southern) part of the ice sheet there was a wide (hundreds of kilometers wide) region of ice with a comparatively small thickness (80 to 100 m at the margin and about 500 m in the rear part of this peripheral region) [3].

The upper diagram of Figure 10.6 shows the Valdai Ice Sheet, the last of the Quaternary ice sheets of Europe, which existed 12,000 to 15,000 years ago. The lower part of the figure shows a cross section of this ice sheet, obtained [3] on the basis of an analysis of geological and geomorphological data. This graph clearly shows the wide, but shallow, peripheral region, which has no analog in present-day Antarctica.

It would be interesting to try to reconstruct the form and the main features of the Quaternary ice sheets of Europe and North America, without resorting to the principle of actualism as it is usually formulated in geomorphology. Instead, let us consider a Quaternary glacier to be an object obeying only the fundamental laws which have been verified by observations of present-day ice sheets. First we will consider the form of the ice-sheet surface.

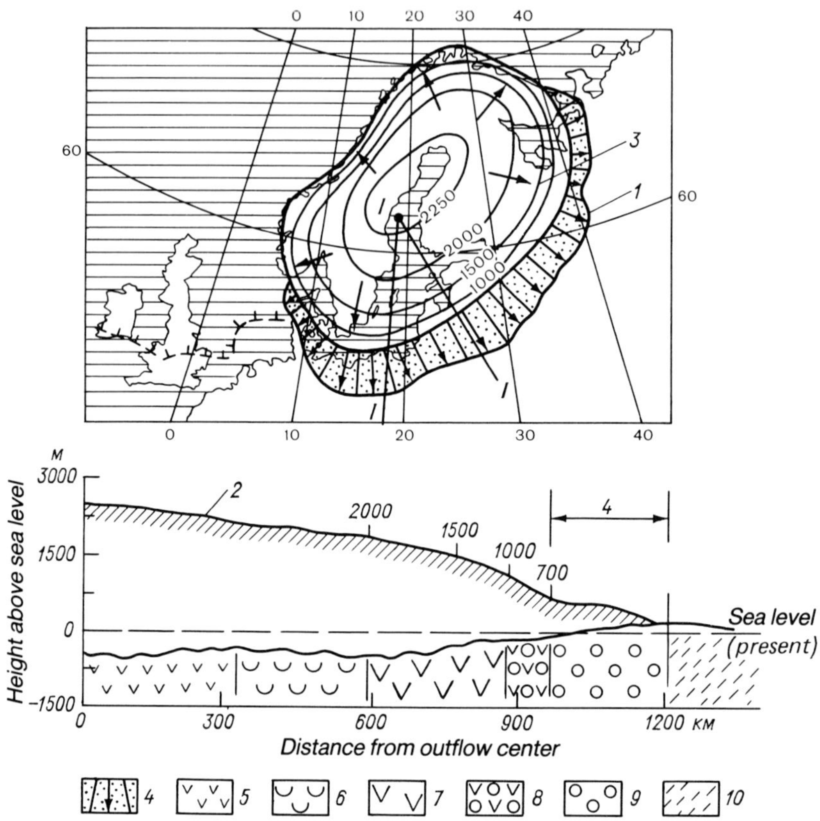

Fig. 10.6. Boundaries of Valdai Ice Sheet and its cross section for region I-I (after [3]).

(1) outer boundary of peripheral (shallow) part of ancient glacier; (2) shape of central part of glacier, reconstructed using geomorphological data; (3) isolines of surface heights of reconstructed glacier; (4) tapered peripheral part of glacier; (5) zone of weak erosion; (6) zone of enhanced erosion; (7) zone of appreciable, but nonuniform, erosion; (8) transition zone (erosion alternating with accumulation); (9) accumulation region; (10) Quaternary deposits.

10.2.1. The Form of the Surface of an Ice Sheet

While studying the movements inside a glacier, we have to make certain assumptions. Let us assume that the ice moves very slowly, that is, we neglect the inertial terms of the equations. Also, the temperature in each vertical cross section of the glacier is taken to be constant over the thickness and equal to some mean value typical of that section. We also assume that the density of the ice in the glacier is constant. All these assumptions are usually made, even when analyzing the conditions of movement of the large present-day ice sheets. However, then it is also assumed that the glacier is steady-state, and that its thickness, rate of nourishment, and temperature are constant in time. The latter assumption is quite valid for the ice sheets of Antarctica and Greenland, which have remained nearly the same size for the past several million years; however, it does not hold true for the ice sheet of Europe, which more than once has appeared and disappeared during much shorter periods of time.

Now let us consider the assumption of a steady state. We will try to arrive at general expressions for the form and movement of the ice which are not limited to the steady state. We will find out if these general expressions can be used to analyze the state of a glacier, by comparing the calculations with observations of present-day glaciers. Finally, we will use these expressions to ascertain specific features of the ice sheets of the past.

First of all, we note that the ice sheets of the past, when viewed from above, were very elongated, so that the outflow of ice southward, for instance, can be considered a movement for which the flow lines on the map are parallel for a very large region of the ice sheet. The same can be said of present-day ice sheets. Thus we will consider the two-dimensional problem, keeping in mind that the calculations can easily be repeated for an axisymmetric ice sheet as well (radial ice outflow).

It will also be assumed [14] that the tangential stress in the glacier interior varies linearly from zero at the surface to a maximum τ_0 at the bed. The stress at the bed depends on the thickness of the glacier H and the slope of the surface (to the horizontal) in the direction of motion of the ice. For a glacier with a flat bed this stress is given by Equation (6.41). The linear variation of the tangential stress from τ_0 at the bed to zero at the surface can be represented as

$$\tau = \rho g H \frac{dH}{dx}\left(\frac{H-y}{H}\right). \tag{10.7}$$

The y axis is directed vertically upward from the origin at the bed. Now we note that the rate of angular shear (γ) in Glen's law (3.24) is none other than the gradient of the horizontal ice velocity over the glacier thickness du/dy. Then Equation (10.7) enables us to determine the variation in the ice velocity over the thickness for each point located a distance y from the bed:

$$\frac{\partial u}{\partial y} B\tau^n = B\left[\rho g H \frac{dH}{dx}\left(\frac{H-y}{H}\right)\right]^n. \tag{10.8}$$

Let us assume that at the glacier bed the ice can slide with a velocity u_{sl} and let us integrate formula (10.8) from zero to y. This gives us an expression for the velocity of horizontal ice movement at any distance y from the surface:

$$u = u_{sl} + \frac{B\left(\rho g \frac{dH}{dx}\right)^n}{n+1}\left(H^{n+1} - y_1^{n+1}\right), \tag{10.9}$$

where y_1 is the distance from the glacier surface to the given point: $y_1 = H-y$.

Accordingly, the mean ice velocity u_{av} over the glacier thickness is found from the condition $u_{av}H = \int_0^H u dy$ to be

$$u_{av} = u_{sl} + \frac{BH^{1+n}}{n+2}\left(\rho g \frac{dH}{dx}\right)^n.$$

In this case the ice discharge M through the vertical ice section being considered, which has a thickness H and unit width, is

$$M = u_{av}H = u_{sl}H + \frac{BH^{2+n}}{n+2}\left(\rho g \frac{dH}{dx}\right)^n. \tag{10.10}$$

The first term on the right-hand side of (10.10) is the ice discharge due to sliding (the Blockschollen component of the discharge), and the second term is the discharge due to the increase in the ice velocity from the bed to the surface. In view of the indeterminacy of u_{sl}, let us designate as $\bar{m}$ the part of the discharge due to Blockschollen movement relative to the discharge given by the second term in (10.10). Then we have

$$M = \frac{B(1-\bar{m})}{n+2} H^{2+n}\left(\rho g \frac{dH}{dx}\right)^n. \tag{10.11}$$

It should be recalled that formula (10.11) gives the ice discharge through a surface with a width of unity and a height equal to the glacier thickness H at a distance x from the margin, caused by the inclination dH/dx of the glacier surface to the horizontal.

Now let us reconsider the conditions of mass exchange at the glacier surface. If A is the rate of nourishment (plus sign) or ablation (minus sign) at the glacier surface at a distance x from the margin, then the total amount of substance precipitating per unit time onto the entire space above section x, right up to the ice divide, will be

$$G(x) = \int_x^L A\, dx. \tag{10.12}$$

The difference between the total amount G(x) of input of substance above section x and the discharge of substance M due to the ice outflow caused by the inclination of the glacier surface at this point will then be the amount of ice consumed to vary the mass of the glacier in the volume from the ice divide to section x. This difference (the difference between (10.12) and (10.11)) can be written as

$$\int_x^L A\, dx - \frac{B(1+\bar{m})}{n+2} H^{2+n}\left(\rho g \frac{\partial H}{\partial x}\right)^n = \int_x^L \frac{\partial H}{\partial t}\, dx. \tag{10.13}$$

Formula (10.13) is the fundamental equation determining the form of ancient unsteady-state ice sheets and their rates of change, for glaciers in which the ice obeys Glen's law. For a present-day ice sheet of Antarctica or Greenland, which is under steady-state conditions ($\partial H/\partial t = 0$), formula (10.13) simplifies to

$$\int_x^L A\, dx = \frac{B(1 + \bar{m})}{n+2} H^{2+n} \left(\rho g \frac{\partial H}{\partial x}\right)^n . \tag{10.14}$$

10.2.2. The Form of the Surface of a Steady-State Ice Sheet

In [14] Equation (10.14) is applied to a glacier which does not slide over its bed ($\bar{m} = 0$), for various values of parameter n in Glen's law and various profiles of the accumulation variation with the distance from the glacier margin. Figure 10.7 shows some of the profiles for a

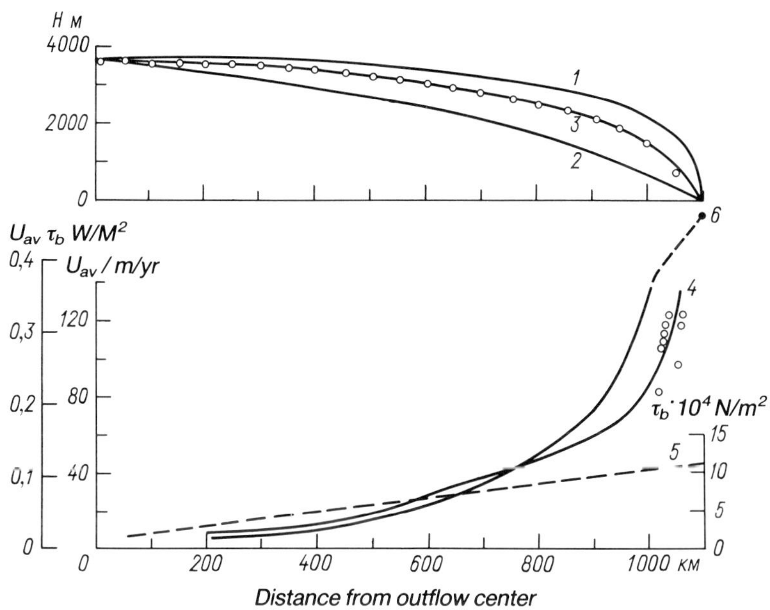

Fig. 10.7. Configuration of present-day ice sheets and ice movement in them.

(1) form of surface for viscous flow [14]; (2) form of surface for ice flow as plastic body (parabola) [14]; (3) viscoplastic motion (Glen's law); (4) ice velocity (average over glacier thickness); (5) shear stress at bed τ_b; (6) work done against bed friction $u_{av}\, \tau_b$. Circles (points) correspond to observed thicknesses and velocities in East Antarctica, on profile I-I in Figure 3.5.

steady-state ice sheet, obtained on the basis of the solutions of [14]. Curve 1 is the profile of a glacier moving according to the law for an ideal viscous liquid (n=0), while curve 2 corresponds to ideal plasticity. The points on the graph delineate the surface of the Antarctic Ice Sheet on a profile from Mirny Station to Vostok Station [14]. Curve 3 gives the profile of the ice sheet for ice obeying Glen's law with n=2, under conditions of constant accumulation over the whole surface at a rate of 0.25 m/yr. Such a distribution corresponds, as a first approximation, to the actual distribution of precipitation on this profile. The value of a corresponding to curve 3 was found to be close to 7.27×10^{-19} m^4 °C/(N^2 s).

It should be kept in mind that the value obtained for B corresponds to conditions in the Antarctic Ice Sheet for mean englacial temperatures of about -40 °C. This value, taking into account its temperature dependence according to Rouillon's formula (Chapter 3), agrees well with the value of B determined by Glen under laboratory conditions. Consequently, Equation (10.14) describes satisfactorily the form of the surface of an ice sheet.

10.2.3. The Form of the Surface of an Ice Sheet with an Ablation Zone

Now let us return to our study of ancient ice sheets, using Equation (10.14). Once again we note, as at the beginning of this section, that, unlike the large present-day ice sheets, which are usually bounded by a coastal barrier, the ancient ice sheets were largely inland formations which terminated on land. In this case the only limitation to their horizontal extent was the boundary of the accumulation zone. The region between this boundary and the glacier margin lay in the ablation zone, which was just wide enough to ensure the discharge of all the ice formed in the nourishment zone and moving horizontally into the ablation zone. The presence of this ablation region at the margin of an ancient ice sheet must have influenced its form considerably. This influence manifested itself as a flattening out of the marginal part of the ice sheet, since the marginal ablation kept the horizontal ice velocities at the edge of the glacier low, and thus the quantities H and dH/dx determining these velocities must have been relatively low as well (expression (10.11)). In addition, the ice in the ablation region had, on the average, a temperature considerably higher than the ice in the nourishment region, so that its fluidity must have been higher as well. This temperature difference had two causes: (1) the mean surface temperatures in the nourishment region were, over the years, lower; (2) the bulk of the ice in the nourishment zone had a temperature close to the surface temperature, since in this zone the particles of cold surface ice moved downward, cooling the lower layers of ice, whereas in the ablation zone the warm ice particles near the bottom ascended and warmed the glacier interior there (cf. Figures 4.2, 4.3, 4.4, 4.5). Let L_1 be the length of the region of accumulation at a rate A_1 which is constant with varying x, and let us assume that at points in the glacier which are further from the ice divide the ice melts at a rate A_2. Under the simplest, that is, steady-state, conditions the width of the ablation region is defined as the width at which all the ice arriving from the nourishment zone melts. Then the distance L from the ice divide to the glacier terminus will be

$$L = \left(1 - \frac{A_1}{A_2}\right) L_1. \tag{10.15}$$

The conditions under which an ancient ice sheet attained its maximum development, just prior to the start of its degradation, can be assumed to be close to steady-state (the glacier then neither grew nor diminished in size). Therefore, Equation (10.14) applies to this state. Calculation formulas for the equilibrium profile of the ice sheet should be derived in this case for two regions.

The first region is the central part of the ice sheet (nourishment zone) $L > x > (L - L_1)$. If the glacier thickness in the middle of this zone (at the ice divide) is H_L, the thickness at a distance L_1 from the margin (that is, at the snowline) is H_{L_1}, and the mean value of B for zone I is B_1, then in Equation (10.14), after separation of variables, we can integrate from $L - L_1$ to x:

$$\int_{L-L_1}^{x} \left[\int_x^L A(x)\,dx\right]^{1/n} dx = \left[\frac{B_1(1+\bar{m})}{n+2}\right]^{1/n} \rho g \int_{H_{L_1}}^{H} H^{(2+n)/n}\,dH \tag{10.16}$$

It should be noted that for $A(x) = A_1 = \text{const}$ we have

$$\int_x^L A(x)\,dx = A_1(L - x).$$

Integration of Equation (10.16) taking into account the indicated relation then gives the following expression for the form of the surface of the ice sheet in the region $L > x > (L - L_1)$:

$$H^{(2n+2)/n} - H_{L_1}^{(2n+2)/n} = \frac{2A_1^{1/n}}{\rho g} \left[\frac{n+2}{(1+\bar{m})B_1}\right]^{1/n} \left[L_1^{(1+n)/n} - (L - x)^{(1+n)/n}\right]. \tag{10.17}$$

Consequently, the glacier thickness at the ice divide is related as follows to the thickness at the snowline:

$$H_L^{(2n+2)/n} = \frac{2A_1^{1/n}}{\rho g} \left[\frac{n+2}{(1+\bar{m})B_1}\right]^{1/n} L_1 + H_{L_1}^{(2n+2)/n}. \tag{10.18}$$

The second region is the peripheral part of the glacier (in the ablation zone) $(L - L_1) > x > 0$. The ice thickness in this region varies from H_{L_1} to zero at the margin (where $x = 0$). If B_2 is the value of B in this zone, and if the ice discharge $\int_x^L A(x)\,dx$ through each section of this zone is $A_1L_1 + A_2(L - L_1 - x)$, then integration of an equation of the type of (10.16) gives the following expressions for the form of the ice sheet in the marginal zone of ablation and the glacier thickness at the snowline:

$$H^{(2n+2)/n} = \frac{-A_2^{1/n}}{\rho g}\left[\frac{n+2}{(1+\bar{m})B_2}\right]^{1/n} x^{(1+n)/n}; \quad H_{L_1}^{(2n+2)/n} =$$

$$= \frac{-A_2^{1/n}}{\rho g}\left[\frac{n+2}{(1+\bar{m})B_2}\right]^{1/n} (L-L_1)^{(1+n)/n}. \tag{10.19}$$

Earlier we showed that for a large ice sheet like the Antarctic Ice Sheet it is advisable to take $n = 2$. Quantities B_1 and B_2 should be expressed in terms of Rouillon's formula: $B_{1,2} = K_R/1 + T_{av1,2}$ [77]. Then the value of B obtained by us for the Antarctic Ice Sheet (mean temperature of Interior of about -40°C) gives the following value for constant K_R 2.9×10^{-19} m^4 °C / (N^2 s). Now we can substitute into Equations (10.17), (10.18), and (10.19) the values of B in terms of K_R according to Rouillon's formula and, for $n = 2$ and $K_R = 2.9 \times 10^{-19}$ m^4 °C / (N^2 s), we then get the following calculation formulas for the form of an ancient ice sheet and such of its fundamental parameters as the ice thickness at the ice divide and at the snowline. The thickness of the glacier at its center (along the ice-divide line $x = L$) is

$$H_L^3 = \frac{2}{\rho g}\sqrt{\frac{4(1 + |T_{av1}|)}{(1+\bar{m})\ K_R}}\, A_1L_1^{3/2} +$$

$$+ \frac{2}{\rho g}\sqrt{\frac{4(1 + |T_{av2}|)}{(1+\bar{m})\ K_R}}\,(-A_2)(L - L_1)^{3/2}. \tag{10.20}$$

The form of the glacier above the snowline $(L > x > (L - L_1))$ is represented as

$$H^3 = H_L^3 - \frac{2}{\rho g}\sqrt{\frac{4(1 + |T_{av1}|)}{(1+\bar{m})\ K_R}}\, A_1\ (L - x)^{3/2}. \tag{10.21}$$

The glacier thickness at the snowline $(x = (L - L_1))$ is

$$H_{L_1} = \sqrt[6]{\frac{4(1 + |T_{av2}|)}{(1+\bar{m})\ K_R}}\,(-A_2)\ \sqrt[3]{\frac{2}{\rho g}}\ \sqrt{L - L_1}. \tag{10.22}$$

Finally, the form of the glacier in the ablation region $(L - L_1) > x > 0)$ is

$$H = \sqrt[6]{\frac{4(1 + |T_{av2}|)}{(1+\bar{m})\; K_R} (-A_2)} \sqrt[3]{\frac{2}{\rho g}} \sqrt{x} \qquad (10.23)$$

Here $T_{av\ 1}$ and $T_{av\ 2}$ are the mean values of the temperature inside the glacier in the nourishment zone and the ablation zone, respectively. L is defined in terms of L_1 by Equation (10.15).

10.2.4. The Surface of the Last Quaternary Ice Sheet of Europe

Figure 10.8 shows a reconstruction of the last ice sheet of Europe along line I - I (Figure 10.6). Curve 1 in Figure 10.8 corresponds to the steady-state form of an ice sheet with a nourishment zone 800 km wide, a nourishment rate of 0.25 m/yr, and an ablation rate A_2 of 0.5 m/yr. In this case the ablation zone is half as long as the nourishment zone, giving a total glacier length of 1200 km. The calculation was carried out for $K_R = 2.9 \times 10^{-19}$ m^4 °C/ $(N^2$ s) and mean englacial temperatures of -30 °C in the nourishment zone and -2 °C in the ablation zone.

This temperature difference exists because in the central part of the glacier there is vertically downward ice movement, which has a cooling effect, while in the marginal ablation zone the ice temperature throughout the glacier is very close to the temperature of the bed, because of the warming effect of the upward ice movement (see Chapter 4). This will be considered in more detail below. The rate of sliding at the bed is taken to be the same for both regions and equal to unity. Curve 2 in Figure 10.8 shows the calculated form of the ice sheet, constructed by analogy with a present-day (antarctic) ice sheet for $K_R = 2.9 \times 10^{-19}$ m^4 °C/$(N^2$ s) and a mean temperature of -30 °C with $\bar{m} = 1$. It is seen that, for the assumptions made, the marginal region of an ancient ice sheet is considerably less steep than that of a present-day ice sheet. It should be noted, however, that the calculated form of the ancient ice sheet does not correspond to paleoreconstuctions. The calculated thickness in the marginal zone turned out to be too great (cf. Figure 10.8).

The calculations give a glacier thickness in the marginal zone of about 500 to 600 m at distances of 300 to 400 km from the edge, only provided that the role of basal sliding in this zone (the ablation region) is taken to be extremely great (100 times as great as the viscous component of the motion). Curve 3 in Figure 10.8 shows the form of this ice sheet.

Therefore, to summarize the foregoing, it must be concluded that the form of an ancient ice sheet during its steady-state period (transition from advance to degradation) differs considerably from the elliptical form suggested by the principle of actualism. An ancient ice sheet typically had a very flattened out marginal ablation zone (Figure 10.8). However, the decrease in the marginal thickness of the ice sheet indicated by the calculational data was not as appreciable as the decrease implied by paleogeographical reconstructions (see Figure 10.6).

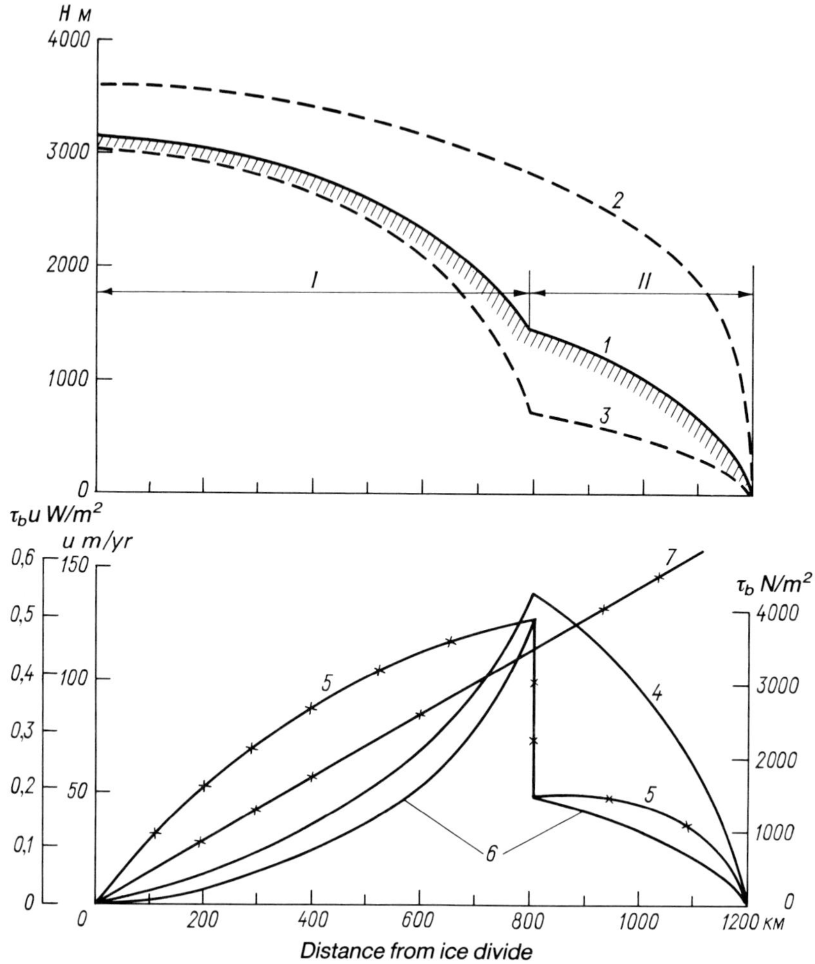

Fig. 10.8. Form of surface of ancient ice sheet and ice movement in it.

(1) surface form for viscoplastic movement with ablation zone (accumulation 0.25 m/yr, ablation 0.5 m/yr, ice temperature in ablation zone -2 °C, ice temperature in nourishment zone -30 °C); (2) form of viscoplastic outflow of ice sheet by analogy with Antarctic Ice Sheet (principle of actualism); (3) form of viscoplastic outflow under conditions of curve 1 with parameter of basal sliding equal to 100; (4) ice velocity (mean value u_{av} over glacier thickness); (5) strèss at bed τ_b; (6) frictional work at bed for Quaternary ice sheet with ablation zone; (7) frictional work at bed for present-day Antarctic Ice Sheet.

Now let us apply Equation (10.13) to the period of advance of the ice sheet. If we assume that the rate of change in glacier thickness W_t is the same at different points, then we can write Equation (10.13) for zone I (central nourishment zone) as follows:

$$\int_x^L (A_1 - W_t)\,dx = B\,\frac{1+\bar{m}}{n+2}\,H^{2+n}\left(\rho g\,\frac{dH}{dx}\right)^n .$$

By separating variables and integrating, we get the rate of change in glacier thickness W_t:

$$W_t = A_1 - \frac{\left(H_L^{2n+2} - H_{L_1}^{2n+2}\right)(\rho g)^n\,(1+\bar{m})\,B_1}{2^n(n+2)\,L_1^{1+n}}, \qquad (10.24)$$

or for n = 2, taking Rouillon's formula into account and noting that $H_{L_1}^3$ is negligible compared to H_L^3, we have the simplified expression:

$$W_t = A_1 - \frac{(\rho g)^2\,K_R}{16}\,\frac{H_{L_{st}}^6}{L_1^3}\left(\frac{H_L}{H_{L_{st}}}\right)^6 . \qquad (10.25)$$

where $H_{L_{st}}$ is the glacier thickness corresponding to steady-state conditions for a given value of L_1. It is easy to see that the quantity

$$\frac{(\rho g)^2\,K_R}{16}\,\frac{H_{L_{st}}^6}{L^3}$$

is the accúmulation rate A_1. But in this case the latter expression leads to the following formula for the rate of growth of the glacier thickness as a function of the ratio of the true ice thickness at the ice divide H_L to the thickness corresponding to a steady state for the given L_1:

$$W_t = A_1\left[1 - \left(\frac{H_L}{H_{L_{st}}}\right)^6\right]. \qquad (10.26)$$

Here and in the following, W_t will only be the rate of increase in the glacier thickness at the ice divide.

Formula (10.26) is extremely interesting, since it shows that only for a glacier thickness close to the steady-state value will the ice buildup due to accumulation go mainly to outflow in the horizontal direction, and also that the rate of increase in ice thickness is small compared to the rate of accumulation of precipitation. However, for considerably smaller ice thicknesses $H_{L_{st}}$, the horizontal outflow drops off sharply, and all the ice gain due to the accumulation of precipitation goes to increase the ice thickness. Figure 10.9 shows the ratio W_t/A_1 as a function of the ratio $H_L/H_{L\ st}$, which we will call the instability parameter. This plot shows, for instance, that for a glacier thickness half as great as the steady-state thickness only about 1% of the ice gain due to accumulation goes to the horizontal outflow, leaving 99% to augment the glacier thickness.

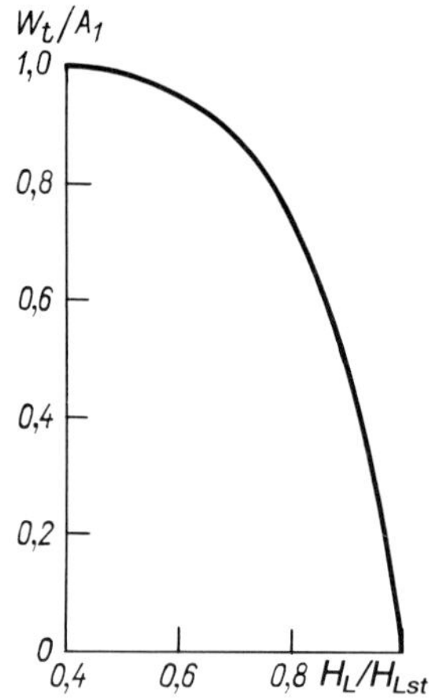

Fig. 10.9. Relative rate of increase in glacier thickness as function of instability parameter $H_L/H_{L\ st}$.

Expressions (10.25) and (10.26) imply that the growth of an ancient ice sheet within its accumulation zone (central region I) took place mainly under conditions when the glacier thickness was much less than the equilibrium (steady-state) value, so that all the accumulated ice went just to increase the thickness of the ice sheet. Under these conditions the horizontal outflow was very sluggish, so that the ablation region (II) during the period of growth was apparently much less developed than indicated in Figure 10.8 for steady-state conditions. The form of the ice sheet in zone I during the entire growth period was less steep than shown in Figure 10.8. Accordingly, the time needed for the ice to attain its maximum equilibrium thickness t_m can be estimated with the aid of the simple formula

$$t_m = \frac{H_{L_{st}}}{A_1^*}, \qquad (10.27)$$

where A_1^* is the time average of the accumulation rate. Thus, for an accumulation rate of 0.25 m/yr an ice sheet 3000 m thick can form in a little more than 12,000 years.

10.2.5. The Thermal Regime of the Last European Ice Sheet

As noted earlier, the temperatures in the interior of the Antarctic Ice Sheet and the Greenland Ice Sheet are not the same. The calculated temperature at the bed of the Greenland Ice Sheet in its central region is everywhere below the melting point of the ice, whereas in the Antarctic Ice Sheet we know that melting takes place at the bed in many places, central as well as peripheral. The presence of meltwater at the bed has been verified by drilling near Byrd Station in the central part of West Antarctica, as well as by the discovery of subglacial lakes. Therefore, the theoretical method described above has been checked sufficiently by a comparison with observations. Let us now use this method to analyze the thermal regime in an ancient ice sheet.

Figure 10.10 shows a section through a melting ice sheet with a rectified bed, reduced to sea level, as well as the mean temperatures (over many years) at the ice-sheet surface, taking into account the marginal reduction of the ice thickness, and also the previously assumed distribution of the accumulation and ablation rates. The dashed curve is a plot of the critical thicknesses. The region within which the true glacier thickness is greater than critical corresponds to continual melting at the bottom of the ice sheet, at the interface between the ice and the underlying bedrock. It is seen that such melting takes place over almost the whole profile of the ice sheet shown in the figure.

The previously mentioned equation for the heat balance at the bed in the presence of melting can be used to determine the melting for each point on the profile. The curve depicting the variation of the melting rate over the profile is also shown in Figure 10.10. Melting at the bed is seen to be absent in the center of the ice sheet, and it increases as the ablation region is approached, reaching a maximum at the firnline. Close to the periphery the melting rate becomes lower because of the reduced velocity of the glacier movement there, but it does not drop to zero.

Therefore, the temperature conditions at the bed of an ancient ice sheet were very similar to those obtaining now in Antarctica. Just as in Antarctica, in a number of places the heat released at the bed due to the glacier movement was many times higher than the geothermal flux, so that it was the ice movement which caused the melting at the bed, and thus the sliding over the bed and the low resistance to the ice motion as well. Under these conditions the viscous local reduction of the ice velocity makes the heat flux of the motion less, and it may drop so low that the bed temperature is below the melting point, causing the lower ice layers to freeze onto the bedrock, bringing a halt to the sliding. Consequently, regions where the melting temperature at the bed is mainly a result of the heat of ice motion will be regions of differentiated ice movement, that is, a situation involving rapidly moving regions of ice, within which the bottom melting keeps the resistance to motion low, alternating with regions of stagnant ice having bed temperatures below the melting point.

In Antarctica the region of bottom melting in the marginal zone is bounded by a very narrow strip, and it is apparently one of the reasons for the presence of a large number of peripheral outlet glaciers.

In the ancient ice sheet of Europe, which had a melting region hundreds of kilometers wide (see Figure 10.8), this differentiation into rapid and slow flows was much more pronounced. The enormous

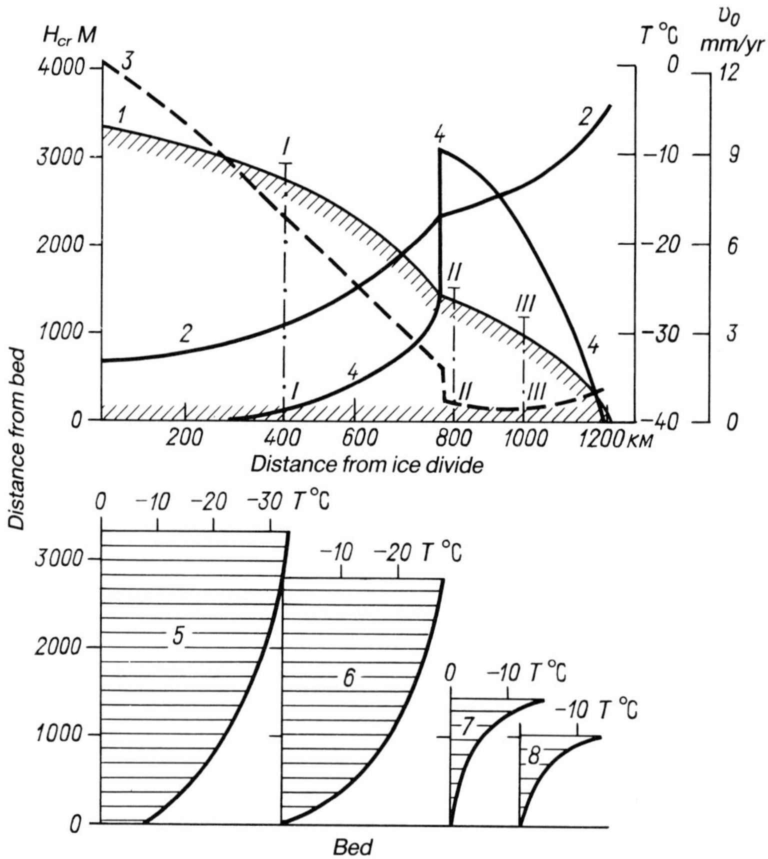

Fig. 10.10. Temperature regime and melting at bed of ancient ice sheet.

(1) form of surface of ice sheet H; (2) mean many-year temperatures T at upper surface; (3) critical glacier thickness H_{cr}; (4) melting rate at bed v_0; (5) temperature distribution inside ice sheet (at ice divide); (6) temperature distribution inside ice sheet at section I-I; (7) temperature distribution at section II-II; (8) temperature distribution at section III-III.

drainage lobes detected during studies of marginal seas [16] are apparently a result of such differentiation. The existence of smaller marginal outlet flows, traces of which have also been found [3], is a less striking manifestation of this process.

In the central region of the ice sheet the temperature gradually rises with increasing distance from the glacier surface into the upper layers, after which it rapidly rises to the melting point in the lower layers of the ice sheet. Such a temperature distribution does not differ much from the temperature field in one of the large present-day ice sheets (Greenland or Antarctica).

The situation was different, however, in the ablation zone. In the peripheral region the temperature distribution was shaped by the heating action of the upward moving ice particles. The lower ice layers were very close to the melting temperature, and only in the upper levels did the temperature drop rapidly to the value determined by the mean many-year temperature of the surface. In Figure 10.10 the typical temperature distributions in these zones are depicted by the corresponding curves.

The difference in the direction of ice motion inside the glacier, from downward and cooling in the accumulation region to upward and heating in the ablation region, is the initial cause of the difference in temperatures and viscous properties of the ice and it provides an explanation of the thin peripheral tongue of the glacier.

10.2.6. Erosion and Accumulation Activity

If the efficiency of the erosion-transportation activity is assumed to be the same over the length of the glacier, then as a first approximation the erosive-transporting effect of the glacier at each point of the bed will be proportional to the work of motion of a unit column of ice above this point. In this case, if this work is constant over the entire length of the glacier, the work will have just a transporting effect. Then a kind of "mobile equilibrium" will exist in the moraine-bearing layer. It consists in a descent to the bed of some of the moraine material brought from the central regions, but the same amount of material is entrained and carried away to take the place of the descending material. The erosion regime corresponds to a work of ice motion per unit time in the direction of ice flow which increases with the downward movement over the glacier.

Figure 10.8 includes a plot of the tangential stress at the bed τ_b and the thermal equivalent of the work of ice movement. The work of ice movement in the glacier being considered is seen to increase at first with increasing distance from the center, and then to decrease. But the zone in which this work increases in the direction of motion is the zone of erosive action of the glacier. The zone where this work decreases is the zone of accumulation of morainal material at the bed. Consequently, the peripheral ablation zone of an ancient ice sheet was simultaneously a zone in which moraine arriving from the central regions was deposited at the bed.

This type of relief-forming activity of an ancient ice sheet was quite different from the activity which is apparently going on now at the bed of the present-day ice sheets of Antarctica and Greenland.

REFERENCES *

[1] Aver'yanov, V.G., Central Antarctica. *Trudy Sov. Antarkt. Eksp.*, Vol. 30, 1963.

[2] Avsyuk, G.A., The temperature of ice in glaciers. *Trudy Inst. Geografii Akad. Nauk SSSR*, No. 67, 1956.

[3] Aseev, A.A., Ancient continental glaciations of Europe. Nauka, Moscow, 1974.

[4] Atlas of Antarctica, Vol. I. Glav. Uprav. Geodez. Kart., MG, 1966.

[5] Budd, W.F., The dynamics of ice masses. *Australian National Antarctic Research Exped. Scient. Report* 108, Melbourne, 1969 [in English].

[6] Budd, W.F., Jenssen, D., and Young, N.W., Calculation of the temperature profile from borehole data obtained at Vostok Station. *Inform. Byull. Sov. Antarkt. Eksp.*, No. 90, 1975.

[7] Barkov, N.I. Drilling deep boreholes in an ice sheet by the melting method. *Inform. Byull. Sov. Antarkt. Eksp.*, No. 40, 1963.

[8] Barkov, N.I., Gordienko, F.G., Korotkevich, E.S., and Kotlyakov, V.M., The isotope analysis of ice cores from Vostok Station (Antarctica) to the depth of 950 m. *Publ. Int. Assoc. Hyd. Sci.* 118, 1977 [English translation].

[9] Barbash, V.R., The thermal regime of the Antarctic Ice Sheet along a flow line from the ice divide via Vostok Station to Byrd Glacier. *Materialy Glyatsiol. Issled. Khronika, Obsuzhdeniya*, No. 36, 1979.

[10] Brichkin, A.V., Mikheev, S.V., and Boev, A.V., Flame drilling of glaciers under mountainous conditions. *Izv. Vses. Geogr. Obshchestva*, Vol. 99, No. 2, 1967.

[11] Veinberg, B.P., Ice. The properties, appearance, and disappearance of ice. Gostekhizdat, Moscow-Leningrad, 1940.

[12] Voitkovskii, K.F., The mechanical properties of ice. Izd. Akad. Nauk SSSR, Moscow, 1960.

[13] Vostretsov, R.N., and Dmitriev, D.N., Results of thermometry in boreholes at Vostok Station. *Inform. Byull. Sov. Antarkt. Eksp.*, No. 99, 1979.

[14] Vyalov, S.S., Toward a theory of viscoplastic flow of ice sheets. *Trudy Sov. Antarkt. Eksp.*, No. 10, 1960.

[15] Vyalov, S.S., Regularities of the deformation of ice. *Trudy Sov. Antarkt. Eksp.*, Vol. 10, 1960.

[16] Gerasimov, I.P., and Markov, K.K., The Ice Age in the USSR. *Trudy Inst. Geografii*, No. 33, 1939.

[17] Golubev, G.N., The hydrology of glaciers. Gidrometeoizdat, Leningrad, 1976.

[18] Golubev, G.N., Sukhanov, L.A., and Khromov, R.S., Coreless thermo-

* The first 81 references are in Cyrillic-alphabetical order. Unless otherwise indicated, these are all in Russian (Translator).

electric drilling and its use to study the structure of Dzhanidatt Glacier. *Materialy Glyatsiol. Issled. Khronika, Obsuzhdeniya*, No. 28, 1976.

[19] Dergunov, I.D., and Gorozhankii, I.P., Measurement of temperatures in the upper strata of the Earth's crust. *Izv. Akad. Nauk SSSR, Ser. Geofiz.*, No. 4, 1954.

[20] Dikikh, A.N., and Blagoobrazov, V.A., The artificial intensification of the melting of some Tien-Shan glaciers. Tezisy Dokladov 5-i Nauch. Konf. Tyan'-Shan'skoi Fiz.-Geogr. Stantsii, Frunze, 1963.

[21] Dolgushin, L.D., Field studies on the artificial intensification of the melting of glaciers of Soviet Central Asia. *Materialy Glyatsiol. Issled. Khronika, Obsuzhdeniya*, No. 11, 1965.

[22] Dubrovin, L.I., and Petrov, V.M., The thermal effect of an air flow on thermometer measurements in the snow-firn interior of a borehole. *Inform. Byull. Sov. Antarkt. Eksp.*, No. 39, 1963.

[23] Evteev, S.A., The geological activity of the ice sheet of East Antarctica, Nauka, Moscow, 1964.

[24] Zagorodnov, V.S., *et al.*, Thermal drilling of Obruchev Glacier. *Materialy Glyatsiol. Issled. Khronika, Obsuzhdeniya*, No. 28, 1976.

[25] Zotikov, I.A., Measurements of the geothermal heat flux in Antarctica. *Inform. Byull. Sov. Antarkt. Eksp.*, No. 29, 1961.

[26] Zotikov, I.A., The thermal regime of the glacier of Central Antarctica. *Inform. Byull. Sov. Antarkt. Eksp.*, No. 28, 1961.

[27] Zotikov, I.A., The thermal regime of the glacier of Central Antarctica. In: Antarctica. Reports of the 1961 Committee. Moscow, 1962. [English translation: Israel Prog. Scient. Trans. (Keter), Jerusalem, 1966].

[28] Zotikov, I.A., The thermal regime of large glaciers. *Materialy Glyatsiol. Issled. Khronika, Obsuzhdeniya*, No. 8, 1963.

[29] Zotikov, I.A., Temperatures in the interior of antarctic glaciers. In: Antarctica. Reports of the 1963 Committee. Moscow, 1964.

[30] Zotikov, I.A., Bottom melting in the central region of the Antarctic Ice Sheet and its effect on the present-day ice balance. *Materialy Glyatsiol. Issled. Khronika, Obsuzhdeniya*, No. 10, 1964.

[31] Zotikov, I.A., A model of the European Ice Sheet, constructed on the basis of various ancient and recent glaciations. In: The paleogeography of Europe in the Late Pleistocene. Reconstructions and models. Inst. Geografii Akad. Nauk SSSR, Moscow, 1973.

[32] Zotikov, I.A., The thermal regime of the Antarctic Ice Sheet. Gidrometeoizdat, Leningrad, 1977.

[33] Zotikov, I.A., Zagorodnov, V.S., and Raikovskii, Yu.V., The discovery of freezing at the bottom of the Ross Ice Shelf (Antarctica). *Dokl. Akad. Nauk SSSR*, Vol. 246, No. 6, 1979.

[34] Zotikov, I.A., and Zagorodnov, V.S., The thermal regime of the Ross Sea beneath the Ross Ice Shelf. In: Antarctica. Reports of the 1980 Committee, No. 19.

[35] Zotikov, I.A., Kapitsa, A.P., Kudryavtsev, E.V., and Sukhanov, A.A., Thermal drilling of glaciers. In: Antarctica. Reports of the 1972 Committee, No. 11.

[36] Zotikov, I.A., Kapitsa, A.P., and Sorokhtin, O.G., The thermal regime of the ice sheet of Central Antarctica. *Inform. Byull. Sov. Antarkt. Eksp.*, No. 51, 1965.

[37] Zotikov, I.A., and Razumeiko, N.G., The temperature distribution in the ice caps of Franz Josef Land as a reflection of climatic fluctuations. *Materialy Glyatsiol. Issled. Khronika, Obsuzhdeniya*, No. 12, 1966.

[38] Zubov, N.N., The maximum thickness of sea ice and continental ice. *Meteorologiya i Gidrologiya*, No. 2, 1959.

[39] Ignatov, V.S., The thermal drilling of boreholes at Vostok Station. *Inform. Byull. Sov. Antarkt. Eksp.*, No. 22, 1960.

[40] Kapitsa, A.P., The dynamics and morphology of the ice sheet in the central sector of East Antarctica. *Trudy Sov. Antarkt. Eksp.*, No. 19, 1961.

[41] Kapitsa, A.P., The subglacial topography of Antarctica. Nauka, Moscow, 1968.

[42] Kotlyakov, V.M., The Earth's snow cover and glaciers. Gidrometeoizdat, Leningrad, 1968.

[43] Kotlyakov, V.M., The glaciology of Antarctica at the beginning of the 1980s. In: Antarctica. Reports of the 1980 Committee, No. 17.

[44] Kotlyakov, V.M., Gordienko, F.G., Barkov, N.I., and Korotkevich, E.S., Isotope studies of cores from Vostok Station and their paleoglaciological interpretation. In: Antarctica. Reports of the 1980 Committee, No. 19.

[45] Kochin, N.E., Kibel', I.A., and Roze, N.V., Theoretical hydrodynamics. Vol. 1. Gostekhizdat, 1955.

[46] Kreiter, A.A., The use of graded dust for the artificial contamination of snow. *Materialy Glyatsiol. Issled. Khronika, Obsuzhdeniya*, No. 11, 1965.

[47] Kreiter, A.A., and Tikhanovskaya, A.A., Variations in the heat and moisture exchange at a glacier surface caused by artificial contamination. *Materialy Glyatsiol. Issled. Khronika, Obsuzhdeniya*, No. 10, 1964.

[48] Kruchinin, Yu. A., The ice shelves of Queen Maud Land. Gidrometeoizdat, Leningrad, 1965.

[49] Kudryashov, B.B., *et al.*, The theory and practice of drilling-melting in the Antarctic. *Materialy Glyatsiol. Issled. Khronika, Obsuzhdeniya*, No. 22, 1973.

[50] Kutasov, I.M., The region of disturbance of the thermal state of rocks due to borehole drilling. *Izv. Akad. Nauk SSSR, Seriya Geofiz.*, No. 10, 1960.

[51] Kutasov, I.M., Determination of the overheating temperature of thermistor sensors. In: Heat and mass exchange in the frozen interior of the Earth's crust. Izd. Akad. Nauk SSSR, Moscow, 1963.

[52] Kutasov, I.M., Determination of the rate of convective heat flow in boreholes. In: Heat and mass exchange in the frozen interior of the Earth's crust. Izd. Akad. Nauk SSSR, Moscow, 1963.

[53] Lebedev, V.L., Heat exchange between antarctic icebergs and glaciers. In: Antarctica. Reports of the 1963 Committee, Moscow, 1964.

[54] Ledenev, V.G., Currents in the vicinity of the Lazarev Ice Shelf. *Inform. Byull. Sov. Antarkt. Eksp.*, No. 45, 1964.

[55] Loitsyanskii, L.G., The aerodynamics of a boundary layer. Fizmatgiz, Moscow, 1962.

[56] Lyubimova, E.A., Starikova, G.N., and Shushpanov, A.A., Thermophysical studies of rocks. In: Geothermal studies. Nauka, Moscow, 1964.

[57] Lyubimova, E.A., Shelyagin, V.A., and Shushpanov, A.P., Equipment for measuring a deep heat flux. In: Problems of a deep heat flux. Nauka, Moscow. 1966.

[58] Lyubimova, E.A., *et al.*, Determination of a heat flux from the Earth's interior and measurement results. In: Geothermal studies. Izd. Akad. Nauk SSSR, Moscow, 1964.

[59] Mikheev, M.A., The fundamentals of heat transfer. Moscow-Leningrad, 1957.

[60] Mikheev, S.V., Ice drilling using the flame method. *Materialy Glyatsiol. Issled. Khronika, Obsuzhdeniya*, No. 18, 1971.

[61] Morev, V.A., Ice drilling using an electrothermal method at Mirny. *Inform. Byull. Sov. Antarkt. Eksp.*, No. 56, 1961.

[62] Morev, V.A., Electrothermal drills for drilling boreholes in an ice sheet. *Materialy Glyatsiol. Issled. Khronika, Obsuzhdeniya*, No. 28, 1976.

[63] Paterson, W.S.B., The physics of glaciers, 2nd ed., Pergamon, 1981 [in English].

[64] Peschanskii, I.S., Ice physics and ice engineering. Gidrometeoizdat, Leningrad, 1967.

[65] Pekhovich, A.I., and Zhidkikh, V.M., Calculations of the thermal regime of solids. Energiya, Leningrad, 1968.

[66] Savatyugin, L.M., Glaciological studies on the Shackleton Ice Shelf (January-April 1978). *Inform. Byull. Sov. Antarkt. Eksp.*, No. 100, 1980.

[67] Silin, S.S., Calculation of temperature fields during the thermal drilling of ice layers. *Inzhen. Fiz. Zhurn.*, Vol. 8, No. 1, 1965.

[68] Sukhanov, L.A., Morev, V.A., and Zotikov, I.A., Portable thermoelectric ice drills. *Materialy Glyatsiol. Issled. Khronika, Obsuzhdeniya*, No. 23, 1974.

[69] Treshnikov, A.F., Features of the ice regime of the Antarctic Ocean. *Trudy Sov. Antarkt. Eksp.*, No. 21, 1963.

[70] Tsykin, E.N., and Tsykina, G.A., Laboratory experiments on the artificial intensification of melting by blackening an ice surface. *Materialy Glyatsiol. Issled. Khronika, Obsuzhdeniya*, No. 14, 1968.

[71] Tsykina, G.A., The use of thermistors as temperature sensors in glaciological experiments. In: Studies of glaciers and glacial regions, Vol. 1, Ser. Rezul'taty MGG. Izd. Akad. Nauk SSSR, 1961.

[72] Tsyferov, M.I., A method for drilling boreholes. *Byull. Izobretenii*, No. 19, 1964.

[73] Schlichting, H., Boundary layer theory (translated into English by J. Kestin). McGraw-Hill, New York, 1968.

[74] Schneider, P.J., Conduction heat transfer. Addison-Wesley, Reading, Mass., 1955 [in English].

[75] Shul'ts, V.L., The advisability of artificially intensifying the melting of snow in the mountains of Soviet Central Asia. *Meteorologiya i Gidrologiya*, No. 5, 1963.

[76] Shumskiy, P.A., and Zotikov, I.A., The bottom melting of ice shelves. In: Antarctica. Reports of the 1962 Committee, Moscow, 1963.

[77] Shumskiy, P.A., Dynamic glaciology. Amerind Publ. Co. New Delhi, 1978. [English translation].

[78] Eckert, E.R., and Drake, R.M., Analysis of heat mass transfer. McGraw-Hill, New York, 1972 [in English].

[79] Yagunov, A.V., The destruction mechanism for flame drilling. *Izv. Vyssh. Ucheb. Zaved. (Gornyi Zhurnal)*, No. 2, 1963.

[80] Yagunov, A.V., Pokrovskii, M.A., Vasil'ev, A.P., and Varich, M.S., Flame drilling of blast boreholes. Gostoptekhizdat, Moscow, 1962.

[81] Yanke, E., and Emde, F., Tables with formulas and curves. Fizmatgiz, Moscow, 1959.

[82] Aamot, H.W.C., Self-contained thermal probes for remove measurements within an ice sheet. Intern. Symp. Antarct. Glaciol. Exploration, Hanover, USA, 1970.

[83] Arnold, K.C., An investigation into methods of accelerating the melting of ice and snow by artificial dusting. *Geology of Arctic*, No. 2, 1960.

[84] Benfield, A.E., A problem of temperature distribution in a moving medium. *Quart. J. Appl. Math.*, No. 6, 1949.

[85] Browning, J.A., Flame drilling through the Ross Ice Shelf. *Northern Eng.*, Vol. 10, No. 1, 1978.

[86] Budd, W.F., Jenssen, D., and Radok, U., Derived physical characteristics of the Antarctic Ice Sheet. Meteorology Department University of Melbourne, 1971.

[87] Bullard, E.C., The time necessary for a borehole to attain temperature equilibrium. *Month. Not. Astr. Soc. Geophys. Sup.*, Vol. 5, No. 5, 1947.

[88] Clough, G.W., Hansen, B.L., The Ross Ice Shelf Project. *Science*, Vol. 203, No. 4379, 1979.

[89] Cooper, L.R, and Jones, C., The determination of virgin strata temperatures from observations in deep survey boreholes. *Geophys. J. R. Astr. Soc.*, Vol. 2, No. 2, 1959.

[90] Crary, A.P., Geophysical studies along Northern Ellesmere Island. *Arctic*, Vol. 9, No. 3, 1956.

[91] Crary, A.P., Arctic ice island and ice shelf studies. Part II. *Arctic*, Vol. 13, No. 1, 1960.

[92] Crary, A.P., Presidential address. International symposium on Antarctic Glaciological Exploration. IASH Publ., No. 98, 1970.

[93] Crary, A.P., Robinson, E.S., and Boyd, W.J., Glaciological studies of the Ross Ice Shelf, Antarctica, 1957-1960. IGY Glaciol. Rep., No. 6, 1962.

[94] Debenham, F., A new mode of transportation by ice: The raised marine muds of South Victoria Land. *Quart. J. Geol. Soc. London*, Vol. 75, pt. 2, 1920.

[95] Frolich, K., Tauch, E.I., Hogan, I.I., and Peer, A.A., Solubilities of gases in liquids of high pressure. *Industr. and Eng. Chemistry*, Vol. 23, No. 4, 1931.

[96] Gilmour, A.E., Hydrological heat and mass transport across the boundary of the ice shelf in McMurdo Sound, Antarctica. *New Zealand J. Geology and Geophysics*, Vol. 6, No. 3, 1963.

[97] Gilmour, A.E., Ross Ice Shelf sea temperature. *Science*, Vol. 203, No. 4379, 1979.

[98] Gilmour, A.E., MacDonald, W.J.P., and Hoeven, E., Winter measurements of sea currents in McMurdo Sound. *New Zealand J. Geology and Geophysics*, Vol. 5, No. 5, 1962.

[99] Glen, J.W., Experiments on the deformation of ice. *J. Glaciol.* Vol. 2, No. 12, 1952.

[100] Gow, A.J., Deep bore hole measurements at Byrd Station, Antarctica. *J. Glaciol.*, Vol. 4, No. 36, 1963.

[101] Gow, A.J., *et al.*, New light on the mode of uplift of the fish and fossiliferous moraines of the McMurdo Ice Shelf, Antarctica. *J. Glaciol.*, Vol. 5, No. 42, 1965.
[102] Jacobs, S.S., Gordon, A.L., and Ardai, J.L. Jr., Circulation and melting beneath the Ross Ice Shelf. *Science*, Vol. 203, No. 4379, 1979.
[103] Jenssen, D., and Radok, U., Transient temperature distributions in ice caps and ice shelves. Union Geod. et Geoph. Intern. Assemblee Generale de Helsinki, 1961. (IASH, Publ. 55, 1961).
[104] Jenssen, D., and Radok, U., Heat conduction in thinning ice sheets. *J. Glaciol.*, Vol. 4, No. 34, 1963.
[105] Lee, W.H.K., Heat flow data analyses. *Rev. Geophysics*, No. 1, 1963.
[106] Loewe, F., On melting of fresh-water ice in sea water. *J. Glaciol.*, Vol. 3, No. 30, 1961.
[107] Morgan, V.I., Oxygen isotope evidence for bottom freezing on the Amery Ice Shelf. *Nature*, Vol. 238, No. 5364, 1972.
[108] Nye, J.F., The flow of glaciers and ice sheets as a problem in plasticity. *Proc. Roy. Soc., Ser. A.*, Vol. 207, No. 1091, 1951.
[109] Nye, J.F., The motion of ice sheets and glaciers. *J. Glaciol.*, Vol. 3, No. 26, 1959.
[110] Nye, J.F., The mechanics of glacier flow. *J. Glaciol.*, Vol. 2, No. 12, 1952.
[111] Oswald, G.K.A., and Robin, G. de Q., Lakes beneath the Antarctic ice sheet. *Nature*, Vol. 245, No. 5423, 1973.
[112] Paterson, W.S.B., *et al.*, An oxygen-isotope climate record from the Devon Island ice cap, Arctic Canada. *Nature*, Vol. 266, No. 5602, 1977.
[113] Philberth, K., On the temperature response in ice sheets to radioactive waste deposits. *J. Glaciol.* , Vol. 16, No. 74, 1976.
[114] Radok, U., Temperatures in polar ice caps. *Nature*, Vol. 184, No. 4692, 1959.
[115] Robin, G. de Q., Ice movement and temperature distribution in glaciers and ice sheets. *J. Glaciol.*, No. 18, 1955.
[116] Robin, G. de Q., Polar ice sheets: a review. *Polar Record*, Vol. 16, No. 100, 1972.
[117] Robin, G. de Q., Is the basal ice of a temperate glacier at the pressure melting point? *J. Glaciol.*, Vol. 16, No. 74, 1976.
[118] Robin, G. de Q., Reconciliation of temperature - depth profiles in polar ice sheets with past surface temperatures deduced from oxygen-isotope profiles. *J. Glaciol.*, Vol. 16, No. 74, 1976.
[119] Rosenthal, D., The theory of moving sources of heat and its application to metal treatments. *Trans. ASME*, Vol. 68, No. 8, 1946.
[120] Rubin, M.J., Atmospheric advection and the antarctic mass and heat budget. *Geophys. Monogr.*, No. 7, Publ. Amer. Geophys. Union, Washington 1962.
[121] Shreve, R.L., Theory of performance of isothermal solid-nosed hot points boring in temperate ice. *J. Glaciol.*, Vol. 4, No. 32, 1962.
[122] Schumskiy, P.A., and Zotikov, I.A., On the bottom melting of the Antarctic Ice Shelf. Union Geod. et Geoph. Intern. Assemblee Generale de Berkeley, 1963.
[123] Steinemann, S.,Results of preliminary experiments on the plasticity of ice crystals. *J. Glaciol.*, Vol. 2, No. 16, 1954.
[124] Ueda, H.T., and Garfield, D.E., Deep core drilling program at Byrd Station (1967-1968). *Antarct. J. US*, Vol. 3, No. 4, 1968.

[125] Untersteiner, N., and Badglay, F.I., Preliminary results of thermal budget studies of Arctic pack ice during summer and autumn. - Arctic sea ice, Washington D.C. National Acad. of Sciences, Nat. Res. Council Publ., 1958.

[126] Van Der Merwe, J.H., The influence of convection on measured borehole temperatures. *South Afr. J.S.*, 47, 253, 1951.

[127] Weertman, J., On the sliding of glaciers. *J. Glaciol.*, Vol. 3, No. 21, 1957.

[128] Weertman, J., The unsolved general glacier sliding problem. *J. Glaciol.*, Vol. 23, No. 89, 1979.

[129] Weertman, J., Sibert, J., Weeks, W.F., and Sternig, J., Radioactive wastes on ice. *Science Public Affairs*, Vol. 30, No. 1, Jan. 1974.

[130] Winding, C.C., and Cheney, A.J., Mass and heat transfer in tube banks. *Industr. and Eng. Chemistry*, Vol. 40, 1948.

[131] Yin-Chao Jen, Chi-Tien, Laminar heat transfer over a melting plate, the modified Lévêque problem. *J. Geophys. Res.*, Vol. 68, No. 12, 1963.

[132] Zeller, E., Saunders, D.F., and Angono, E.E., Putting radioactive wastes on ice: A proposal for an international radionuclide depository in Antarctica. Science Public Affairs. Bull. Atom. Sci. Vol. 29, No. 1, 1973.

[133] Zotikov, I.A., Bottom melting in the central zone of the Antarctic Ice Sheet and its influence on the present balance of the ice mass. *IASH Bull.* 8, No. 1, 1963.

[134] Zotikov, I.A., Antifreeze-thermodrilling for core through the central part of the Ross Ice Shelf (J-9 Camp), Antarctica. CRREL Report 79-24. Cold Regions Res. and Eng. Lab., Hanover, New Hampshire, U.S.A., 1979.

[135] Zotikov, I.A., and Gow, A.J., The thermal and compositional structure of the Koettlits Ice Tongue, McMurdo Sound, Antarctica. Proc. Intern. Conf. on Low Temperature Science. Sapporo, Japan, 1967.

[136] Zotikov, I.A., and Moiseeva, G.P., A theoretical study of ice surface dusting influence on melting intensity. Proc. of Intern. Symposium on the Role of Snow and Ice in Hydrology Banff. Canada, 1973.

[137] Zotikov, I.A., Zagorodnov, V.S., and Raikovsky, J.V., Core drilling through the Ross Ice Shelf (Antarctica) confirmed basal freezing. *Science*, Vol. 207, p.1463, 1980.

[138] Zumberge, J.H., Horizontal strain and absolute movement of the Ross Ice Shelf between Ross Island and Roosevelt Island, Antarctica, 1958-1963. Antarctic Snow and Ice Studies, Washington D.C., Geophys. Union, Antarctic Research Series, Vol. 2, 1964.

[139] Zumberge, J.H., Ross Ice Shelf Project. Drilling in and below ice will reveal physical, chemical, biological features. *Antarct. J. US*, Vol. 6, No. 6, 1970.

INDEX